FOOD PROCESS ENGINEERING

Emerging Trends in Research
and Their Applications

Innovations in Agricultural and Biological Engineering

FOOD PROCESS ENGINEERING

Emerging Trends in Research and Their Applications

Edited by

Murlidhar Meghwal, PhD
Megh R. Goyal, PhD, PE

APPLE
ACADEMIC
PRESS

Apple Academic Press Inc. | Apple Academic Press Inc.
3333 Mistwell Crescent | 9 Spinnaker Way
Oakville, ON L6L 0A2 | Waretown, NJ 08758
Canada | USA

First issued in paperback 2021

Exclusive worldwide distribution by CRC Press, a member of Taylor & Francis Group
No claim to original U.S. Government works

ISBN 13: 978-1-77463-703-6 (pbk)
ISBN 13: 978-1-77188-402-0 (hbk)

Library and Archives Canada Cataloguing in Publication

Food process engineering : emerging trends in research and their applications / edited by Murlidhar Meghwal, PhD, Megh R. Goyal, PhD, PE.

(Innovations in agricultural and biological engineering)
Includes bibliographical references and index.
Issued in print and electronic formats.
ISBN 978-1-77188-402-0 (hardcover).--ISBN 978-1-77188-403-7 (pdf)
1. Food industry and trade--Research. 2. Food industry and trade--Technological innova-tions. I. Meghwal, Murlidhar, editor II. Goyal, Megh Raj, editor III. Series: Innovations in agricultural and biological engineering
TP370.F68 2016   664   C2016-906081-0   C2016-906082-9

Library of Congress Cataloging-in-Publication Data

Names: Meghwal, Murlidhar, editor. | Goyal, Megh Raj, editor.
Title: Food process engineering : emerging trends in research and their applications / editors, Murlidhar Meghwal, PhD, Megh R. Goyal, PhD, PE.
Description: Toronto : Apple Academic Press, 2017. | Includes bibliographical references and index.
Identifiers: LCCN 2016046209 (print) | LCCN 2016046474 (ebook) | ISBN 9781771884020 (hardcover : alk. paper) | ISBN 9781771884037 (ebook)
Subjects: LCSH: Food industry and trade. | Food--Preservation. | Processed foods. | Food han-dling.
Classification: LCC TP370 .F62527 2017 (print) | LCC TP370 (ebook) | DDC 664/.02--dc23
LC record available at https://lccn.loc.gov/2016046209

Apple Academic Press also publishes its books in a variety of electronic formats. Some content that ap-pears in print may not be available in electronic format. For information about Apple Academic Press products, visit our website at **www.appleacademicpress.com** and the CRC Press website at **www.crc-press.com**

CONTENTS

LIST OF CONTRIBUTORS

Azad Ahmad Ahanger, PhD
Associate Professor, Division of Pharmacology and Toxicology, Sher-e-Kashmir University of Agricultural Sciences and Technology, Kashmir, Shuhama, Alustang, Srinagar–190006, India. E-mail: azadpharm@rediffmail.com

Asaad Rehman Saeed Al-Hilphy, PhD
Assistant Professor, Department of Food Science, College of Agriculture, University of Basrah, Basra City, Iraq. Mobile: +00-96 47702696458. E-mail: aalhilphy@yahoo.co.uk

A. Beltran, PhD
Researcher, Analytical Chemistry, Nutrition and Food Sciences Department, University of Alicante, 03080, Alicante, Spain. E-mail: ana.beltrán@ua.es

Shraddha Bhatt, PhD
Assistant Professor, Department of Biotechnology, Junagadh Agricultural University, Junagadh, Gujarat, India

Sudhanshi Billoria, PhD Student
Research Scholar, Department of Agricultural and Food Engineering, Indian Institute of Technology, Kharagpur – 721302, West Bengal, INDIA. Mobile: +91-8768126479. E-mail: sudharihant@gmail.com

Satyanarayana Bonthu, MTech Student
Research Scholar, Agricultural and Food Engineering Department, Indian Institute of Technology Kharagpur, West Bengal, Kharagpur – 721302, India. Mobile: +91-7059972916, E-mail: satyanarayanabonthu@yahoo.com

A. Canals, PhD
Professor, Analytical Chemistry, Nutrition and Food Sciences Department, University of Alicante, 03080, Alicante, Spain. E-mail: a.canals@ua.es

Rupesh Chavan, PhD
Quality Assessment Analyst, Mother Dairy, Gandhinagar, Gujarat. Mailing Address: B 501 Nobel Platinum, Rayjibaug Near Motibaug Circle, Junagadh 362001, Gujarat, India. Mobile: +91-9992525438, E-mail: rschavanb_tech@rediffmail.com

D. V. Chidan, PhD
Assistant Professor, Indian Institute of Crop Processing Technology, Thanjavur – 613005. Phone: +91-9750968417; E-mail: chidanand@iicpt.edu.in

Madhusweta Das, PhD
Associate Professor, Agricultural and Food Engineering Department, Indian Institute of Technology, Kharagpur – 721302, West Bengal, India. Tel.: +91-32222-83108; E-mail: madhu@agfe.iitkgp.ernet.in

Navneet Singh Deora, PhD
Research Scientist, Prasan Solutions (India) Pvt. Limited (Nalanda R&D Center), Cochin – 682021, Kerala. Mobile: +91-7042307007, E-mail: navneetsinghdeora@gmail.com

Aastha Deswal, PhD
Research Scientist, Prasan Solutions (India) Pvt. Limited (Nalanda R&D Center), Cochin – 682021, Kerala. Mobile: +91-8137892690, E-mail: deswalad@gmail.com

Deepika Goswami, MSc
Scientist, Food Grains and Oilseeds Processing Division, ICAR-CIPHET, Ludhiana – 141004, Punjab, India. Phone: +91-9592317693; E-mail: deepikagoswami@rediffmail.com

Megh R. Goyal, PhD, PE
Retired Faculty in Agricultural and Biomedical Engineering from General Engineering Department, University of Puerto Rico – Mayaguez Campus; and Senior Technical Editor-in-Chief in Agriculture Sciences and Biomedical Engineering, Apple Academic Press Inc., USA. E-mail: goyalmegh@gmail.com

N. Grane, PhD
Professor. Analytical Chemistry, Nutrition and Food Sciences Department, University of Alicante, 03080, Alicante, Spain. E-mail: nuria.grane@ua.es

Raj K. Keservani, M.Pharm.
Researcher, School of Pharmaceutical Sciences, Rajiv Gandhi Proudyogiki Vishwavidyalaya, Bhopal (MP), 462036, India. Mobile: +91-7897803904; E-mail: rajksops@gmail.com

Rajesh K. Kesharwani, PhD
Assistant Professor, Department of Biotechnology, NIET, NIMS University, Shobha Nagar, Jaipur, Rajasthan, 303121, India. E-mail: rajiiita06@gmail.com

Anit Kumar, PhD Scholar
Department of Food Science and Technology, National Institute of Food Technology Entrepreneurship and Management (NIFTEM), Kundli, Sonepat, Haryana, India

Deepak Kumar, PhD
Postdoctoral Research Associate, ADM Institute for the Prevention of Postharvest Loss, University of Illinois at Urbana–Champaign, Phone: (541) 7408711; E-mail: kumard@illinois.edu

Manish Kumar, PhD
Assistant Professor, Veterinary and Animal Sciences, Institute of Agricultural Sciences, Banaras Hindu University, Varanasi – 221005 (UP), India. E-mail: manish.vet82@gmail.com

Priya Ranjan Kumar, PhD
Assistant Professor, Veterinary and Animal Sciences, Institute of Agricultural Sciences, Banaras Hindu University, Varanasi – 221005 (UP), India. E-mail: dr.pranjan007@gmail.com

Sanjith Madhavan, MSc
Technical Director, Prasan Solutions (India) Pvt. Limited (Nalanda R&D Center), Cochin – 682021, Kerala. Mobile: +91-8137892690, E-mail: sanjith@prasansolutions.com

Archana Mahapatra, M.Tech.
Research Scholar, Agricultural and Food Engineering Department, Indian Institute of Technology, Kharagpur, Kharagpur–721302, West Bengal, India

Harshad M. Mandge, MSc
Assistant Professor (PHT), College of Horticulture, Banda University of Agriculture and Technology, Banda – 210001, Uttar Pradesh, India, E-mail: mandgeharshad@gmail.com

M. L. Martin, PhD
Professor, Analytical Chemistry, Nutrition and Food Sciences Department, University of Alicante, 03080, Alicante, Spain. E-mail: mluisa.martin@ua.es

Murlidhar Meghwal, PhD
Assistant Professor, Food Technology, Centre for Emerging Technologies, Jain University, Jain Global Campus, Jakkasandra – 562112, Kanakapura Main Road, Dist. Ramanagara, Karnataka, India. Mobile: +91-9739204027; E-mail: murli.murthi@gmail.com

Tanmay Nalawade, PhD
Assistant Professor, Department of Bachelor of Food Technology and Management, CNCVCW, CSIBER, Kolhapur, India

Alaa Kareem Niamah, PhD
Assistant Professor, Department of Food Science, College of Agriculture, University of Basrah, Basra City, Iraq. Mobile: +00-96 47709042069. E-mail: alaakareem2002@hotmail.com

Ramadevi Nimannapalli, PhD
Professor, Veterinary and Animal Sciences, Institute of Agricultural Sciences, Banaras Hindu University, Varanasi – 221005 (UP), India. E-mail: ramadevi.nimannapalli@gmail.com

Shruti Pandey, PhD
Research Scientist, Department of Grain Science & Technology, CSIR – Central Food Technological Research Institute, Mysore, 570020, India. Mobile: +91-9449859778, E-mail: shruti@cftri.res.in

Akash Pare, PhD
Assistant Professor, Indian Institute of Crop Processing Technology, Thanjavur – 613005. Phone: +91-9750968416; E-mail: akashpare@iicpt.edu.in

Shahid Prawez, PhD
Associate Professor, Veterinary and Animal Sciences, Institute of Agricultural Sciences, Banaras Hindu University, Varanasi – 221005 (UP), India. Mobile: +91-9170770003; E-mail: Shahidprawez@gmail.com

Hradesh Rajput, MSc
Senior Research Fellow, Department of Food Science and Technology, Punjab Agricultural University, Ludhiana – 141004, Punjab, India, E-mail: hrdesh802@gmail.com

Marina Ramos, PhD
Researcher, Analytical Chemistry, Nutrition and Food Sciences Department, University of Alicante, 03080, Alicante, Spain. E-mail: marina.ramos@ua.es

B. V. Sathyendra Rao, PhD
Senior Principal Scientist, Department of Grain Science & Technology, CSIR – Central Food Technological Research Institute, Mysore, 570020, India. Mobile: +91-9986846780, E-mail: sathyendra@cftri.res.in

Ashish Rawson, PhD
Assistant Professor, Indian Institute of Crop Processing Technology, Thanjavur – 613005. Phone: +91-7373068426; E-mail: ashish.rawson@iicpt.edu.in

Anirban Ray, PhD
Post-Doctoral Research Fellow, Department of Molecular biology and Biotechnology, University of Kalyani, West Bengal, Kalyani – 741235, India. Mobile: +91-9231695435; E-mail: anirbanrayiitkgp@gmail.com

Jagbir Rehal, MSc
Assistant Fruit and Vegetable Technologist, Department of Food Science and Technology, Punjab Agricultural University, Ludhiana – 141004, Punjab, India, E-mail: kaur_jagbir@yahoo.com

P. Roman, PhD
Researcher, Analytical Chemistry, Nutrition and Food Sciences Department, University of Alicante, 03080, Alicante, Spain. E-mail: ivan.roman@ua.es

Chandani Sen, MS
PhD Research Scholar, Agricultural and Food Engineering Department, Indian Institute of Technology, Kharagpur – 721302, West Bengal, India, E-mail: chandani.sen@agfe.iitkgp.ernet.in

Anil K. Sharma, MPharm
Researcher, Department of Pharmaceutics, Delhi Institute of Pharmaceutical Sciences and Research, New Delhi, 110017, India. E-mail: sharmarahul2004@gmail.com

Prem Prakash Srivastav, PhD
Associate Professor, Department of Agricultural and Food Engineering, Indian Institute of Technology, Kharagpur – 721302, West Bengal, India. Mobile: +91-9434043426. Telephone: +91-3222-283134, Fax: +91-3222-282224. E-mail: pps@agfe.iitkgp.ernet.in

C. K. Sunil, PhD
Assistant Professor, Indian Institute of Crop Processing Technology, Thanjavur – 613005. Phone: +91-9750968423; E-mail: sunil.ck@iicpt.edu.in

Somya Tewari, PhD
Senior Research Fellow, Department of Food Process Engineering, National Institute of Food Technology Entrepreneurship and Management (NIFTEM), Kundli, Sonepat, Haryana, India

P. Punam Tripathy, PhD
Assistant Professor, Agricultural and Food Engineering Department, Indian Institute of Technology, Kharagpur, Kharagpur – 721302, West Bengal, India, E-mail: punam@agfe.iitkgp.ernet.in

Deepak Kumar Verma, MSc
PhD Research Scholar, Department of Agricultural and Food Engineering, Indian Institute of Technology, Kharagpur – 721302, West Bengal, India. Mobile: +91-74071-70260, +91-93359-93005. Telephone: +91-32222-81673, Fax: +91-3222-282224. E-mail: deepak.verma@agfe.iitkgp.ernet.in, rajadkv@rediffmail.com

Vaibhav Vyas, BTech
M. Tech. Scholar, Department of Food Science and Technology, National Institute of Food Technology Entrepreneurship and Management (NIFTEM), Kundli, Sonepat, Haryana, India

LIST OF ABBREVIATIONS

AA	ascorbic acid
AD	air drying
C16:1	palmitoleic acid
C18:0	stearic acid
C18:1	oleic acid
C18:2	linoleic acid
C18:3	linolenic acid
CHD	cardiac health disease
CVD	cardiovascular disease
DHA	docosahexaenoic acid
EPA	eicosapentaenoic acid
FAME	fatty acid methyl esters
FIR	far-infrared
GC-MS	gas chromatography mass spectrometry
GHz	Giga Hertz
H2SO4	sulphuric acid
HDPE	high-density PE
IR	infrared
LDL	low-density-lipoprotein
LDL-C	low density lipoprotein cholesterol
LDPE	low-density PE
LLDPE	linear low-density PE
LPSSD	low-pressure superheated steam drying
M	molar concentration (mol L–1)
MeOH	methanol
MFD	microwave-finish drying
MHz	Mega Hertz
mm	millimeter
MRPs	Maillard reaction products
MW	molecular weight
MWSB	microwave and spouted bed

NaCl	sodium chloride
NaOH	sodium hydroxide
PID	proportional-integral-derivative controller
ppm	parts per million
PUS	power ultrasound
RH	relative humidity
RW	refractance window
STP	standard temperature and pressure
TD	tray drying
TPs	total phenolic
US	ultrasonication
VLDPE	very low density PE
VMD	vacuum microwave drying
μm	micrometer

LIST OF SYMBOLS

A_c	collector surface area, m^2
A_e	area of edges, m^2
C_p	specific heat of air, kJ/kg K
d	distance between electrodes
d_0	penetration depth, m
E	voltage
f	frequency of electromagnetic field
F'	collector efficiency factor
F_R	collector heat removal factor
$h_{c,\,p-g}$	convective heat transfer coefficient from absorber to glass cover, W/m^2 K
$h_{r'p-amb}$	radiative heat transfer coefficient from absorber to ambient, W/m^2 K
$h_{r'p-g}$	radiative heat transfer coefficient from absorber to glass cover, W/m^2 K
h_w	wind heat transfer coefficient, W/m^2 K
I	incident solar radiation, W/m^2
K_{in}	thermal conductivity of insulation, W/m K
L_{in}	insulation thickness, m
m	mass flow rate of air, kg/s
MT	million ton
P	power
Q_u	useful energy gain of collector, W/m^2
T_{amb}	temperature of ambient air, °C
$\tan \delta$	loss tangent
T_{in}	temperature of inlet air, °C

T_{out}	temperature of outlet air, °C
U_b	bottom heat loss coefficient, W/m^2 K
U_e	edge heat loss coefficient, W/m^2 K
U_L	overall heat loss coefficient, W/m^2 K
U_t	top heat loss coefficient, W/m^2 K
ε'	dielectric constant of the material
ε''	dielectric loss
η	efficiency of collector, %
λ_0	wavelength, m
σ	Stefan–Boltzmann constant, W/m^2 K^4
τ	transmissivity

PREFACE 1
By Murlidhar Meghwal

Food processing engineering is very important because everybody has to eat to survive. This sector consists of a wide range of activities. Some of its specific activities are research and development of new foods; development of new food-related pharmaceutical products; design and installation of food processes; development and operation of food manufacturing, packaging and distributing systems for food products; and marketing and technical support for food manufacturing plants. It is the responsibility of food engineers and food scientists to provide the food technological knowledge required to have food products and services be cost-effective in production and commercialization. They are employed in food processing, food machinery, food packaging, ingredient food manufacturing, instrumentation, and control for food.

Food Process Engineering: Emerging Trends in Research and Their Applications provides a global perspective of the present-age frontiers in food process engineering research, innovation, and emerging trends. It includes selected recent emerging trends and issues of food engineering. The book volume explores topics of food engineering, food technology, food science and food process engineering, that can help to provide solutions for the different issues, problems and complexity related to food crises all over globe. with the help of limited resources and technology.

Part I: Emerging Trends and Technologies in Food Processing includes trends in food packaging technology; emerging technology-based drying for food and feed products; application of emerging technologies for freezing and thawing of foods; principles of novel freezing and thawing technologies for foods application; and overview of applications of dryers for foods including: industrial, solar, novel, and infrared methods.

Part II: Ultrasonic Treatment of Foods include chapters on ultrasonic-assisted derivatization of fatty acids from edible oils and determination by GC-MS and principles of ultrasonic technology for treatment of milk and milk products. *Part III: Foods for Specific Needs* covers new research on natural food colors: a technical insight; potential use of pseudo cereals:

buckwheat, quinoa and amaranth; and nutraceutical and functional foods for cardiovascular health. *Part VI: Food Preservation* includes natural antioxidants during frying: food industry perspective. *Part V: Food Hazards and Their Controls* includes chapters on the Hazard Analyses Critical Control Point Program and antibiotics in food producing animals and resistance hazards.

The targeted audience for this book includes practicing food process engineers, food technologists, researchers, lecturers, teachers, professors, food industry professionals, students of these fields and all those who have inclination for food processing sector. The book not only covers the practical aspect but also provides a lot of basic information. It is also instructive. Therefore students in undergraduate, graduate courses, and postgraduate and post-doctoral researchers will also find it informative. In order for the book to be useful to engineers, coverage of each topic is comprehensive enough to serve as an overview of the most recent and relevant research and technology. Numerous references are included at the end of each chapter.

The editors wish to acknowledge all individuals who have contributed to this book.

—*Murlidhar Meghwal, PhD*
December 2015

PREFACE 2
By Megh R. Goyal

The discovery of a new dish (processed food) does more for the hap-
piness of the human race than the discovery of a star.
—Anthelme Brillat Savarin

https://en.wikipedia.org/wiki/Food_processing indicates: *"Food pro-*
cessing is the transformation of raw ingredients, by physical or chemical
means into food, or of food into other forms. Food processing combines
raw food ingredients to produce marketable food products that can be
easily prepared and served by the consumer. Food processing typically
involves activities such as mincing and macerating, liquefaction, emulsi-
fication, and cooking (such as boiling, broiling, frying, or grilling); pick-
ling, pasteurization, and many other kinds of preservation; and canning or
other packaging. Primary processing such as dicing or slicing, freezing or
drying when leading to secondary products are also included."

When designing processes for the food industry, the following performance parameters may be taken into account: Hygiene, energy efficiency,
minimization of waste, labor used, minimization of cleaning stops measured, and reduction of fat content in final product.

Prehistoric people knew such food-processing technology as sun-dry-
ing, preserving with salt, and various types of cooking (such as roasting,
smoking, steaming, and oven baking) for using processed foods in his
daily life. Evidence for the existence of these methods can be found in the
writings of the ancient Greek, Chaldean, Egyptian and Roman civiliza-
tions as well as archaeological evidence from Europe, North and South
America, and Asia. These tried and tested processing techniques remained
essentially the same until the advent of the industrial revolution. Examples of ready-meals also date back to before the preindustrial revolution.
Modern food processing technology was developed in large part to serve
military needs. Although initially expensive and somewhat hazardous
due to the lead used in cans, canned goods would later become a staple
around the world. Pasteurization improved the quality of preserved foods

and also introduced wine, beer, and milk preservation. In the 20th century, World War II, the space race, and the rising consumer society contributed to the growth of food processing with such advances as spray drying, juice concentrates, freeze drying, and the introduction of artificial sweeteners, coloring agents, and such preservatives as sodium benzoate. In the late 20th century, products such as dried instant soups, reconstituted fruits and juices, and self-cooking meals such as MRE food rations were developed. Processors utilized the perceived value of time to appeal to the postwar population, and this same appeal contributes to the success of convenience foods today.

Benefits of food processing include toxin removal; preservation; easing of marketing and distribution tasks; food consistency; yearly availability of many foods; enabling transportation of delicate perishable foods across long distances from the source to the consumer; reducing the incidence of foodborne disease; allowing more free time and improving the quality of life for people with allergies, diabetics, and other people who cannot consume some common food elements. Food processing can also add extra nutrients such as vitamins.

Any processing of food can affect its nutritional density. The amount of nutrients lost depends on the food and processing method. For example the heat destroys vitamin C. Therefore, canned fruits possess less vitamin C than their fresh alternatives. The USDA study in 2004 indicates that in the majority of foods, processing reduces nutrients by a minimal amount. On average this process reduces any given nutrient by as little as 5–20%. Abundant food processing (not fermentation of foods) endangers that environment. Using food additives (e.g., sweeteners, preservatives, and stabilizers) may represent another safety concern. Certain additives can also result in an addiction to a particular food item. The mixing, grinding, chopping and emulsifying equipment in the production process may introduce a number of contamination risks.

Who does not know how to cook food, boil milk or prepare omelet? I learned food-processing skills when I was in the seventh grade. I knew how to dry red chillies in the open sun, prepare mango pickles, cook rice, and prepare evaporated milk with sugar. When we got married in February of 1970, my wife knew almost zero processing and cooking Indian foods. I taught her the culinary skills. Now she does not let me enter into the kitchen. At the first Mango Festival at the Agricultural Experiment Station – University of Puerto Rico in Jana Diaz, I prepared my own recipes (mango shake/ice cream/yogurt/chutney/pickles/cookies/jam or jelly, etc.) for sale at the festival. Many of my recipes were promoted in the local newspapers. My three grandchildren enjoy *Indian Potato-Filled Cooked Bread (Parantha)*. What I want to emphasize that each one you has processed food for personal use at least once in your life.

At the 49th annual meeting of the Indian Society of Agricultural Engineers at Punjab Agricultural University (PAU) during February 22–25 of 2015, a group of ABEs and FEs convinced me that there is a dire need to publish book volumes on focus areas of agricultural and biological engineering (ABE). This is how the idea was born for new book series titled "Innovations in Agricultural and Biological Engineering." This book, *Food Process Engineering: Emerging Trends in Research and Their Applications*, is the fifth volume under this book series, and it contributes to the ocean of knowledge on food engineering.

The contributions by all cooperating authors to this book volume have been most valuable in the compilation. Their names are mentioned in each chapter and in the list of contributors. I appreciate the authors for having patience with my editorial skills. This book would not have been written without the valuable cooperation of these investigators, many of whom are renowned scientists who have worked in the field of food engineering throughout their professional careers.

I am glad to introduce Dr. Murlidhar Meghwal, who is an Assistant Professor in the Food Technology, Center for Emerging Technologies at Jain University – Jain Global Campus in District Karnataka, India. With several awards and recognitions, including from the President of India, Dr. Meghwal brings his expertise and innovative ideas to this book series. Without his support, and leadership qualities as editor of the book volume and his extraordinary work on food engineering applications, readers will not have this quality publication.

I will like to thank editorial staff, Sandy Jones Sickels, Vice President, and Ashish Kumar, Publisher and President at Apple Academic Press, Inc., for making every effort to publish the book when the diminishing water and food resources are a major issue worldwide. Special thanks are due to the AAP Production Staff for the quality production of this book.

I request that the reader offer your constructive suggestions that may help to improve the next edition.

I express my deep admiration to my family and colleagues for their understanding and collaboration during the preparation of this book volume. Can anyone live without food or water? Who has escaped from processed food today? As an educator, there is a piece of advice to one and all in the world: *"Permit that our almighty God, our Creator, provider of all and excellent teacher, feed our life with Healthy Food Products and His Grace—and Get married to your profession."*

—Megh R. Goyal, PhD, PE
Senior Editor-in-Chief
December 31, 2015

WARNING/DISCLAIMER

READ CAREFULLY

The goal of this book volume on *Food Process Engineering: Emerging Trends in Research and Their Applications* is to guide the world community on how to manage efficiently the technology available for different processes in food engineering. The reader must be aware that the dedication, commitment, honesty, and sincerity are most important factors in a dynamic manner for complete success. It is not a one-time reading of this compendium. Read and follow every time, it is needed.

The editors, the contributing authors, the publisher and the printer have made every effort to make this book as complete and as accurate as possible. However, there still may be grammatical errors or mistakes in the content or typography. Therefore, the contents in this book should be considered as a general guide and not a complete solution to address any specific situation in food engineering. For example, one type of food process technology does not fit all case studies in dairy engineering/science/technology.

The editors, the contributing authors, the publisher and the printer shall have neither liability nor responsibility to any person, any organization or entity with respect to any loss or damage caused, or alleged to have caused, directly or indirectly, by information or advice contained in this book. Therefore, the purchaser/reader must assume full responsibility for the use of the book or the information therein.

The mention of commercial brands and trade names is only for technical purposes. This does not mean that a particular product is endorsed over to another product or equipment not mentioned.

All weblinks that are mentioned in this book were active on June 30, 2015. The editors, the contributing authors, the publisher and the printing company shall have neither liability nor responsibility, if any of the weblinks is inactive at the time of reading of this book.

ABOUT THE LEAD EDITOR

 Murlidhar Meghwal, PhD, is a distinguished researcher, engineer, teacher and professor at the Food Technology, Centre for Emerging Technology, Jain Global Campus, Jain University, Bangalore, India. He received his BTech degree (Agricultural Engineering) in 2008 from the College of Agricultural Engineering Bapatla, Acharya N. G. Ranga Agricultural University, Hyderabad, India; his MTech degree (Dairy and Food Engineering) in 2010 and PhD degree (Food Process Engineering) in 2014 from the Indian Institute of Technology Kharagpur, West Bengal, India.

He worked for one year as research associate at INDUS Kolkata for the development of a quicker and industrial-level parboiling system for paddy and rice milling. In his PhD research, he worked on ambient and cryogenic grinding of fenugreek and black pepper by using different grinders to select a suitable grinder.

Currently, Dr. Meghwal is working on developing inexpensive, disposable and biodegradable food containers using agricultural wastes; quality improvement, quality attribute optimization and storage study of kokum (*Garcinia indica Choisy*); and freeze drying of milk. At present, he is actively involved in research and is the course coordinator for MTech (Food Technology) courses. He is also teaching at the Food Science and Technology Division, Jain University Bangalore, India. He has written two books and many research publications on food process engineering. He has attended many national and international seminars and conferences. He is reviewer and member of editorial boards of reputed journals.

He is recipient of Bharat Scout Award from the President of India as well as the Bharat Scouts Award from the Governor. He received a meritorious Foundation for Academic Excellence and Access (FAEA-New Delhi) Scholarship for his full undergraduate studies from 2004 to 2008. He also received a Senior Research Fellowship awarded by the Ministry of

Human Resources Development (MHRD), Government of India, during 2011–2014; and a Scholarship of Ministry of Human Resources Development (MHRD), Government of India research during 2008–2010.

He is a good sports person, mentor, social activist, critical reviewer, thinker, fluent writer and well-wishing friend to all.

Readers may contact him at: murli.murthi@gmail.com.

ABOUT THE SENIOR EDITOR-IN-CHIEF

 Megh R. Goyal, PhD, PE, is a Retired Professor in Agricultural and Biomedical Engineering from the General Engineering Department in the College of Engineering at University of Puerto Rico – Mayaguez Campus; and Senior Acquisitions Editor and Senior Technical Editor-in-Chief in Agricultural and Biomedical Engineering for Apple Academic Press Inc.

He received his BSc degree in Engineering in 1971 from Punjab Agricultural University, Ludhiana, India; his MSc degree in 1977 and PhD degree in 1979 from the Ohio State University, Columbus; his Master of Divinity degree in 2001 from Puerto Rico Evangelical Seminary, Hato Rey, Puerto Rico, USA.

Since 1971, he has worked as Soil Conservation Inspector (1971); Research Assistant at Haryana Agricultural University (1972–1975) and the Ohio State University (1975–1979); Research Agricultural Engineer/Professor at Department of Agricultural Engineering of UPRM (1979–1997); and Professor in Agricultural and Biomedical Engineering at General Engineering Department of UPRM (1997–2012). He spent one-year sabbatical leave in 2002–2003 at Biomedical Engineering Department, Florida International University, Miami, USA.

He was first agricultural engineer to receive the professional license in Agricultural Engineering in 1986 from College of Engineers and Surveyors of Puerto Rico. On September 16, 2005, he was proclaimed as "Father of Irrigation Engineering in Puerto Rico for the twentieth century" by the ASABE, Puerto Rico Section, for his pioneer work on micro irrigation, evapotranspiration, agroclimatology, and soil and water engineering. During his professional career of 45 years, he has received awards such as: Scientist of the Year, Blue Ribbon Extension Award, Research Paper Award, Nolan Mitchell Young Extension Worker Award, Agricultural Engineer of the Year, Citations by Mayors of Juana Diaz and Ponce, Membership Grand

Prize for ASAE Campaign, Felix Castro Rodriguez Academic Excellence, RashtryaRatan Award and Bharat Excellence Award and Gold Medal, Domingo Marrero Navarro Prize, Adopted son of Moca, Irrigation Protagonist of UPRM, Man of Drip Irrigation by Mayor of Municipalities of Mayaguez/Caguas/Ponce and Senate/Secretary of Agriculture of ELA, Puerto Rico.

He has authored more than 200 journal articles and textbooks on: "Elements of Agroclimatology (Spanish) by UNISARC, Colombia"; two bibliographies on "Drip Irrigation." Apple Academic Press Inc. (AAP) has published his books, namely: "Management of Drip/Trickle or Micro Irrigation," and "Evapotranspiration: Principles and Applications for Water Management."

During 2014–2015, AAP has published his ten-volume set in *Research Advances in Sustainable Micro Irrigation.* During 2016–2017, AAP will be publishing book volumes on emerging technologies/issues/challenges under book series, *"Innovations and Challenges in Micro Irrigation,"* and *"Innovations in Agricultural and Biological Engineering."* Readers may contact him at: goyalmegh@gmail.com.

BOOK ENDORSEMENTS

Here is a well-written book on recent advances in food process engineering that will be useful for food process engineering professionals, industrialists, undergraduate and postgraduate students.

—Deepak Kumar Garg, PhD
Postdoctoral Research Associate
ADM Institute for the Prevention of Postharvest Loss
University of Illinois at Urbana–Champaign, USA

My heartiest congratulations to Professors Murlidhar Meghwal and Megh R. Goyal for carrying out this book, which covers most of the topics relevant to food process engineering. This will be a great help for all the academicians and non-academicians interested in food systems.

—Abhinav Mishra, PhD
Research Scholar, Department of Nutrition and Food Science and
Center for Food Safety and Security Systems
University of Maryland, College Park, Maryland, USA

Food Process Engineering: Emerging Trends in Research and Their Applications is a well-arranged and well-written book on recent advances in food process engineering, food technology and food technology related topics. This book will be very useful for the fraternity of educators.

—Tridib Kumar Goswami, PhD
Professor, Agricultural and Food engineering Department
Indian Institute of Technology, Kharagpur, West Bengal, India

This book provides a comprehensive coverage of the various aspects of food engineering. Topics including emerging trends and technologies in food processing, ultrasonic treatment of foods, foods for specific needs, food preservation, food hazards and their controls and health-related aspects will be very useful to students and professionals in food process engineering. Increasing awareness on food processing and preservation and the growing processed food markets make this book an excellent source for reference in these areas.

—Narendra Reddy, PhD
Professor and Ramalingaswami Fellow
Centre for Emerging Technologies, Jain University
Jain Global Campus, Jakkasandra Post, Bangalore

The topics in the book have been wisely selected. They cover not only the entire aspect of food engineering and processing, but also the health benefits. This book would be useful to the students, researchers and professionals in the field of food process engineering.

—Soumitra Banerjee, PhD
Professor, Food Technology
Centre for Emerging Technologies, Jain University
Jakkasandra, Ramanagara, Karnataka

The book (*Food Engineering Emerging Issues, Modeling, and Applications*, Editors: Murlidhar Meghwal, PhD, Megh R. Goyal, PhD) is nicely written and surely will be very useful for researchers, industry people and students.

—Kacoli Banerjee, PhD
Assistant Professor, Department of Zoology
Maharaja Sayajirao University of Baroda, Baroda, India

This book is a nice piece of work on food process engineering for professionals, industrialists, and others.

—Winny Routray, PhD
Assistant Professor, Marine Bioprocessing Facility
Centre of Aquaculture and Seafood Development
Marine Institute of Memorial University of New Foundland
St. John's, Newfoundland, Canada

OTHER BOOKS ON AGRICULTURAL AND BIOLOGICAL ENGINEERING BY APPLE ACADEMIC PRESS, INC.

Management of Drip/Trickle or Micro Irrigation
Megh R. Goyal, PhD, PE, Senior Editor-in-Chief

Evapotranspiration: Principles and Applications for Water Management
Megh R. Goyal, PhD, PE, and Eric W. Harmsen, Editors

Book Series: Research Advances in Sustainable Micro Irrigation
Senior Editor-in-Chief: Megh R. Goyal, PhD, PE

Volume 1: Sustainable Micro Irrigation: Principles and Practices
Volume 2: Sustainable Practices in Surface and Subsurface Micro Irrigation
Volume 3: Sustainable Micro Irrigation Management for Trees and Vines
Volume 4: Management, Performance, and Applications of Micro Irrigation Systems
Volume 5: Applications of Furrow and Micro Irrigation in Arid and Semi-Arid Regions
Volume 6: Best Management Practices for Drip Irrigated Crops
Volume 7: Closed Circuit Micro Irrigation Design: Theory and Applications
Volume 8: Wastewater Management for Irrigation: Principles and Practices
Volume 9: Water and Fertigation Management in Micro Irrigation
Volume 10: Innovation in Micro Irrigation Technology

Book Series: Innovations and Challenges in Micro Irrigation
Senior Editor-in-Chief: Megh R. Goyal, PhD, PE
  Volume 1: Principles and Management of Clogging in Micro Irrigation
  Volume 2: Sustainable Micro Irrigation Design Systems for Agricultural Crops: Methods and Practices
  Volume 3: Performance Evaluation of Micro Irrigation Management: Principles and Practices
  Volume 4: Potential of Solar Energy and Emerging Technologies in Sustainable Micro Irrigation
  Volume 5: Micro Irrigation Management: Technological Advances and Their Applications
  Volume 6: Micro Irrigation Engineering for Horticultural Crops: Policy Options, Scheduling, and Design
  Volume 7: Micro Irrigation Scheduling and Practices

Book Series: Innovations in Agricultural and Biological Engineering
Senior Editor-in-Chief: Megh R. Goyal, PhD, PE
  • Modeling Methods and Practices in Soil and Water Engineering
  • Food Engineering: Modeling, Emerging issues and Applications.
  • Emerging Technologies in Agricultural Engineering
  • Dairy Engineering: Advanced Technologies and their Applications
  • Food Process Engineering: Emerging Trends in Research and Their Applications
  • Soil and Water Engineering: Principles and Applications of Modeling
  • Soil Salinity Management in Agriculture: Technological Advances and Applications
  • Developing Technologies in Food Science: Status, Applications, and Challenges
  • Engineering Practices for Agricultural Production and Water Conservation: An Interdisciplinary Approach
  • Flood Assessment: Modeling and Parameterization
  • Food Technology: Applied Research and Production Techniques
  • Processing Technologies for Milk and Milk Products: Methods, Applications, and Energy Usage
  • Engineering Interventions in Agricultural Processing
  • Technological Interventions in the Processing of Fruits and Vegetables
  • Technological Interventions in Management of Irrigated Agriculture

EDITORIAL

Apple Academic Press Inc., (AAP) will be publishing various book volumes on the focus areas under book series titled *Innovations in Agricultural and Biological Engineering*. Over a span of 8 to 10 years, Apple Academic Press Inc., will publish subsequent volumes in the specialty areas defined by American Society of Agricultural and Biological Engineers (http://asabe.org).

The mission of this series is to provide knowledge and techniques for agricultural and biological engineers (ABEs). The series aims to offer high-quality reference and academic content in Agricultural and Biological Engineering (ABE) that is accessible to academicians, researchers, scientists, university faculty, and university-level students and professionals around the world. The following material has been edited/ modified and reproduced below [From: *"Goyal, Megh R., 2006. Agricultural and biomedical engineering: Scope and opportunities. Paper Edu_47 Presentation at the Fourth LACCEI International Latin American and Caribbean Conference for Engineering and Technology (LACCEI' 2006): Breaking Frontiers and Barriers in Engineering: Education and Research by LACCEI University of Puerto Rico – Mayaguez Campus, Mayaguez, Puerto Rico, June 21–23"*]:

WHAT IS AGRICULTURAL AND BIOLOGICAL ENGINEERING (ABE)?

"Agricultural Engineering (AE) involves application of engineering to production, processing, preservation and handling of food, fiber, and shelter. It also includes transfer of technology for the development and welfare of rural communities," according to http://isae.in. *"ABE is the discipline of engineering that applies engineering principles and the fundamental concepts of biology to agricultural and biological systems and tools, for the safe, efficient and environmentally sensitive production, processing, and management of agricultural, biological, food, and natural resources*

systems," according to http://asabe.org. *"AE is the branch of engineering involved with the design of farm machinery, with soil management, land development, and mechanization and automation of livestock farming, and with the efficient planting, harvesting, storage, and processing of farm commodities,"* the definition by http://dictionary.reference.com/browse/agricultural+engineering.

"AE incorporates many science disciplines and technology practices to the efficient production and processing of food, feed, fiber and fuels. It involves disciplines like mechanical engineering (agricultural machinery and automated machine systems), soil science (crop nutrient and fertilization, etc.), environmental sciences (drainage and irrigation), plant biology (seeding and plant growth management), animal science (farm animals and housing) etc.," as indicated by http://www.ABE.ncsu.edu/academic/agricultural-engineering.php.

According to https://en.wikipedia.org/wiki/Biological_engineering: *"BE (Biological engineering) is a science-based discipline that applies concepts and methods of biology to solve real-world problems related to the life sciences or the application thereof. In this context, while traditional engineering applies physical and mathematical sciences to analyze, design and manufacture inanimate tools, structures and processes, biological engineering uses biology to study and advance applications of living systems."*

SPECIALTY AREAS OF ABE

Agricultural and Biological Engineers (ABEs) ensure that the world has the necessities of life including safe and plentiful food, clean air and water, renewable fuel and energy, safe working conditions, and a healthy environment by employing knowledge and expertise of sciences, both pure and applied, and engineering principles. Biological engineering applies engineering practices to problems and opportunities presented by living things and the natural environment in agriculture. BA engineers understand the interrelationships between technology and living systems, have available a wide variety of employment options. The http://asabe.org indicates that *"ABE embraces a variety of following specialty areas."* As new technology and information emerge, specialty areas are created, and many overlap with one or more other areas.

1. **Aqua Cultural Engineering**: ABEs help design farm systems for raising fish and shellfish, as well as ornamental and bait fish. They specialize in water quality, biotechnology, machinery, natural resources, feeding and ventilation systems, and sanitation. They seek ways to reduce pollution from aqua cultural discharges, to reduce excess water use, and to improve farm systems. They also work with aquatic animal harvesting, sorting, and processing.

2. **Biological Engineering** applies engineering practices to problems and opportunities presented by living things and the natural environment.

3. **Energy:** ABEs identify and develop viable energy sources—biomass, methane, and vegetable oil, to name a few—and to make these and other systems cleaner and more efficient. These specialists also develop energy conservation strategies to reduce costs and protect the environment, and they design traditional and alternative energy systems to meet the needs of agricultural operations.

4. **Farm Machinery and Power Engineering**: ABEs in this specialty focus on designing advanced equipment, making it more efficient and less demanding of our natural resources. They develop equipment for food processing, highly precise crop spraying, agricultural commodity and waste transport, and turf and landscape maintenance, as well as equipment for such specialized tasks as removing seaweed from beaches. This is in addition to the tractors, tillage equipment, irrigation equipment, and harvest equipment that have done so much to reduce the drudgery of farming.

5. **Food and Process Engineering:** Food and process engineers combine design expertise with manufacturing methods to develop economical and responsible processing solutions for industry. Also food and process engineers look for ways to reduce waste by devising alternatives for treatment, disposal and utilization.

6. **Forest Engineering**: ABEs apply engineering to solve natural resource and environment problems in forest production systems and related manufacturing industries. Engineering skills and expertise are needed to address problems related to equipment design and manufacturing, forest access systems design and construction; machine-soil interaction and erosion control; forest operations

analysis and improvement; decision modeling; and wood product design and manufacturing.

7. **Information and Electrical Technologies Engineering** is one of the most versatile areas of the ABE specialty areas, because it is applied to virtually all the others, from machinery design to soil testing to food quality and safety control. Geographic information systems, global positioning systems, machine instrumentation and controls, electromagnetics, bioinformatics, biorobotics, machine vision, sensors, spectroscopy: These are some of the exciting information and electrical technologies being used today and being developed for the future.

8. **Natural Resources:** ABEs with environmental expertise work to better understand the complex mechanics of these resources, so that they can be used efficiently and without degradation. ABEs determine crop water requirements and design irrigation systems. They are experts in agricultural hydrology principles, such as controlling drainage, and they implement ways to control soil erosion and study the environmental effects of sediment on stream quality. Natural resources engineers design, build, operate and maintain water control structures for reservoirs, floodways and channels. They also work on water treatment systems, wetlands protection, and other water issues.

9. **Nursery and Greenhouse Engineering**: In many ways, nursery and greenhouse operations are microcosms of large-scale production agriculture, with many similar needs—irrigation, mechanization, disease and pest control, and nutrient application. However, other engineering needs also present themselves in nursery and greenhouse operations: equipment for transplantation; control systems for temperature, humidity, and ventilation; and plant biology issues, such as hydroponics, tissue culture, and seedling propagation methods. And sometimes the challenges are extraterrestrial: ABEs at NASA are designing greenhouse systems to support a manned expedition to Mars!

10. **Safety and Health:** ABEs analyze health and injury data, the use and possible misuse of machines, and equipment compliance with standards and regulation. They constantly look for ways in which the safety of equipment, materials and agricultural practices can be improved and for ways in which safety and health issues can be communicated to the public.

11. **Structures and Environment:** ABEs with expertise in structures and environment design animal housing, storage structures, and greenhouses, with ventilation systems, temperature and humidity controls, and structural strength appropriate for their climate and purpose. They also devise better practices and systems for storing, recovering, reusing, and transporting waste products.

CAREER IN AGRICULTURAL AND BIOLOGICAL ENGINEERING

One will find that university ABE programs have many names, such as biological systems engineering, bioresources engineering, environmental engineering, forest engineering, or food and process engineering. Whatever the title, the typical curriculum begins with courses in writing, social sciences, and economics, along with mathematics (calculus and statistics), chemistry, physics, and biology. Student gains a fundamental knowledge of the life sciences and how biological systems interact with their environment. One also takes engineering courses, such as thermodynamics, mechanics, instrumentation and controls, electronics and electrical circuits, and engineering design. Then student adds courses related to particular interests, perhaps including mechanization, soil and water resource management, food and process engineering, industrial microbiology, biological engineering or pest management. As seniors, engineering students work in a team to design, build, and test new processes or products.

For more information on this series, readers may contact:

Ashish Kumar, Publisher
and President
Sandy Sickels, Vice President
Apple Academic Press, Inc.
Fax: 866-222-9549
E-mail: ashish@appleacademicpress.com
http://www.appleacademicpress.com/publishwithus.php

Megh R. Goyal, PhD, PE
Senior Editor-in-Chief
Innovations in Agricultural and Biological Engineering
E-mail: goyalmegh@gmail.com

PART I

EMERGING TRENDS AND TECHNOLOGIES IN FOOD PROCESSING

CHAPTER 1

TRENDS IN FOOD PACKAGING TECHNOLOGY

CHANDANI SEN[1] and MADHUSWETA DAS[2]

[1]*PhD Research Scholar, Agricultural and Food Engineering Department, Indian Institute of Technology, Kharagpur – 721302, West Bengal, India, E-mail: chandani.sen@agfe.iitkgp.ernet.in*

[2]*Associate Professor, Agricultural and Food Engineering Department, Indian Institute of Technology, Kharagpur – 721302, West Bengal, India. Tel.: +91-32222-83108; E-mail: madhu@agfe.iitkgp.ernet.in*

CONTENTS

1.1   INTRODUCTION

Packaging is one of the most important processes to maintain the quality of food products for storage, transportation and consumption. It prevents quality deterioration and facilitates distribution and marketing. The basic

TABLE 1.1 Trends in the Evaluation of Food Packaging

Period	Functions and issues		References
1960	Convenience		[12, 28, 37]
1970	Light weight, source reduction, energy saving	Efficiency of packaging material	
1980	Tamper evidence		
1990 onwards	In addition to efficiency of packaging material its environmental impact also came into picture.		

functions of packaging are protection, containment, information and convenience. Apart from preservation, packaging also has secondary functions- such as selling and sales promotion, which contributes significantly to a business profit [12].

Food industry uses a lot of packaging materials, and thus even a small reduction in the amount of material used for each package would result in a significant cost reduction, and may improve solid waste problems. Packaging technology has attempted to reduce the volume and/or weight of materials in efforts to minimize resources and costs. Trends of food packaging can be summarized as presented in Table 1.1.

The food-packaging sector has a huge marketing potential. Food and beverage packaging comprises about 65–70% of the global food packaging sales [5] which has been projected to rise by 3% in real terms to $797 billion (approx. Rs. 500,211 million) in 2013 [2] and are expected to grow at an annual rate of 4% to 2018 [2] fetching about $284 billion (approx. = 170,892 million Rs.), of which drinks packaging shares around 50% [5].

This chapter summarizes the brief history of food packaging, basic mechanisms of mass transfer followed by the innovative packaging trends like, modified atmosphere packaging (MAP), active packaging (AP) intelligent packaging (IP) and biodegradable packaging (BDP).

1.2 BRIEF HISTORY OF FOOD PACKAGING

It took over 150 years for food packaging to undergo several steps and finally evolve into the current form. A brief review of the most popular packaging developments is described in this chapter.

1.2.1 PAPER

Paper (derived from Greek word *papyrus*), invented in ancient China during 206 BC to 220 AD, is the oldest form of "flexible packaging" [3]. During the next fifteen hundred years, the papermaking technique was refined and transported to the Middle East, Europe and USA. Paperboard was first used to manufacture folding cartons in the early 1800s [27]. Corrugated boxes that today are widely used as a shipping container to hold a number of smaller packages were developed in the 1850s [27].

1.2.2 PLASTICS

Plastics including cellulose nitrate, styrene, and vinyl chloride were discovered in the 1800s. Polyethylene was one of the first plastics used widely for food packaging. There are several types of polyethylene in use today including low-density (LDPE), high-density (HDPE), linear low-density (LLDPE), and very low density (VLDPE). LDPE was the first to be developed by Imperial Chemical Industries in 1933 by compressing ethylene gas and heating it to a high temperature [19].

Isotactic polypropylene was discovered by Professor Giulio Natta in 1954 [13]. The film is often oriented after the casting or forming process by first stretching the material in the machine direction and then stretching it in the crosswise direction to give oriented polypropylene (OPP). This stretching aligns the molecules, making a film with a better moisture vapor barrier, better clarity, and more stiffness.

One process that is used to improve barriers even further is metallization. In this process, an aluminum wire is heated to 1700°C in a large vacuum chamber [13]. This vaporizes the aluminum, which deposits on the surface of the film as it is run through the chamber. In the case of a 50 gauge polyester film, metallizing improves the *moisture vapor transmission rate* (MVTR) from 2.0 g/(100 in.^{2}; 24 h. 90% RH) to 0.05, i.e., a 40-fold improvement [13]. Oriented polypropylene and polyethylene terephthalate (PET) are the most common films used for metallization followed by nylon, polyethylene and cast polypropylene [24].

One process that has improved overall properties of plastic films is co extrusion, developed in 1964 by Hercules [27]. In this process, a film with two or more layers of different types of plastic can be made in one step, without any adhesive and eliminating the use of solvents. Multiple-layer films offer better protection for products as some films are better moisture barriers and others offer better barriers to gases. One example is polyester film, which provides a better gas barrier, whereas polypropylene and *ethylene vinyl alcohol* (EvOH) films are better moisture barriers. These three can be combined readily in one structure to give protection from both moisture and oxygen permeation [27].

In addition to broad developments in materials, there have been a number of specific packages that have both created new food categories and changed the way to deliver a product to the consumer. Polyethylene naphthalene (PEN) (approved for food contact by FDA), polyethylene terephthalate (PET) and aluminum cans, which currently have a huge market potential for carbonated beverages-are among some of the most popular categories.

1.3   MASS TRANSFER THROUGH PACKAGING MATERIALS

In all flexible packaging, permeability plays a significant role. For food powder packaging, water vapor permeability of the film should be less to maintain free flow character, essential for long shelf life. For aromatic components like tea leaves or spices, the packaging material should have proper aroma barrier property for better aroma retention. In case of MAP of fresh fruits and vegetables, where the principle is to extend the shelf life by controlling reaction rate, the oxygen/carbon di oxide gas permeability of the film is the decisive factor. So, permeability is one of the major properties of packaging films on which its application is dependent. To know what permeability is, basic knowledge of mass transfer is necessary.

Under steady state condition, gas will diffuse through film at a constant rate if a constant pressure difference is maintained across the barrier (Figure 1.1). The diffusive flux, J, of a permeant can be defined as the amount passing through a plane (surface) of unit area normal to the direction of flow in unit time:

$$J = Q/At \qquad (1)$$

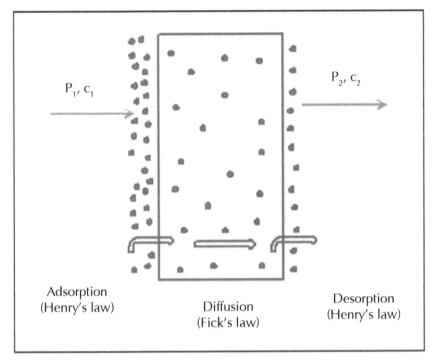

FIGURE 1.1 Basic mechanism of gas and vapor permeation through a packaging film.

where, Q is the total amount of permeant, which has passed through area A in time t.

The rate of permeation and the concentration gradient is directly proportional to each other and embodied in Fick's first law:

$$J = -D\frac{\partial c}{\partial X} \qquad (2)$$

where, J is the flux per unit area of permeant through the polymer film, kg mol/m²; D is the diffusion coefficient, m²/h; c is the concentration of the permeant, kg mol/m³; $\delta c/\delta x$ is the concentration gradient of the permeant across a thickness δX, m.

In steady state condition, when J = constant, integrating Eq. (2), we get:

$$JX = D (c_1 - c_2) \qquad (3)$$

Substituting for J using Eq. (1), the quantity of permeant diffusing through a film of area A in time t can be calculated:

$$Q = \frac{D.(c_1 - c_2).A.t}{X} \tag{4}$$

When the permeant is a gas, it is more convenient to measure the vapor pressure p rather than the actual concentration. Henry's law applies at low concentrations and c can be expressed as:

$$C = Sp \tag{5}$$

where, S is the solubility coefficient of the permeant in the polymer (it reflects the amount of permeant in the polymer).

Combining Eqs. (4) and (5),

$$Q = \frac{D.S.(p_1 - p_2).A.t}{X} \tag{6}$$

The product of D and S is referred to as the permeability, and is represented by the symbol P. Thus,

$$P = \frac{Q.X}{A.t.(p_1 - p_2)} \tag{7}$$

Hence, permeability (P) is the proportionality constant between the flow of the penetrant gas per unit film area per unit time and the driving force (partial pressure difference) per unit film thickness. The amount of gas penetrating through the film is expressed in terms of either moles per unit time (flux) or weight or volume of the gas at STP. Commonly, it is expressed in terms of volume.

1.3.1 TEMPERATURE QUOTIENT FOR PERMEABILITY

The influence of temperature on permeability of polymeric films was quantified with the value, which is the permeability increase for a 10°C rise in temperature and is given by the following equation:

$$Q_{10}^P = \left(\frac{P2}{P1}\right)^{10/(T2-T1)} \tag{8}$$

where, Q_{10}^P, the temperature quotient for permeability, and P_1 and P_2 are the permeabilities at temperatures T_1 and T_2, respectively (Table 1.2).

TABLE 1.2 Gas and Water Vapor Permeability of Different Commercial Packaging Materials [20, 28]

Material	$P \times 10^{11}$ [ml(STP)cm cm^{-2} s^{-1} (cm Hg)$^{-1}$]			
Polymer	O_2	CO_2	N_2	H_2O at 90% RH
Low density polyethylene	30–69	130–280	1.9–3.1	800
High density polyethylene	6–11	45	3.3	180
Polypropylene	9–15	92	4.4	680
Polyethylene Terephthalate	0.3–0.75	1.6–3.0	0.04–0.06	1300
Polystyrene	15–27	105	7.8	12,000–18,000

1.4 INNOVATIVE FOOD PACKAGING

Traditional food packages are passive barriers designed to delay the adverse effects of the environment on the food product. Advance technologies, like modified atmosphere packaging, active packaging, intelligent packaging, and biodegradable packaging, however, are needed to allow packages to take care of the food and environment as well [6, 18].

1.4.1 MODIFIED ATMOSPHERE PACKAGING (MAP)

Modified Atmosphere Packaging (MAP), usually used for fresh produces, is a package in which the atmosphere inside the package is modified or altered to provide an optimum atmosphere for increasing shelf life. Modification of the atmosphere may be achieved either actively or passively. Active modification involves displacing the air with a controlled, desired mixture of gases, and is generally referred to as gas flushing. Passive modification occurs as a consequence of the respiration/metabolism of the enclosed commodity, which changes the gaseous concentrations inside the package.

The normal composition of air by volume is 78.08% nitrogen, 20.95% oxygen, 0.93% argon, 0.03% carbon dioxide, and traces of other nine gases. The three main gases used in active MAP are O_2, CO_2 and N_2, either singly or in combination. The choices of suitable packaging materials for the MAP of respiring produce such as fruits and vegetables are complex due to the dynamic nature of the product. The main characteristics to be considered when selecting packaging materials for MAP are the package permeability to gases and the respiration characteristics of the commodity. Different types of MAP are shown in Figure 1.2.

As discussed above, the optimum gas/temperature combination is different for different commodities. Table 1.3 summarizes the temperature/gas combination for various MAP conditions.

1.4.2 ACTIVE PACKAGING (AP)

The popularity of Active Packaging (AP) has signified a major paradigm shift in packaging during the past 2 decades. The protection function of

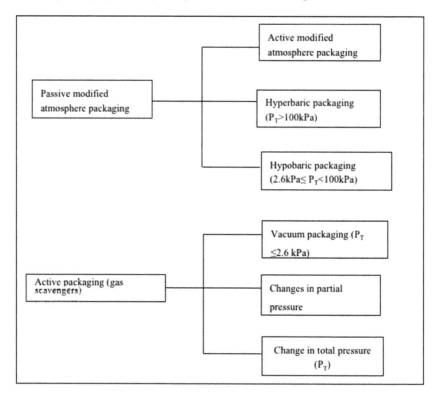

FIGURE 1.2   Types of Modified Atmosphere Packaging [30, 42].

TABLE 1.3   Recommended MAP Conditions for Various Products [10, 14, 26]

Commodity	Temperature range (°C)	O_2 (%)	CO_2 (%)
Baked products			
Bread (sliced)	Ambient	0	100
Pizza (crust)	Ambient	0	90
Cheeses			
Hard cheeses	1–4	0	100
Soft cheeses	1–4	0	20–40
Fish			
Lean	0–2	30	40
Oily and smoked	0–2	0	60
Fruits			
Apple (whole)	0–5	2–3	1–5
Apple (sliced)	0–5	10–12	8–11
Avocado	5–13	2–5	3–10
Banana	12–15	2–5	2–5
Kiwi fruit	0–5	2	5
Mango	10–15	5	5
Pineapple	10–15	5	10
Strawberry	0–5	10	15–20
Meats			
Beef	−1 to 2	60–80	20–40
Pork	−1 to 2	30	30
Poultry	−1 to 2	0	25–35
Processed foods			
Dried foods	Ambient	0	0–100
Low moisture foods	Ambient	0	0–100
Fats and oils	Ambient	0	0
Vegetables			
Asparagus	0–5	20	5–10
Broccoli	0–5	1–2	5–10
Cabbage	0–5	3–5	5–7
Lettuce (head)	0–5	2–5	0
Lettuce (shredded)	0–5	1–2	10–12
Mushrooms	0–5	21	10–15
Spinach	0–5	21	10–20
Tomatoes (mature)	12–20	3–5	0

packaging has been shifted from passive to active. Previously, primary packaging materials were considered as "passive," meaning that they functioned only as an inert barrier to protect the product against oxygen and moisture. Recently, a host of new packaging materials have been developed to provide "active" protection for the product. AP has been defined as a system in which the product, the package, and the environment interact in a positive way to extend shelf life or to achieve some characteristics that cannot be obtained otherwise [22]. It has also been defined as a packaging system that actively changes the condition of the package to extend shelf life or improve food safety or sensory properties, while maintaining the quality of the food [38].

Food packaging materials have traditionally been chosen to avoid unwanted interactions with the food. During the past two decades a wide variety of packaging materials have been devised or developed to interact with the food. These packaging materials, which are designed to perform some desired role other than to provide an inert barrier to outside influences, are termed 'active packaging.' The benefits of active packaging are based on both chemical and physical effects [29].

Active packaging elements can be divided into three categories: Absorber, releasing system and other system. For any fresh fruits, absorbing system is used as active packaging components to remove undesired gases and substances (oxygen, carbon dioxide, moisture, ethylene, and taints) in order to extend the shelf life [6, 18]. Table 1.4 lists active packaging system for fresh fruits and vegetables [31].

1.4.3   INTELLIGENT PACKAGING (IP)

Intelligent packaging is defined as a packaging system that is capable of carrying out intelligent functions (such as detecting, sensing, recording, tracing, communicating, and applying scientific logic) to facilitate decision making to extend shelf life, enhance safety, improve quality, provide information, and warn about possible problems [43]. Comparing AP with IP, the later one is a provider of enhanced communication, whereas, AP is a provider of enhanced protection. Thus, in the total packaging system, IP is the component responsible for sensing the environment and processing information, and AP is the component responsible for taking some action

TABLE 1.4 Active Packaging System for Fresh Fruits and Vegetables [31]

Active packaging components	Scavengers/absorbers	Working principle	Purpose	References
O_2 absorbers (sachet, labels, films, corks)	Ferro-compound (iron powders), ascorbic acid, metal salt, glucose oxidase, alcohol oxidase.	The most successful oxygen scavengers of sachet form on commercial scale are based on iron. Powdered iron is contained alone or with other catalysts in oxygen permeable film pouch. The basic oxidation reaction for absorbing oxygen is: $$4\,Fe + 3O_2 + 6\,H_2O \rightarrow 4\,Fe\,(OH)_3$$ Rough estimation of iron's capacity to absorb oxygen is around 300 ml O_2 per gram of iron.	Reducing respiration rate, mould, yeast and aerobic bacteria growth, prevention oxidation of fats, oil, vitamins, and colors. Prevention damage by worms, insects and insect eggs.	[17, 20]
CO_2 absorbers (sachets)	$Ca(OH)_2$ and NaOH or KOH, CaO, MgO, activated charcoal and silica gel	The most versatile commercial CO_2 absorber is calcium hydroxide, which reacts with carbon dioxide to produce calcium carbonate. Sodium carbonate can also absorb CO_2 under high humidity condition: $$Na_2CO_3 + CO_2 + H_2O \rightarrow 2NaHCO_3.$$ Active charcoal and zeolite acts as physical adsorbents of carbon dioxide.	Removing excess CO_2 formed during storage to prevent fruit damage and bursting of package.	[16, 20, 32]

TABLE 1.4 (Continued)

Active packaging components	Scavengers/absorbers	Working principle	Purpose	References
Ethylene absorbers (sachets/ films)	Aluminum oxide and potassium permanganate (sachets), activated hydrocarbon (squalane, apiezon) + metal catalyst (sachets), Builder- clay powders (films), zeolite films, japaneseoya stone (films) and other compound like silicones (phenyl- methyl silicone)	The most popular method is oxidation of ethylene by potassium permanganate ($KMnO_4$) adsorbed on an inert carrier with large surface area such as silica gel, alumina, and activated carbon.	Prevention fast ripening and softening.	[17, 38, 45]
Humidity absorbers (drip absorbent sheets, films, sachets)	Silica gel (sachets), clays (sachets), sucrose, xylitol, sorbitol, potassium chloride, calcium chloride and sodium chloride.	Desiccants of silica gel, calcium chloride, and calcium oxide are most widely. Silica gel removes moisture by physical adsorption mechanism, which can be reversible by temperature change. Calcium oxide reacts to remove water irreversibly as: $CaO + H_2O \rightarrow Ca(OH)_2$	Excess moisture control in packed produce. Water activity reduction on food surface to check moulds, yeast and spoilage bacteria	[33]

(e.g., release of an antimicrobial) to protect the food product. It may be noted that the terms IP and AP are not mutually exclusive; some packaging systems may be classified either as IP or AP or both, but this situation does not detract the usefulness of these terms. In appropriate situations, functions of IP, AP, and the traditional packaging work synergistically to provide a desirable solution [43].

Intelligent package devices are small, labels or tags that are attached onto primary packaging (e.g., pouches, trays, and bottles), or more often onto secondary packaging (e.g., shipping containers), to facilitate communication throughout the supply chain so that appropriate actions may be taken to achieve desired benefits in food quality and safety enhancement. There are two basic types of smart package devices: data carriers (such as barcode labels and radio frequency identification [RFID] tags) that are used to store and transmit data, and package indicators (such as, time-temperature indicators, gas indicators, biosensors) that are used to monitor the external environment and, whenever appropriate, issue warnings.

1.4.3.1 Barcodes

Barcodes are the least expensive and most popular form of data carriers. The UPC (Universal Product Code) barcode is a linear symbology consisting of a pattern of bars and spaces to represent 12 digits of data to store limited information such as manufacturer identification number and item number. To address the growing demand for encoding more data in a smaller space, a new family of barcode symbologies called the Reduced Space Symbology (RSS) is recently being introduced. The RSS-14 Stacked Omni-directional barcode encodes the full 14-digit Global Trade Item Number (GTIN), and it may be used for loose produce items such as apples or oranges. The RSS Expanded Barcode (also available in stacked format) encodes up to 74 alphanumeric characters, and it may be used for variable measure products.

1.4.3.2 Radio Frequency Identification Tags

The RFID tag is an advanced form of data carrier for automatic product identification and traceability. In a typical RFID system, a reader focus radio waves to capture data from an RFID tag, and the data is then passed onto a host computer (which may be connected to a local network or to the Internet) for analysis and decision making [40]. Inside the RFID tag is a minuscule microchip connected to a tiny antenna. RFID tags may be classified into 2 types: passive tags that have no battery and are powered by the energy supplied by the reader, and active tags that have their own battery for powering the microchip's circuitry and broadcasting signals to the reader. The more expensive active tags have a reading range of 30 m or more, while the less expensive passive tags have a reading range of up to 4.5 m.

1.4.3.3   Time-Temperature Indicators

Temperature is usually the most important environmental factor influencing the kinetics of physical and chemical deteriorations, as well as microbial growth in food products. Time-temperature indicators (TTIs) are typically small self-adhesive labels attached onto shipping containers or individual consumer packages. These labels provide visual indications of temperature history during distribution and storage, which is particularly useful for warning of temperature abuse for chilled or frozen food products. They are also used as "freshness indicators" for estimating the remaining shelf life of perishable products. There are 3 basic types of commercially available TTIs: critical temperature indicators, partial history indicators, and full history indicators [34].

1.4.3.4   Gas Indicators

The gas composition in the package headspace often changes as a result of the activity of the food product, the nature of the package, or the environmental conditions. Gas indicators in the form of a package label or printed on packaging films can monitor changes in the gas composition, thereby providing a means of monitoring the quality and safety of food products.

1.4.3.5   Biosensors

A biosensor is a compact analytical device that detects, records, and transmits information related to biochemical reactions. This smart device consists of two primary components: a bio receptor that recognizes a target analyte and a transducer that converts biochemical signals into a quantifiable electrical response. The bio receptor is an organic or biological material such as an enzyme, antigen, microbe, hormone, or nucleic acid. The transducer can assume many forms (such as electrochemical, optical, acoustic) depending on the parameters being measured. These can be used for rapid, accurate, on-line sensing for in situ analysis of pollutants, detection and identification of pathogens, and monitoring of post-processing food quality parameters

1.4.4   BIODEGRADABLE PACKAGING (BDP)

In recent days, use of synthetic polymer has to be restricted because they are not totally recyclable and/or biodegradable and packages developed from homo polymer can be recycled a limited number of times, and show degraded properties after further persuasion and pose serious ecological problems [35]. Concerned to synthetic multilayer plastic packaging, recycling these material is impracticable and most of the times economically not convenient. Incineration of any plastic put carbon foot prints in atmosphere. As a consequence, several thousands of tons of plastic packages are landfilled, increasing the problem of municipal waste disposal [15]. The growing environmental awareness imposes to packaging films and process possessing both user-friendly and eco-friendly attributes. As a consequence biodegradability is not only a functional requirement but an important environmental attribute. Different sources of biodegradable polymers used for development of films are summarized in Figure 1.3.

1.4.4.1   Limitations and Modifications of Biopolymers Based Films

Films from natural polymers are often associated with poor mechanical and barrier properties and low thermal stability. For this reason different modification strategies are needed to improve their properties. Some strategies are based on blends with synthetic polymers [36], by increasing hydrophobicity [23] or by blending with other biodegradable polymers

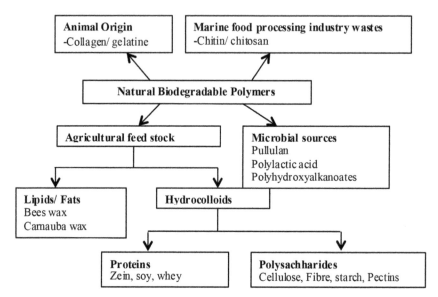

FIGURE 1.3   Animal and plant sources for natural biodegradable polymers.

with different properties [1]. The use of plasticizers and chemical modifiers like cross-linking agents are few among the much needed advanced technologies [4, 11].

1.4.4.2   Plasticizer and Cross-Linking Agent to Modify Film Properties

A plasticizer is a substance that is incorporated into biopolymer to increase its flexibility, workability, and dispensability. Addition of plasticizer produces a film, which is less likely to be, and is more flexible. Thus, its use in starch film is an invariable part to avoid stiffness. The content of plasticizer necessarily varies from 10–60% (dry basis) according to the nature and type of film and its application [44]. Among various available plasticizers, glycerol is the most widely used. The effectiveness of glycerol in biodegradable blend films is most likely due to its small size, which allows it to be more readily inserted between the polymer chains [41].

Cross-linking is a key technique for modifying the properties of starches and can be achieved by adding intra- and inter-molecular bonds at random locations [41]. Inter chain cross-linking tends to limit the contact of the free OH-groups with surrounding water [9]. Starch cross-linking is normally performed by treating starches with reagents (e.g., Glutaraldehyde) capable of forming either ether or ester linkages between hydroxyl (–OH) groups. Manoj and Rizvi [21] explained that the increase of mechanical property by cross-linking is due to reinforcing the structure of starch and limiting its water absorption, thereby restricting the mobility of the starch chain in the amorphous region.

1.4.4.3 Incorporation of Natural Antimicrobials in Biodegradable Films

Development of biopolymer films includes incorporation of antimicrobials as these films are usually susceptible to be infected with air-borne microbes, particularly in humid condition. So, irrespective of nature of uses, inclusion of antimicrobial preservative(s) in the film may become essential for its stability [7].

Weak organic acids and their salts such as, propionic acid, potassium sorbate and benzoic acid, and some essential oils (cinnamon, oregano, clove etc.) which are commonly used as antimicrobial preservatives in food systems, may be incorporated into biodegradable films to inhibit the outgrowth of both bacterial and fungal cells. Being GRAS certified, they do not affect the food quality, or the eco system when discarded after use. Furthermore, it is also claimed that these impart some additional cross-linking effect that gives more structural stability to the system [25]. However, enough experimental verification to the above fact is not available in literature. Also, more research is needed to evaluate the effect of antimicrobial on various film properties.

1.5 CONCLUSIONS

The food industry has seen great advances in the packaging sector since its inception in the 18th century with most innovations occurring during the past century. These advances have led to improved food quality and safety. However, excessive use of petroleum based synthetic packaging materials leads to ecological imbalance, global warming and continuous depletion of limited petroleum resources. Although films made from natural bio-polymer usher promising solution, to arrive at commercial utilization, lot of research is needed in this field.

1.6   SUMMARY

In this chapter, the different role of packaging materials along with the total market share of food packaging and brief history of food packaging have been discussed. Basic mechanisms of mass transfer followed by the innovative packaging trends like, MAP, AP, IP and BDP have also been discussed in this chapter.

KEYWORDS

- Absorber
- Active packaging
- Antimicrobial packaging
- Biodegradable packaging
- Diffusion
- Fick's law
- Flexible packaging
- HDPE
- Henry's law
- Intelligent packaging
- LDPE
- LLDPE
- MAP
- Metal cans
- Metallization
- Packaging history
- Permeability

- **Permeant**
- **Plastic**
- **Polypropylene**
- **Preservation**
- **Releaser**

REFERENCES

1. Abugoch, L. E., Tapia, C., Villamán, M. C., Yazdani-Pedram, M., & Díaz-Dosque, M. (2011). Characterization of quinoa protein-chitosan blend edible films. *Food Hydrocolloids*, 25, 879–886.
2. Anonymous, (2013). Global packaging market to reach $975 billion by 2018. Available from: http://www.smitherspira.com/news/2013/december/global-packaging-industry-market-growth-to-2018 Accessed on 30th July, 2015.
3. Anonymous, (2015). History of paper. Available from: https://en.wikipedia.org/wiki/History_of_paper. Accessed on 5th August, 2015.
4. Bigi, A., Cojazzi, G., Panzavolta, S., & Rubini, K. (2001). Mechanical and thermal properties of gelatin films at different degrees of glutaraldehyde cross-linking. *Biomaterials*, 22, 763–768.
5. Brody, A. (2008). Packaging by the numbers. *Food Technology*, 62(2), 89–91.
6. Brody, A., Strupinsky, E. R., & Kline, L. R. (2001). Odor removers. In: *Active Packaging for Food Applications. Lancaster*. Brody, A., Strupinsky, E. R., Kline, L. R. (Ed.). Technomic Publishing Company. pp. 107–117.
7. Chowdhury, T., & Das, M. (2010). Moisture sorption isotherm and isosteric heat of sorption characteristics of starch based edible films containing antimicrobial preservative. *International Food Research Journal*, 17, 601–614.
8. Dagoon, J. D. (1989). *Applied Nutrition and Food Technology*. Rex Bookstore, Inc., pp. 471.
9. El-Tahlawy, K., El-Nagar, K., & Elhendawy, A. G. (2007). Cyclodextrin-4 hydroxybenzophenone inclusion complex for UV protective cotton fabric. *Textile Institute*, 98(5), 453–462.
10. Floros, J. D., & Matsos, K. I. (2005). Introduction to modified atmosphere packaging. In: *Han H J (Ed). Innovations in Food Packaging*. Elsevier, Academic Press. 1, 159–172.
11. González, A., & Igarzabal, A. C. I. (2013). Soy protein e Poly (lactic acid) bilayer films as biodegradable material for active food packaging. *Food Hydrocolloids*, 33, 289–296.
12. Han, H. J. (2005). New technologies in food packaging. In: *Han H J (Ed). Innovations in Food Packaging.* Elsevier, Academic Press. 1, 3–11.

13. Hanlon, J. F. (1992). *Handbook of Package Engineering*. Technomic Publishing: Lancaster, PA, 3, 59–103.
14. Kader, A. A., Zagory, D., Kerbel, E. L., & Wang, C. Y. (1989). Modified atmosphere packaging of fruits and vegetables. *Critical Reviews in Food Science and Nutrition*, 28(1), 1–30.
15. Kirwan, M. J., & Strawbridge, J. W. (2003). Plastics in food packaging. *Food Packaging Technology*, 174–240.
16. Lee, D. S., Shin, D. H., Lee, D. U., Kim, J. C., & Cheigh H. S. (2001). The use of physical carbon dioxide absorbents to control pressure build up and volume expansion of kimchi packages. *Journal of Food Engineering*, 48, 183–188.
17. Lee, D. S., Yam, K. L., & Piergiovanni, L. (2010). Active and intelligent packaging. In: *Food Packaging Science and Technology*, CRC Press, Taylor & Francis Group. 1, 445–473.
18. Lopez-Rubio, A., Almenar, E., Hernandez-Munoz, P., Lagaron, J. M., Catala, R., & Gavara, R. (2004). Overview of active polymer-based packaging technologies for food applications. *Food Reviews International*, 20(4), 357–387.
19. Malpass, D. (2010). *Introduction to Industrial Polyethylene: Properties, Catalysts, and Processes*. John Wiley and Sons. pp. 1–3.
20. Mangaraj, S., & Goswami, T. K. (2009). Modified atmosphere packaging - An ideal food preservation technique. *Journal of Food Science and Technology*, 46(5), 399–410.
21. Manoj, K., & Rizvi, S. S. H. (2010). Physicochemical characteristics of phosphorylated cross-linked starch produced by reactive supercritical fluid extrusion. *Carbohydrate Polymers*, 81(4), 687–694.
22. Miltz, J., Passy, N., & Mannheim, C. H. (1995). Trends and applications of active packaging systems. *Food and Packaging Materials – Chemical Interaction*. The Royal Society of Chemistry, 201–210.
23. Monedero, F. M., Fabra, M. J., Talens, P., & Chiralt, A. (2010). Effect of oleic acid beeswax mixtures on mechanical, optical and water barrier properties of soy protein isolate based films. *Journal of Food Engineering*, 91, 509–515.
24. Mount, E. (2004). Metallized films for food packaging. *Converting Magazine*, 3.
25. Ojagh, S. M., Rezaei, M., Razavi, S. H., Hosseini, S., & Mohamad, H. (2010). Development and evaluation of a novel biodegradable film made from chitosan and cinnamon essential oil with low affinity toward water. *Food Chemistry*, 122, 161–166.
26. Parry, R. T., ed. (1993). *Principles and Applications of Modified Atmosphere Packaging of Food*. Glasgow, UK, Blackie. pp. 1–18.
27. Risch, S. J. (2009). Food packaging history and innovations. *Journal of Agricultural and Food Chemistry*, 57, 8089–8092.
28. Robertson, G. L. (2006). Food packaging. In: *Food Science and Technology*. Campbell-Platt G. (Ed). Wiley-Blackwell publishing Ltd. 1, 279–298.
29. Rooney, M. L. (1995). Overview of active food packaging. In: *Active Food Packaging*. Rooney M L (Ed.). Blackie Academic and Professional, New York. pp. 1–2.
30. Sen, C., Mishra, H. N., & Srivastav, P. P. (2012). Modified atmosphere packaging and active packaging of banana (Musa spp.): A review on control of ripening and extension of shelf life. *Journal of Stored Product and Postharvest Research*, 3(9), 122–132.

31. Sen, C., Mishra, H. N., & Srivastav, P. P. (2014). *Active packaging—an approach to minimize post-harvest loss of banana.* Lambert Academic Publishing. pp. 13–14.

32. Shin, D. H., Cheigh, H. S., & Lee, D. S. (2002). The use of Na_2CO_3-based CO_2 absorbent systems to alleviate pressure build up and volume expansion of kimchi packages. *Journal of Food Engineering*, 53, 229–235.

33. Shirazi, A., & Cameron, A. C. (1992). Controlling relative humidity in modified atmosphere packages of tomato fruit. *Hortscience*, 27, 336–339.

34. Singh, J., Kaur, L., & McCarthy, O. J. (2007). Factors influencing the physico-chemical, morphological, thermal and rheological properties of some chemically modified starches for food applications—A review. *Food Hydrocolloids*, 21(1), 1–22.

35. Singh, R. P. (2000). Scientific principles of shelf-life evaluation. In: *Shelf-life Evaluation of Food.* Man, D., & Jones, A. (Ed). Gaithersburg, Md., Aspen Publishers. 2, 3–22.

36. Sorrentino, A., Gorrasi, G., & Vittoria, V. (2007). Potential perspectives of bio-nanocomposites for food packaging applications. *Trends in Food Science & Technology*, 18, 84–95.

37. Stilwell, E. J., Canty, R. C., Kopf, P. W., Montrone, A. M., & Arthur, D. L. (1991). *Packaging for the Environment—A Partnership for Progress.* AMACOM, American Management Association. New York.

38. Tian, H., Wang, Y., Zhang, L., Quan, C., & Zhang, X. (2010). Improved flexibility and water resistance of soy protein thermoplastics containing waterborne polyurethane. *Industrial Crops and Products*, 32, 13–20.

39. Vermeiren, L., Heirlings, L., Devlieghere, F., & Devevere, J. (2003). Oxygen, ethylene and other scavengers. In: *Novel Food Packaging Techniques.* Ahvenainen, R. (Ed). Cambridge-UK, Woodhead Publishing. pp. 22–49.

40. Want, R. (2004). Enabling ubiquitous sensing with RFID. *Computer,* 37(4), 84–86.

41. Wittaya, T. (2012). Rice starch-based biodegradable films: properties enhancement. *Structure and Function of Food Engineering,* 51, 103–134.

42. Yahia, E. M., & Singh, S. P. (2009). Tropical fruits. In: *Modified and Controlled Atmospheres for the Storage, Transportation, and Packaging of Horticultural Commodities.* Yahia, E. M. (Ed). CRC Press, Taylor & Francis Group. 1, 397–444.

43. Yam, K. L., Takhistov, P. T., & Miltz, J. (2005). Intelligent packaging: concepts and applications. *Journal of Food Science*, 70(1), R1–R10.

44. Yang, L., Paulson, A. T. (2000). Effects of lipids on mechanical and moisture barrier properties of edible gellan film. *Food Research International*, 33(7), 571–578.

45. Zagory, D. (1995). Ethylene-removing packaging. In: *Active Food Packaging.* Rooney, M. L. (Ed.). Blackie Academic & Professional. London, pp. 38–54.

CHAPTER 2

EMERGING TECHNOLOGY-BASED DRYING: FOOD AND FEED PRODUCTS

SATYANARAYANA BONTHU[1] and ANIRBAN RAY[2]

[1]*Research Scholar, Agricultural and Food Engineering Department, Indian Institute of Technology Kharagpur, West Bengal, Kharagpur – 721302, India. Mobile: +91-7059972916, E-mail: satyanarayanabonthu@yahoo.com*

[2]*Post-Doctoral Research Fellow, Department of Molecular Biology and Biotechnology, University of Kalyani, West Bengal, Kalyani – 741235, India. Mobile: +91-9231695435; E-mail: anirbanrayiitkgp@gmail.com*

CONTENTS

2.1   INTRODUCTION

The purpose of processing is to prevent undesirable changes in the plant materials in a cost effective way, which control growth of microorganisms, reduce chemical, physical and physiological changes of an undesirable nature, obviate contamination and thus prolonging the shelf life of the concerned product. The development of modern methods of processing and preserving the raw material of medicinal plants helps to maintain their quality for a longer time [23]. To obtain high quality products attention should be given towards cultivation, post harvesting handling, processing and comprehensive quality control. The processing of plant materials can be accomplished by chemical and physical methods. Chemical processing involves the addition to plant material of such substances as sugars, salts, antioxidants, stabilizers or acids, or exposure of plant material to chemicals, such as smoke or fumigants.

Physical approaches to processing the plant material include temporary increases in the product's energy level (heating, irradiation), controlled reduction of the product's temperature (chilling, freezing), controlled reduction in the product's water content (drying) by concentration, air dehydration, freeze drying, irradiation, and the use of protective packages [12]. Due to need for emerging trends, physical methods of processing are used extensively in developed countries of the world and they are likely to become more common world-wide.

This chapter encompasses the fundamental concepts of drying, their types and influence on the physical and chemical properties of plant products with special reference to aloe gel.

2.2   DIFFERENT DRYING TECHNIQUES AND BASIC PRINCIPLES BEHIND THE ACTIVITY

Drying is a type of physical method used to process plant materials. The main principal of drying is the decrease in the water activity in the product by decreasing its water content, inhibiting the development of microorganisms and decreasing spoilage reactions, thus resulted in prolonged shelf life of the product. An important advantage of dehydrated products is the reduced cost of packing, storage and transportation due to the smaller volume and mass of the dried product than that of the unprocessed one [23]. In addition, products with low moisture content can be stored for longer periods at room temperature [12].

The process of drying happens by effecting vaporization of the liquid by supplying heat to the wet feedstock. Heat may be supplied by convection through direct dryers, by conduction with contact or indirect dryers, radiation or volumetrically by placing the wet material in a microwave or radio frequency electromagnetic field. Over 85% of industrial dryers are of convective type with hot air or direct combustion gases as the drying medium. Over 99% of the drying applications involve removal of water from the native substances. Different types of drying techniques involved in drying of plant materials are discussed in this section.

2.2.1 FREEZE DRYING

Lyophilization or freeze-drying is an operation where water is removed from the wet materials by transfer from the solid and icy state to the gaseous state (water vapor), and this operation referred as sublimation can only be accomplished when the vapor pressure and temperature of the ice surface at which the sublimation takes place are below those at triple point, i.e., 4.58 torr at a temperature of approximately 0°C [9].

Freeze drying requires a high-energy supply at reduced pressure. The advantages of freeze-drying are minimum physical and chemical damage to thermal sensitive materials so that the dehydrated product can be rapidly and completely rehydrated. But initial investment for freeze drying, operational cost and time being are major constraints of this dehydration process [27], though it is the best method of water removal with final products with highest quality as compared to other methods of drying [7, 25]. Freeze drying is mainly based on the dehydration by sublimation of a frozen product. Due to the absence of liquid water and the low temperatures required for the process, most of deterioration and microbiological reactions are halted, which gives a final product of excellent quality. The solid state of water during freeze-drying protects the structure and the shape of the products with minimal reduction of volume. Despite of many advantages, freeze-drying has been considered to be the most extensive process for manufacturing of any dehydrated product [36].

The freeze dryer consists of a drying chamber, a condenser, a vacuum pump, and a heat source. The drying chamber, in which the sample is placed and heating/cooling takes place, must be air tight and with temperature-controlled shelves. The condenser must have sufficient condensing surface and cooling capacity to collect water vapor released by the product. As vapors contact the condensing surface, they give up their heat energy and turn into ice crystals that will be removed from the system. A condenser temperature of −65°C is typical for most commercial freeze-dryers [8]. The vacuum pump removes non-condensable gases to achieve high vacuum levels (below 4 mm Hg pressure) in the chamber and condenser. The heating source provides the latent heat of sublimation, and its temperature may vary from −30° to 150°C [5].

2.2.2 DEHUMIDIFIED AIR DRYING OR HOT AIR DRYING

Hot air drying is an ancient method used to preserve materials in which the solid to be dried by exposing to a continuously flowing hot stream of air and absorbed moisture evaporates. The phenomenon underlying this process is a complex process involving simultaneous mass and energy transport in a hygroscopic, shrinking system. Conventional drying with hot air offers dehydrated products that can have an extended life of one year. But air-drying is a difficult food processing operation mainly because of undesirable change in qualities of food products through dehydration. Therefore the dried product may be seriously damaged in comparison to the native state. The major disadvantages of air-drying of foods are low energy efficiency and long drying time during the falling rate period. Because of the low thermal conductivity of food materials in this period, heat transfer to inner sections of foods during conventional heating is limited [9]. This may also cause irreversible modifications to active substances, affecting their original structure and may promote important changes in the proposed physiological and pharmacological properties of active components [10, 35].

A conventional air drier consists of a dryer and a dehumidifier. The dryer comprises of an air blower, air heating chamber, a drying chamber, and a plenum chamber and power supply, and control panel. The plenum chamber is a pressurized chamber containing a gas or fluid (typically air) at positive pressure, means pressure higher than surroundings). Air is drawn into the duct through a mesh guard by a motor driven axial flow fan impeller whose speed can be controlled in the duct. The dehumidifier is used to reduce the relative humidity of air. Two mercury in-glass thermometers are placed at the inlet and exhaust end of the dryer to measure the dry bulb and wet bulb temperatures of incoming and outgoing air to the dryer. These two temperatures is used to determine the relative humidity of inlet and exhaust air. The temperature of the drying air is regulated by PID temperature controller and air velocity is controlled by a fan speed controller, and relative humidity of the drying air is calculated by using psychometric chart is also verified with a humidity/temperature instrument.

2.2.3 MICROWAVE DRYING

In microwave drying electromagnetic radiation is generated in special oscillator tubes namely, magnetrons or klystrons, and is radiated into a closed chamber containing the product. The chamber is constructed of highly reflective walls to reflect the radiation back and forth until it is absorbed by the product under treatment. There is a wave deflector located at the point, where the radiation enters the oven to distribute the radiation uniformly in the oven [4, 31].

Wavelengths of microwave range from 1 mm to 1 m, corresponding to a frequency range of 300 MHz to 300 GHz [31]. Since a conventional dryer is limited by heat transfer to the core of product and mass transfer of water out of the material [21], it would be expected that microwave drying would perform more uniformly and faster due to the volumetric heating. In microwave drying, heat is generated by directed transforming electromagnetic energy into kinetic molecular energy, thus the heat is generated within the material. During microwave heating, heat is generated by dielectric materials that absorb microwaves, but materials that are reflectors will not be heated directly. Microwave drying has gained immense popularity as an alternative and convenient drying method in food industry as a rapid and energy efficient technique as compared to conventional hot-air drying [4].

It is distinct from conventional drying, which is driven by the difference in temperature between the outside and inside of the material. Microwave drying is not governed by temperature gradients but the heat arises from the oscillation of molecular dipoles and movement of ionic constituents, respectively in response to alternating electric fields at high frequency. The resulting energy is absorbed throughout the volume of the wet material. The increase in internal pressure drives out the moisture from the interior to the surface of the material [31]. Here, the energy is converted into kinetic energy of water molecules and then into heat, when the water molecules realign in the changing electrical field and interact with the surrounding molecules because of (friction). Dielectric properties are electric properties of basic interest in microwave processing, since the microwave heating mechanism is closely related to these properties. Temperature and moisture profiles of foods under a microwave field are directly related to the electrical properties, which greatly influence the rate of absorption and distribution of energy within the product. Therefore, relating these factors to electrical properties is of considerable importance in predicting heating characteristics and designing efficient food processes [14].

Microwave energy transfer causes a rapid evaporation of water from the sample tissue, hence treatment time is shorter and oxidation is limited with a substantial preservation of color, flavor, and sensory qualities of products. The dryer comprises microwave generator, reflected power absorber, microwave cavity, vacuum pump, fiber optic temperature sensors, electronic scale and computer. Magnetron microwave generator produces microwaves and two sensors are located at the rectangular waveguide to detect input microwave power and reflected microwave power in watts. A vacuum pump is used to create a vacuum in the circular plastic container in which the vacuum level is controlled by an open-ended valve. Desiccators filled with anhydrous calcium sulphate as moisture absorber is located between the container and the vacuum pump to ensure absorption of the moisture vapor. The sample is placed in a vacuum chamber during microwave vacuum studies. A vacuum pump provides sub-atmospheric pressures within the chamber, and is connected in-line with the vacuum hose to remove moisture from the vacuumed air before reaching the vacuum pump. The anhydrous calcium sulfate ($CaSO_4$) is replaced once it becomes saturated. The plate rotates generally at 10 rpm in the cavity during operation to ensure uniformity of the volumetric heating in the sample [14].

2.2.4   INFRARED (IR) DRYING

Energy conservation is one of the fundamental factors determining profitability and success of any unit operation. When infrared radiation is used to dry moist materials, the radiation impinges the exposed material, penetrates it and the energy of radiation converts into heat [8]. Since a material is heated intensely, the temperature gradient in the material reduces within a short period of time. Therefore, energy consumption in infrared drying process is relatively lesser than that of the other methods. IR energy is transferred from the heating element to the product surface without heating surrounding air [13].

IR radiation can be classified into three regions: near-infrared (NIR), mid-infrared (MIR), and far-infrared (FIR) corresponding to the spectral ranges of 0.75 to 1.4, 1.4 to 3, and 3 to 1000 µm, respectively [29]. In general, FIR radiation is advantageous for food processing because most of the food components absorb radiative energy in the FIR region [30]. The use of infrared radiation technology in dehydrating foods has several advantages: Decreased drying time, high energy efficiency, high quality finished products, uniform temperature in the product while drying, and a reduced necessity for air flow across the product [20].

Conventional types of infrared radiators used for the heating process are electric and gas-fired heaters. These two types of IR heaters generally fit into different temperature ranges: 343 to 1100°C for gas and electric IR, and 1100 to 2200°C for electric IR only. The IR temperatures are typically range from 650 to 1200°C to prevent charring of products. The capital cost of gas heaters is higher, while the operating cost is cheaper than that of electric infrared systems. The electrical infrared heaters are popular because of installation controllability, ability to produce prompt heating rate, and cleaner form of heat. Electric infrared emitters also provide flexibility in producing the desired wavelength for a particular application. In general, the operating efficiency of an electric IR heater ranges from 40% to 70%, while that of gas-fired IR heaters ranges from 30% to 50% [11]. The spectral region suitable for industrial process heating varied from 1.17 to 5.4 μm, which corresponds to 260 to 2200°C [32].

The infrared radiation is transmitted through water at short wavelength, whereas at longer wavelengths it is absorbed at the surface [29]. Hence, drying of thin layers seems to be more efficient at the FIR region, while drying of thicker bodies should give better results at the NIR region. The rate of color development by FIR heater is greater with NIR heater, primarily due to a more rapid heating rate on the surface. During infrared radiation, the absorbed energy may induce changes in the electronic, vibrational and rotational states of atoms and molecules of a product as a result water is transported as vapor.

2.2.5 SPRAY DRYING

The spray drying is presently one of the most promising and ideal technology, where the end-product must comply with the precise quality standards for particle size distribution, residual moisture content, bulk density and morphology. Spray drying produces predominately amorphous material due to the almost instantaneous transition between liquid and solid phases; however it can also be used to obtain crystalline products [33]. Dry milk powder, detergents and dyes are just a few instances of spray-dried products. It also provides the advantage of weight and volume reduction than that of the native products. It is the transformation of feed from a fluid state into a dried particulate form by spraying the feed into a hot drying medium.

The spray drying involves evaporation of moisture from an atomized feed by mixing the spray and the drying medium, which is typically air. The drying proceeds until the desired moisture content is reached in the sprayed particles and the product is then separated from the air. The mixture being sprayed can be a solvent, emulsion, suspension or dispersion. The dispersion can be achieved with a pressure nozzle, a two fluid nozzle, a rotary disk atomizer or an ultrasonic nozzle. Different kinds of energy can be used to disperse the liquid body into fine particles. Many Spray drying operations produce spherical particles while others result in non-spherical particles. Particles may be hollow or solid. Pressure spray nozzles can produce particles ranging in size from 20 to 600 microns. Two fluid nozzles generally produce particles with sizes in the range from 10 to 200 microns and larger [34]; and produce more uniform particle sizes as compared to pressure atomizers. Concurrent dryers produce powders with lower bulk densities than that of the counter-current dryers.

2.3   EFFECTS OF DRYING METHODS ON PLANT MATERIALS: AN OVERVIEW

Drying process is also frequently employed to process plant materials. Drying occurs by effecting vaporization of the liquid by supplying heat to the wet feedstock. The main attribute of drying process involves reduction in water content and inhibition of development of microorganisms and the decrease in spoilage reactions, thus prolonging the shelf life of the product [23]. The drying causes irreversible modifications to the active constituents; and affects their physical and biochemical structure, which may promote important changes in physiological properties of these substances. Therefore, proper processing guidelines need to be followed to ensure the biological integrity of the final product [10].

Khraisheh et al. [15] conducted a study to evaluate the quality and structural changes in potatoes during microwave and convective drying or air-drying. The quality aspects of the dehydrated potato samples included: rehydratibility, ascorbic acid (vitamin C) retention and the structure or shrinkage behavior). Ascorbic acid degradation was found to exhibit first order kinetics during microwave and convective drying. The potato samples dried in a microwave field exhibited less shrinkage than those of air-dried samples. The rehydration of potato samples was quantified on the basis of coefficient of rehydration and rehydration ratio. The rehydration properties of the microwave-dried samples were superior to the convective dried samples. The extent of rehydration was also increased with increasing power level. However, at high power levels (i.e., 38 W), starch gelatinization was observed thus reducing degree of rehydration.

Leeratanarak [16] studied the effects of blanching and drying temperature on the drying kinetics as well as various sensory attributes (texture,
color and brown pigment accumulation) of potato slices during both low-
pressure superheated steam drying (LPSSD) and hot air drying. It was
found that LPSSD took shorter time to dry the product to the final desired
moisture content than that required by hot air drying, when the drying
temperatures were higher than 80°C. Hot air drying resulted in higher
browning index than did LPSSD at higher drying temperatures. A higher
degree of non-enzymatic browning occurring during hot air drying might
be due to both Maillard reaction and ascorbic acid oxidation. In the case of
LPSSD, there was no oxygen left in the drying chamber and the main cause
of non-enzymatic browning could be only Maillard reaction. The results of
the browning index were also related to the color changes, especially the
change of redness. The results showed similar trends for both physical and
chemical changes. From the results of color changes and browning index,
it was concluded that hot air drying resulted in more severe chemical damage of potato chips than did LPSSD [16].

The drying characteristics (color and rehydration) of microwave, air
and microwave finishing dried banana samples were studied by Maskan
[18]. Initially the rate of moisture loss was high under hot air drying but at
low moisture content the hot air drying was not advantageous as the diffusion process was slow. The drying rate was increased remarkably with
power output of the microwave oven. Microwave finish drying increased
the drying rate and reduced the drying time. It had little effect on the color
and rehydration capacity of finished products as compared to the hot air
and Microwave drying methods. The drying temperature and time are
important parameters for color change during drying. The lower color degradation of microwave finish dried banana may be due to the substantial
reduction in drying time.

Winnie et al. [37] studied the behavior of 16 volatile compounds of
banana during air drying, freeze drying and combination of air drying and
vacuum microwave drying (VMD). Samples that underwent more VMD
had significantly lower levels of volatile compounds, which are attributed
to the decreased formation of an impermeable solute layer on the surface
of the chips. The optimal process of 90% AD/10% VMD yielded crisper
banana chips with significantly higher volatile levels. Almost all volatile

compounds decreased as the extent of VMD increased, with the least total volatile compounds, esters, and acetates found in VMD banana chips. The freeze dried banana chips retained the most total volatile compounds, esters, and acetates; and 90%AD/10%VMD banana chips retained the next highest concentration of these volatile compounds, and air drying banana chips retained less of these volatiles than the 90%AD/10%VMD banana chips. Retention of volatiles during VMD can be influenced by the physical properties of the particular compound as well as the environmental conditions, drying temperature, and the amount of heat input.

The effects of three common postharvest processing treatments (freezing, freeze drying, and air drying) on the total phenolic (TPs) and ascorbic acid contents of corn were investigated by Danny et al. [3]. The highest levels of TPs and ascorbic acid were consistently found in the extractions of frozen samples, followed by freeze dried and then air dried. In general, air-drying at temperatures >60°C is regarded as unfavorable due to the possibility of inducing oxidative condensation or decomposition of thermos-labile compounds, such as (+)-catechin. Conversely, freeze-drying may lead to higher extraction efficiency of TPs because it can lead to the development of ice crystals within the plant matrix. Ice crystals can result in a greater rupturing of plant cell structure, which may allow for better solvent access and extraction. With air-drying there is little or no cell rupture and there is the added effect of heat, which can cause losses in phenolic and ascorbic acid.

The effect of the drying bed thickness on drying characteristics and quality of rough rice subjected to IR heating was studied by Pan et al. [24]. Samples of freshly harvested medium grain rice (M202 variety) with 20.5% and 23.8% moisture contents were dried with two different radiation intensities (4685 and 5348 W) and exposure times of 15, 30, 40, 60, 90, and 120 seconds for each drying bed thickness. The three tested drying bed thicknesses were single layer, 5 mm, and 10 mm. The heating and drying rates were decreased with the increase of bed thickness. IR heating under tested conditions did not have adverse effects on rice sensory and milling quality, including total rice yield, head rice yield, and degree of milling of the dried rice.

The effect of osmotic pretreatment on the mass transfer kinetics and quality of dried rehydrated Chilean papaya was investigated by Lemus-

Mondaca et al. [17]. Osmotic treatments were sucrose solutions of 40, 50, and 60% w/w and dried at 60°C; non-pretreated samples were dried at different temperatures of 40, 60, and 80°C by hot air drying. Different quality parameters under evaluation were: proximal composition, rehydration ratio, water-holding capacity, color, vitamin C content, firmness, and microstructure. Non-pretreated samples showed a clear turgor loss, color loss and low ascorbic acid retention when rehydrated. The conventional hot air drying of papaya typically reduces the nutritional and commercial quality in the rehydrated product. The firmness and stability of the color of papaya were strongly dependent on the drying temperature of the product. Higher air temperatures produced softer rehydrated samples with higher color loss. The different sucrose concentrations 40, 50, and 60% w/w improved the quality of the rehydrated papayas. The pretreated samples showed higher ascorbic acid retention and yielded best firmness and color quality.

Different drying technologies for retention of physical quality and antioxidants in asparagus (*Asparagus officinalis* L.) were evaluated by Nindo et al. [22]. They used five drying methods such as: tray drying (TD), spouted bed (SB) drying, combined microwave and spouted bed drying (MWSB), refractance window (RW) drying and freeze-drying. MWSB drying produced asparagus particles with good rehydration and color characteristics, and was the fastest among the methods where heated air was used. When using MWSB drying, the power level of 2 W/g and 60°C heated air resulted in highest retention of total antioxidant activity. The antioxidant activity of asparagus was enhanced after RW and freeze-drying. The highest amount of ascorbic acid was retained in the product after RW drying, followed by freeze-drying, MWSB and SB drying. TD resulted in the least retention of ascorbic acid. In the RW drying process, the pureed product heats up very fast, this caused increased release of phenolic compounds bound in the cell matrix. Moisture loss was very intensive during the first one minute of drying, and the partial pressure of oxygen near the product became very low due to high local vapor pressure created by moisture evaporation. This condition prevented the oxidation of phenolic antioxidants in asparagus that were made more available by the heating process. It has recently been shown that thermal processing of sweet corn caused antioxidant activity and total phenolics to increase by 44% and 54%, respectively, although 25% loss of ascorbic acid was observed.

The effects of different drying methods (e.g., hot-air drying, microwave drying and vacuum-freeze drying) on antioxidant activity and antioxidants in sweet potato (*Ipomoea batatas* L. Lam.) tubers were investigated by Yang et al. [39]. The dried sweet potatoes in microwave possessed the highest antioxidant activity, while the lowest activity was observed in hot-air dried samples. The phenolic contents were positively correlated with scavenging activity and reducing power of DPPH$^+$. The microwave-dried samples retained the highest antioxidant activity with the highest content of phenolic compounds in dried sweet potatoes. In their study, the temperature of Microwave drying (95–105°C) was higher than hot-air drying (65°C) and freeze-drying (–20°C). The highest content of polyphenols in microwave-dried samples could be associated with the release of more bound phenolics from breakdown of cellular constituents during thermal treatment. In addition, the increase in phenolic content could be explained, at least partially, by the formation of Maillard reaction products (MRPs) with phenolic type structure during the thermal process [28].

Physico-chemical modifications promoted by heat treatment and dehydration at different temperatures (30–80°C) on acemannan, a bioactive polysaccharide from *Aloe vera* parenchyma [25, 26] were studied by Femenia et al. [6]. Modification of acemannan was particularly significant when dehydration was performed above 60–80°C. Heating promoted marked changes in the average molecular weight (MW) of the bioactive polysaccharide, increasing from 45 kDa, in fresh aloe, to 75 and 81 kDa, for samples dehydrated at 70 and 80°C, respectively. This could be attributed to structural modifications, such as de-acetylation and losses of galactose rich side chains from the mannose backbone. These structural modifications were reflected by the significant changes occurring in the related functional properties, such as swelling, water retention capacity, and fat adsorption capacity, which exhibited a significant decrease as the temperature of dehydration was increased. Further, dehydration also promoted significant modification of the main type of cell wall polysaccharides present within the aloe parenchyma tissues. Pectic polysaccharides from the cell wall matrix were affected by heating, probably due to either β-elimination processes or enzyme catalyzed degradation.

Effects of heat treatments on the stabilities of polysaccharides substances and barbaloin in gel juice from *Aloe vera* were investigated by Xiu
et al. [38]. Thermostatic water bath was used for Aloe gel juice to allow
it to heat at 50, 60, 70, 80 and 90°C, respectively. The content of polysaccharides decreased maximum in the sample dried at 90°C then followed
by 50°C, 60°C, 80°C and 70°C heated samples. Sample at 90°C and 50°C
exhibited the highest losses of polysaccharide. The marked decrease at
90°C may be due to the thermal degradation of the polysaccharide, which
is consistent with Cohen and Yang [2] that heat drying may promote a
breakdown of the cell wall polysaccharides network. The significant
decrease at 50°C and 60°C was probably as a result of the presence of
enzymes which could be responsible for the hydrolysis of large molecular
polysaccharides into lower ones. Heating promoted a remarkable decrease
in barbaloin content with temperature and time. The higher temperature
and the longer period of heat treatment may be more effective on the instability of barbaloin. The most rapid decline was obtained at 90°C, whereas
the least rapid decline was noticed at 60°C.

Changes in physico-chemical and functional properties during convective drying of *Aloe vera* leaves were studied by Gulia et al. [11]. *Aloe vera*
leaves were dried at different temperatures in hot air oven and was pow-
dered. The percent powder yield was 2.60%, 2.60%, 2.55% and 2.52%
at 50, 60, 70 and 80°C, respectively. Powder samples had the pH (1%
solution) of 3.51, 3.53, 3.52 and 3.53 with the rise of drying temperature
in the selected range. Wettability of powder at 70°C was 32 seconds as
compared to 35, 35 and 37 seconds in the samples obtained at 50, 60 and
80°C, respectively. The water absorption capacity of powder at 70°C was
359% as compared to 351%, 354%, and 356% at 50, 60, and 80°C powder
samples. Longer time period of drying at 50 and 60°C and higher temperature at 80°C than 70°C during the drying may be responsible to cause more
structural changes in the various components along with the polysaccharides, which may probably have resulted better water absorption capacity
of the powder at 70°C. The polysaccharides stability was dependent on
time and temperature. The aloin content was decreased from 10.6 to 1.7
ppm as temperature increased from 50°C to 80°C. The decrease in the
aloin content may be due to its heat sensitive property [10].

The effect of air temperature on the physico-chemical and nutritional properties and antioxidant capacity of *Aloe vera* gel was investigated by Miranda [19]. The following parameters were analyzed: proximal composition, water activity (aw), pH, acidity, non-enzymatic browning, surface color, vitamin content (C and E), mineral content, and antioxidant capacity. The drying kinetics of aloe gel was modeled using the Wang–Singh equation, which provided a good fit for the experimental data. Analysis of variance revealed that the drying temperature exerted a clear influence on most of the quality parameters. A drying temperature of 80 and 90°C resulted in significant variation in and/or loss of the physicochemical and nutritional properties of the gel; in addition, the antioxidant capacity of the gel was decreased at these temperatures. These effects were also observed as a result of a lengthy drying period (i.e., 810 min at 50°C). However, minor alterations in the physicochemical and nutritional properties of aloe gel were produced at drying temperatures of 60–70°C, resulting in the production of a high quality gel. The influence of different drying methods on starchy plant material and aloe gel is summarized in Table 2.1.

2.4 CONCLUSIONS

Different drying methods and their effects on physical and chemical properties of starch based plant products are evaluated in this review. Freeze-drying is highly efficient to obtain the product quality but the process is tedious, costliest and volatile compounds may evaporate during long time process, whereas, Microwave drying is simple, cost effective and can be employed with ease, but chances of degradation of polysaccharides and carotenoids is there. Dehumidified air-drying is long time drying process and often causes heat damage and adversely affects texture, color, flavor and nutritional value of products. Since the prevailing synergism among the component compounds of food and feed products potentiates the nutritional attributes present therein, delicacy should be optimum during the dehydration process of the product of interest. Also, always there should be a concerted effort in developing optimized dehydration methodology specific to particular food product.

TABLE 2.1 A Succinct Overview of the Effects of Different Drying Methods on the Quality of Different Plant Products

Plant name	Plant material	Types of drying methods	Parameters studied	Remarks	Reference
Aloe vera L. (*Aloe barbadensis* Miller)	Aloe gel	Oven drying at 50°C, 60°C, 70°C and 80°C	Wettability and Water Absorption capacity and aloin content	Wettability of sample at 70°C was less and water absorption capacity was more and as the temperature increased from 50 to 80 °C, aloin content decreased from 10.6 to 1.7 ppm	Gulia et al. [10]
Aloe barbadensis Miller	Aloe gel	Air drying at different temperatures	Physicochemical and nutritional properties and antioxidant capacity of aloe gel	Loss of physicochemical and nutritional properties and decreased antioxidant capacity at 80 and 90 °C. Minor alterations produced at drying temperatures of 60–70°C	Miranda et al. [19]
Aloe barbadensis Miller	Aloe gel	Thermostatic water bath drying	Polysaccharides and barbaloin	Polysaccharides exhibited a maximal stability at 70°C decreasing either at higher or lower temperatures and the amount of barbaloin decreased with temperature and time	Xiu et al. [38]
Aloe barbadensis Miller	Aloe gel	Hot air drying at different temperatures (30–80°C)	Physico-chemical modifications of acemannan	Modification of acemannan was particularly significant when dehydration was performed above 60–80°C	Femenia et al. [6]

Asparagus officinalis	Asparagus spears	Tray drying, spouted bed (SB) drying, combined microwave and spouted bed drying (MWSB, refractance window (RW) drying and freeze drying.	Rehydration ratio, ascorbic acid and antioxidants	MWSB produced asparagus particles with good rehydration and color characteristics but the highest amount of AA was retained in the product of RW followed by freeze drying, MWSB and SB	Nindo et al. [22]
Banana (Musa acuminate)	Banana slices	air-drying (60°C at 1.45 m/s), microwave and microwave-finish drying (MFD)	Color, Rehydration	MFD had little effect on the color and rehydration capacity of dried products as compared to the air and microwave drying	Maskan. [18]
Banana (Musa acuminate)	Banana Chips	Air drying, vacuum microwave drying (VMD), freeze drying and combination of air drying and VMD	Flavor and Texture	The optimal process of 90%AD/10% VMD yielded banana chips with significantly higher volatile levels. While freeze drying banana chips retained more volatile compounds than air drying and VMD processed chips	Winnie et al. [37]
Papaya (Chilean Papaya)	Papaya fruit	Air drying (40, 60, and 80°C) on pretreated (osmotic pretreatment with 40, 50, and 60% w/w of sucrose solution) and nonpretreated samples.	Rehydration ratio, water-holding capacity, color, vitamin C content, firmness, and microstructure	Non-pretreated samples showed a clear turgor loss, color loss and low ascorbic acid retention and osmotic pretreatment improved the quality of rehydrated papayas, showing higher ascorbic acid retention and best firmness and color.	Lemus-Mondaca et al. [17]

TABLE 2.1 (Continued)

Plant name	Plant material	Types of drying methods	Parameters studied	Remarks	Reference
Potato (*Solanumtuberosum*)	Potato cylinder	Air drying and microwave drying	Rehydration, Shrinkage and Vitamin C	MD samples retained twice the vitamin C content of AD samples and had improved rehydratibility	Khraisheh et al. [14]
Potato (*Solanumtuberosum*)	Potato slices	Low-pressure superheated steam drying (LPSSD) and air drying	Color, texture, and brown pigment accumulation	LPSSD took shorter time to dry the product than air drying when the drying temperatures were higher than 80°C. Longer blanching time and lower drying temperature resulted in better color retention and led to chips of lower browning index	Leeratanarak et al. [16]
Rice (*Oryza sativa*)	Rice	Infrared (IR) radiation heating (4685 and 5348 W) for 15, 30, 40, 60, 90, and 120 s	Moisture and quality of rice	high heating rate, fast drying, and good rice quality achieved by IR heating of rough rice to about 60°C	Pan et al. [24]
Sweet potato (*Ipomoea batatas*)	Sweet potato tubers	Hot-air drying, microwave drying and vacuum-freeze drying	Antioxidant activity	The dried sweet potatoes in microwave possessed the highest antioxidant activity while the lowest activity was observed in hot-air dried samples	Yang et al. [39]

2.5 SUMMARY

Drying may cause irreversible modifications to active substances, affecting their original structure, which may promote important changes in the proposed physiological and pharmacological properties of aromatic and medicinal plant derived substances, and food products. But now a day, it may not be possible to store the perishable food or feed products for future consumption, if these are not processed. In this respect, a proper processing parameters need to be opted to ensure the biological integrity in the final product. Since the quality of a dried food or feed products is strongly implicated on the drying process as well as the processing conditions, it is necessary to study commonly used different drying methods along with working principles. Considering the significance of various drying techniques, this chapter provides a succinct overview of the effects of drying process on different plant materials along with different drying methodologies with special reference to *Aloe vera* gel.

ACKNOWLEDGMENTS

University Grant Commission (UGC), India is acknowledged duly for providing the UGC Dr. D.S. Kothari Post Doctoral Fellowship (Ref. No. F.42/2006 (BSR)/BL/1415/0529) to Dr. A. Ray.

KEYWORDS

- *Aloe vera*
- **Dehumidified air drying**
- **Feed products**
- **Food**
- **Freeze drying**
- **Infrared drying**
- **Microwave drying**

- **Processing**
- **Spray drying**
- **Storage**
- **Water activity**

REFERENCES

1. Bondaruk, J., Markowski, M., & Baszczak, W. (2007). Effect of drying conditions on the quality of vacuum-microwave dried potato cubes. *Journal of Food Engineering*, 81, 306–312.
2. Cohen, J. S., & Yang, T. C. S. (1995). Progress in food dehydration. *Trends in Food Science and Technology*, 6, 20–26.
3. Danny, A., Yun-Jeong, H., Diane, B., & Alyson, M. (2003). Comparison of the total phenolic and ascorbic acid content of freeze-dried and air-dried marion-berry, strawberry, and corn grown using conventional, organic, and sustainable agricultural practices. *Journal of Agricultural and Food Chemistry*, 51, 1237–1241.
4. Decareau, R. V. (1985). Microwaves. In: *The Food Processing Industry*. Academic Press, 14
5. Donsi, G., Ferrari, G., & Matteo, P. (2001). Utilization of combined processes in freeze-drying of shrimps. *Food and Bioproducts Processing*, 79, 152–159.
6. Femenia, A., Garcia-Pascualb, P., Simala, S., & Carmen, R. (2003). Effects of heat treatment and dehydration on bioactive polysaccharide acemannan and cell wall polymers from *Aloe barbadensis* Miller. *Carbohydrate Polymers*, 51, 397–405.
7. Genin, N., Orlando, F. L., & Rene, F. (1995). Analysis of the role of the glass transition in the methods of food preservation. *Journal of Food Engineering*, 26, 391–408.
8. Ginzburg, A. S. (1969). Application of infrared radiation in food processing. *Chemical and Process Engineering Series*, London: Leonard Hill.
9. Goff, H. D. (1992) Low-temperature stability and the glassy state in frozen foods. *Food Research International*, 25, 317–325.
10. Gulia, A., Sharma, H. K., Sarkar, B. C., Upadhyay, A., & Shitandi, A. (2009). Changes in physico-chemical and functional properties during convective drying of *Aloe vera*. *Food and Bioproducts Processing*, 88, 161–164.
11. Hung, J. Y., Wimberger, R. J., & Mujumdar, A. S. (1995). Drying of coated webs. In: *Handbook of Industrial Drying*. 2nd ed. Mujumdar, A. S. (Ed.), New York: Marcel Dekker Inc. pp. 1007–1038.
12. Jayaraman, K. S., Das Gupta, D. K., & Babu Rao, N. (1990). Effect of pre-treatment with salt and sucrose on the quality and stability of dehydrated cauliflower. *International Journal of Food Science and Technology*, 25, 47–60.
13. Jones, W. (1992). *A Place in the Line for Micronizer*. Special Report, Micronizing Co., UK, 1–3.

14. Khraisheh, M. A. M., & Magee, T. R. A. (1997). Microwave and air drying fundamental considerations assumptions and for the simplified thermal calculations of volumetric power absorption. *Journal of Food Engineering*, 33, 207–219.
15. Khraisheh, M. A. M., McMinn, W. A. M., & Magee, T. R. A. (2004). Quality and structural changes in starchy foods during microwave and convective drying. *Food Research International*, 37, 497–503.
16. Leeratanarak, N., Devahastin, S., & Chiewchan, N. (2006). Drying kinetics and quality of potato chips undergoing different drying techniques. *Journal of Food Engineering*, 77, 635–643.
17. Lemus-Mondaca, R., Miranda, M., Andres, Grau, A., Briones, V., Villalobos, R., & Vega-Galvez, A. (2009). Effect of osmotic pretreatment on hot air drying kinetics and quality of chilean papaya (*Caricapubescens*). *Drying Technology*, 27, 1105–1115.
18. Maskan, M. (2000). Microwave/air and microwave finish drying of banana. *Journal of Food Engineering*, 44, 71–78.
19. Miranda, M., Maureira, H., Rodríguez, K., & Vega-Galvez, A. (2009). Influence of temperature on the drying kinetics, physicochemical properties, and antioxidant capacity of *Aloe vera* (*Aloe barbadensis* Miller) gel. *Journal of Food Engineering*, 91, 297–304.
20. Mongpraneet, S., Abe, T., & Tsurusaki, T. (2002). Accelerated drying of welsh onion by far infrared radiation under vacuum conditions. *Journal of Food Engineering*, 55, 147–156.
21. Mujumdar, A. S., & Ratti, C. (1995). Drying of Fruits. In: *Processing Fruits: Science and Technology-Biology, Principle, and Applications*. Technomic Publication, Lancaster: 522.
22. Nindo, C. I., Sunb, T., Wangb, S., Tanga, J., & Powers, J. R. (2003). Evaluation of drying technologies for retention of physical quality and antioxidants in asparagus (*Asparagus officinalis* L.). *Lebensm.-Wiss.-Technology*, 36, 507–516.
23. Okos, M., Narsimhan, G., Singh, R., & Weitnauer, A. (1992). Food dehydration. In: *Handbook of Food Engineering*. D. R. Heldman and D. B. Lund (Eds.), pp. 437–562.
24. Pan, Z., Khir, R., Bett-Garber, K. L., Champagne, E. T., Thompson, J. F., Salim, A., Hartsough, B. R., & Mohamed, S. (2009). Drying characteristics and quality of rough rice under infrared radiation heating. *Journal of Food Processing Engineering*, 54, 203–210.
25. Ray, A., Aswatha, S. M. (2013). An analysis of the influence of growth periods on physical appearance, and acemannan and elemental distribution of *Aloe vera* L. gel. *Industrial Crops and Products*, 48, 36–42.
26. Ray, A., Ghosh, S., Ray, A., Aswatha, S. M. (2015). An analysis of the influence of growth periods on potential functional and biochemical properties and thermal analysis of freeze-dried *Aloe vera* L. gel. 86, 298–305.
27. Rupprecht, H. (1993). Basic physico-chemical principles of freeze-drying-lyophilization. *Pharmacevski Vestnik*, 44, 193–213.
28. Sahin, H., Topuz, A., Pischetsrieder, M., & Ozdemir, F. (2009). Effect of roasting process on phenolic, antioxidant and browning properties of carob powder. *European Food Research and Technology*, 230, 155–161.
29. Sakai, N., & Hanzawa, T. (1994). Application and advances in far-infrared heating in Japan. *Trends in Food Science and Technology*, 5, 357–362.

30. Sandu, C. (1986). Infrared radiative drying in food engineering: A process analysis. *Biotechnology Progress*, 2, 109–119.
31. Sanga, E., Mujumdar, A. S., & Raghavan, G. S. V. (2000). Principles and applications of microwave drying. In: *drying technology in Agricultural and Food Science,* A. S. Mujumdar (ed.). Science publishers, Enfield, USA, 283–289.
32. Sheridan, P., & Shilton, N. (1999). Application of far-infrared radiation to cooking of meat products. *Journal of Food Engineering*, 41, 203–8.
33. Shoyele, S. A., Cawthorne, S. (2006). Particle engineering techniques for inhaled biopharmaceuticals. *Advanced Drug Delivery Reviews*, 58, 1009–1029.
34. Sommerfeld, M., & Blei, S. (2001). Lagrangian modeling of agglomeration during spray drying processes. *Pharmacy Technician*, 6, 171–173.
35. Thibault, J. F., Lahaye, M., & Guillon, F. (1992). Physico-chemical properties of food plant cell walls. In: *Dietary Fiber—A Component of Food: Nutritional Function in Health and Disease*. T. F. Schweizer & C. A. Edwards (Eds.). London: Springer, 21–39.
36. Welti-Chanes, J., Guerrero, J. A., Aguilera, J. M., Vergara, F., & Barbosa-Canovas, G. V. (2004). Glass transition temperature and water activity of dehydrated apple products. *Journal of Food Process Engineering*, 22, 91–101.
37. Winnie, Mui., Timothy, D., & Christine, S. (2002). Flavor and texture of banana chips dried by combinations of hot air, vacuum, and microwave processing. *Journal of Agricultural and Food Chemistry*, 50, 1883–1889.
38. Xiu Lian, C., Changhai, W., Yongmei, F., & Zhaopu, L. (2006). Effects of heat treatments on the stabilities of polysaccharides substances and barbaloin in gel juice from *Aloe vera* Miller. *Journal of Food Engineering*, 75, 245–251.
39. Yang, J., Jin-Feng, C., Yu-Ying, Z., & Lin-Chun, M. (2010). Effects of drying processes on the antioxidant properties in sweet potatoes. *Agricultural Sciences in China*, 10, 1522–1529.

CHAPTER 3

APPLICATION OF EMERGING TECHNOLOGIES FOR FREEZING AND THAWING OF FOODS

C. K. SUNIL, ASHISH RAWSON, D. V. CHIDAN, and AKASH PARE

Assistant Professor, Indian Institute of Crop Processing Technology, Thanjavur – 613005; E-mail: sunil.ck@iicpt.edu.in, ashish.rawson@ iicpt.edu.in, chidanand@iicpt.edu.in, akashpare@iicpt.edu.in

CONTENTS

3.1   INTRODUCTION

Freezing is a widely applied and excellent preservation method that been used for thousands of years because of the high product quality achieved and ability to achieve stability without damaging initial quality. The advantage of the freezing process is of retaining valuable sensory attributes and nutritive value of fresh foods for a longer period of storage. The freezing process slows down the microbial activity, biochemical reactions, changes in color and texture, and loss of nutrients during storage. Freezing attains the preservation objective by combining two factors: concentration of solute and low temperature of storage. The freezing process is different application apart from storage such as freeze concentration, freeze-drying, cryocomminution and along with thawing for texturizing.

The quality of the frozen foods is closely related to the size and distribution of the crystals formed during the process. The freezing process involves two stages: (i) formation of ice crystals; and (ii) increase in crystal size. The rate of freezing affects the size and distribution of the ice crystals in the frozen food, and is critical to the quality of the frozen foods. Rapid freezing produces small and even size crystals, whereas slow freezing forms large crystals which may have effect on the product quality by causing cell rupture thereby causing damage to the tissue. The crystal size can vary with the type of food, even when subjected to same freezing conditions with same dimensions, because of difference in the free water available in the foods. Both rate of freezing and formation of small crystals play an important role in minimizing mechanical damage, drip loss and thus reducing the quality of the food.

The quality of the frozen food is generally related to the freezing and thawing process. Faster the freezing process, better the quality of the product. Thawing process is slower than freezing. During the process of thawing, the product is subjected to temperature, microorganisms, physical and chemical changes. Thawing process involves the increasing of temperature of the frozen product to the melting or unfrozen state. It is important to know that rapid freezing and slower thawing process is required to achieve the best product quality, yield and safety of frozen foods. But with the advent of the new emerging technologies, the desirable process of thawing should be faster at lower temperatures to avoid rise in temperature of the product, which causes dehydration, change in texture or structure of the product and thereby assuring the product quality.

With the objective of preserving the quality of the frozen foods, many researchers have studied the new emerging technologies, which help in rapid or quick freezing and faster thawing. This chapter discusses application of emerging technologies in freezing and thawing of foods.

3.2 FREEZING

Freezing process can be defined as the process of removing sensible and latent heat in order to store the food at temperature of $-18°C$ or below with a primary objective of preservation of the functionality. The water transforms into ice crystals upon freezing, which preserve the food structure [18]. A typical freeze curve is shown in Figure 3.1. The freezing curve of food will be different from those of a pure substance, as food is a multi-component system. The Figure 3.2 shows the freezing process in foods [60]. The freezing process includes following three phases:

a. **Precooling or chilling phase**: In this stage only the sensible heat from the food is removed, product temperature is lowered and crystallization is about to begin.

b. **Phase change period**: Further lowering the temperature, the free water in the food start to crystallize and form ice crystals with removal of latent heat of fusion. This stage is very important to product quality.

c. **Subcooling or Tempering phase**: This happens when most of the freezable water is converted to ice.

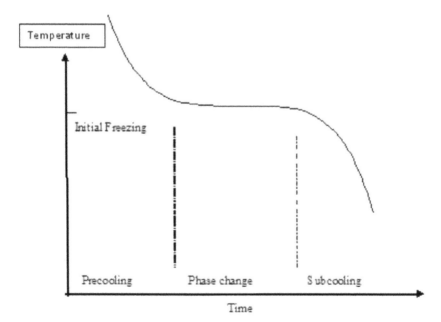

FIGURE 3.1   A typical freezing curve.

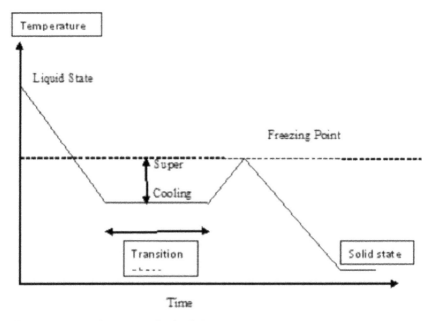

FIGURE 3.2 Freezing process for foods [60].

The freezing process involves two successive processes in ice crystal formation: (i) Formation of ice crystals (nucleation); (ii) Increase in crystal size (growth). Nucleation takes place when temperature of the food is lowered to the initial freezing point. The nucleation takes place in two ways depending on the purity of the water: Homogeneous nucleation (Pure water) and Heterogeneous nucleation (Foods). The growth of the ice crystals depends on the amount of latent heat released and removed. In super-cooling stage, the difference between the actual temperature of the sample and the expected solid–liquid equilibrium temperature at a given pressure is the driving force for ice nucleation and controls the size and number of ice crystals formed [12].

Burke et al. [10] stated that there is an increase in tenfold in the ice nucleation rate with each degree K of super cooling. At rapid cooling rates, there are numerous nucleation sites resulting in formation of many small crystals. The size distribution of the ice crystals is important parameter in frozen foods. Smaller size and even distributions of ice crystals is desired for higher quality of frozen foods.

As mentioned earlier, to achieve the high quality of frozen food, higher freezing rate is desired. Martino et al. [61] mentioned that the size and location of ice crystals formed during freezing depend on the rate of freezing and final temperature and also effects quality parameters such as texture and color of the frozen products. The freezing rate depends on many factors like: temperature between freezing medium and food, effective heat transfer coefficient, food shape and size and thermo-physical properties of food system. The thermal gradient between an interior point in the food product and the surface of the food determines the local freezing rate. Sanz et al. [89] mentioned that the freezing rate decreases towards the center of the product in large volume products.

3.2.1   FREEZING METHODS

Industrial freezers are categorized based on the cooling medium used, methods of contacting food with the cooling medium and methods of conveying foods through the freezer. The freezing methods are classified as: (i) Sharp Freezing; and (ii) Quick/Rapid Freezing. Sharp freezing is batch type method, and slower process of freezing. The time taken varies from 3 to 72 hours. This results in bigger size crystal formation. Quick Freezing is rapid method of freezing, which may take 30 minutes to freeze. The crystals formed are of very small size resulting in better quality of the frozen food. The quick freezing methods are presented in the following subsections.

3.2.1.1   Air Freezing

The foods are frozen by air as cooling medium at temperature of −18° to −40°C. The performance of the air freezer depends on the air velocity in the freezer. The air freezing can be classified as:(i) Air Blast freezing-Batch Type (Tunnel freezing), (ii) Fluidized bed freezing (Continuous type), and (iii) Spiral freezers (Continuous Type).

3.2.1.2   Indirect Contact Freezing

Food products can be frozen by placing them in contact with a metal surface (Often steel plate) cooled either by cold brine or vaporizing refrigerants, such as refrigerant-12, 22 or ammonia. Contact plate freezing is an economical method as it minimizes the problem of product dehydration, defrosting of equipment and package bulging.

3.2.1.3   Immersion Freezing

Foods to be frozen are immersed in the cold liquid media, primary aqueous solutions. Solutions of glycols, glycerol, sodium chloride, calcium chloride, and mixtures of salt and sugars can be used as cooling medium.

3.2.1.4   Cryogenic Freezing

Cryogenic freezing refers to the process in which the food is exposed directly to liquid boiling at a very low temperature or a solid subliming at a very low temperature. The cooling media used for the cryogenic cooling are liquid and solid CO_2, N_2, N_2O, and Freon 12 (CCl_2F_2). This method of freezing has the advantage of achieving very rapid heat transfer.

The problems observed in the conventional freezing are non-uniform crystal development, destruction of food material structure and loss in the food quality; and these have given rise to application of high-pressure freezing, ultrasound freezing, osmo-dehdyofreezing, antifreeze proteins and ice-nucleation proteins.

3.2.2   APPLICATION OF EMERGING TECHNOLOGIES: FREEZING

3.2.2.1   High Pressure Freezing

The high pressure processing as a non-thermal and its application is widely increasing in food processing. Many studies have been conducted dealing with the application of high-pressure effects on ice-water transitions. It is a known fact that water undergoes different phase changes when subjected to high pressure. At a pressure of 207.5 MPa, water remains in a liquid state below 0°C, with a reduction in freezing point to a minimum of –22°C. Because water expands on freezing by Le Chatelier's principle, increase in pressure will cause decrease in freezing point. By observing the effect of pressure on the phase diagram of water, various methods of high-pressure applications are studied.

The phase diagram (Figure 3.3) shows that the ice exists in several phases. The type of the ice formed during the freezing is an important factor, which affects the quality of the product. A total of nine solid ice phases are known. The common ice form, which exists at low pressure, is called Ice I, which has a lower density than that of liquid water. The formation of ice I results in a volume increase of 9% on freezing at 0°C and about 13% at 20°C [41]. The freezing point of water decreases with pressure up to 210 MPa and opposite was observed to other forms of Ice other than Ice I. The increase in volume results in tissue damage during freezing. The increase in pressure, other types of ice is formed as shown in Figure 3.3. The types of ice formed under pressure have higher densities than water and they do not expand during phase transition. So, the tissue damage is reduced when food is subjected to higher pressure.

The effect of pressure P on the freezing temperature T of ice is related to the volume change ΔV and change in enthalpy ΔH, according to the Clausius–Clapeyron equation;

$$\frac{dT}{dP} = \frac{\Delta V.T}{\Delta H} \tag{1}$$

As water expands on freezing, the ΔV is positive in Ice I and ΔH is negative as heat is lost from the water so that the right hand side is negative and the freezing point decreases with increase in pressure. The increase in volume causes the damage in the cells. The volume of increment is negative in liquid-ice III, liquid-ice V or liquid-ice VI phases, resulting in less damage upon freezing (Figure 3.3). Three different types of high-pressure freezing processes can be distinguished based on the occurrence of the phase transition:

1. High-pressure assisted freezing (HPAF): Phase transition under constant pressure.
2. High-pressure shift freezing (HPSF): Phase transition due to pressure release.
3. High-pressure induced freezing (HPIF): Phase transition initiated by a pressure change and continued at constant pressure.

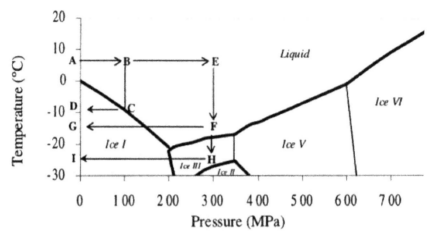

FIGURE 3.3 High pressure freezing and thawing processes on the phase diagram [53]. (a) ABCD: pressure-assisted freezing, (b) DCBA: pressure-assisted thawing, (c) ABEFG: pressure shift freezing, (d) GFEBA: pressure-induced thawing, (e) ABEFHI: freezing to ice III, (f) IHFEBA: thawing to ice III.

The terminologies were first suggested by Knorr et al. [50] and also they proposed six different high-pressure freezing and thawing processes as shown in Figure 3.3.

3.2.2.2   High-Pressure Assisted Freezing (HPAF)

The phase transition occurs under constant pressure, higher than atmospheric pressure and the temperature is lowered to the corresponding freezing point. Reduction in phase transition times can be achieved as the latent heat of crystallization is reduced when pressure increases. In Figure 3.3, HPAF is shown as ABCD. In this process, sample is cooled under pressure upto its phase change temperature at an applied pressure. This method is applied to various products like tofu, carrots, Chinese cabbage, or agar gel, etc. There is a substantial reduction in freezing times in application of HPAF compared with times required at atmospheric pressure [50].

3.2.2.3   High-Pressure Shift Freezing (HPSF)

In this method, the phase transition occurs as a result of pressure change, promoting metastable condition and instant ice production. The pressure can be released slowly over a period of time (minutes) or quickly in 1–2 s, which results in a high super cooling effect and the ice, nucleation is increased [54]. ABEFG region in Figure 3.3 shows HPSF. On expansion of the food, the pressure release occurs instantaneously and a subsequent decrease in its temperature is produced, also resulting in large scale super cooling. The main advantage of HPSF is instantaneous formation of ice initially and homogeneous throughout the product. Therefore, HPSF can be useful for freezing of foods of large dimensions, having effect of freeze cracking because of thermal gradient [61]. The HPSF has been applied in freezing of different products like fruits (Peaches, mangoes), vegetables (Potatoes), pork, lobster, and tofu etc. A model has been developed considering that when water expanded, it does not extend over its melting curve but reaches a metastable state to calculate the amount of ice formed instantaneously after a rapid expansion in HPSF [74].

3.2.2.4   High-Pressure Induced Freezing (HPIF)

In this process phase transition initiated by a pressure change and continued at constant pressure. This was first described by Urrutia Benet et al. [102].

The high pressure freezing process is done in high pressure resistant vessels, where the packed product is immersed in the pressure or cooling medium (Table 3.1). The cooling medium is used as pure or as mixture. The cooling media is selected based on the freezing point of the fluid under pressure, viscosity and other properties like heat capacity, thermal expansion coefficient and specific volume. The different cooling media used are: silicon oil, propylene glycol, glycol/water (62/38 v/v), ethylene glycol/water: (75/25 v/v), ethanol/glycol (20/80 v/v), castor oil/ethanol (15/85, v/v), etc.

TABLE 3.1   Applications of High Pressure Freezing

Food	Conditions	Observations
Carrot [30]	• HPSF • 100–700 MPa • –20°C, later –30°C at 1 atm	• Maintain shape and texture
Chinese cabbage [31]	• HPSF • 100–700 MPa • –20°C, later –30°C at 1 atm	• Texture loss at 100 and 700 MPa
Eggplant [76]	• HPSF	• Reduction in drip volume (20%)
Mangoes [75]	• HPSF	• Maintain texture • Reduction in drip loss • Reduction in freezing time
Norway lobsters [12]	• HPSF • 200 MPa • –18°C	• Undesirable increase in toughness- induced protein denaturation
Peaches [75]	• HPSF	• Maintain texture • Reduction in drip loss • Reduction in freezing time
Porcine and bovine muscle [26]	• HPSF • 200 MPa • –20°C , 30 min	• Fragmentation of myofibrils
Pork [61]	• HPSF • 200 MPa • –20°C	• Low microstructure damage
Potato [51]	• HPSF • 400 MPa • –15°C	• Texture maintained • Color maintained • Less dissolved substances in drip • Less enzymatic browning
Tofu [42]	• HPSF • 100–700 MPa • –20°C	• Maintain initial shape and texture • No drip loss • Very small size crystals

3.2.3 ULTRASOUND

Ultrasound application is increasingly being applied in food processing and preservation. The ultrasound is mainly classified into two fields: high frequency low energy diagnostic ultrasound (5–10 MHz) and low frequency high-energy power ultrasound (20–100 kHz range). The former is used as an analytical technique for quality assurance, non-destructive inspection, process control, to determine food properties, to measure flow rate, to inspect food packages [4, 63, 80]. The latter is used in different processes like crystallization, drying, degassing, extraction, filtration, homogenization, meat tenderization, oxidation, sterilization, etc. [62]. The ultrasound has been applied in various aspects of food freezing such as initiation of nucleation [44, 109] control of ice crystals size [43, 87], acceleration of freezing rate [18, 36, 55, 108] and quality improvement [38, 92, 108]. The acoustic energy can be directly applied to the product (direct immersion of ultrasonic probe) or indirectly from transducer coupling through parts of the processing vessel [1]. Therefore, the type of ultrasonic apparatus varies according to the product and the type of freezers. Zheng and Sun [116] have explained in detail the arrangement of the ultrasonic device in different types of freezers like air blast freezer, immersion freezer, plate freezer, chest freezer and scraped surface freezer to minimize the negative effects. Ultrasonic intensity decreases abruptly as the distance from the radiating surface increases and is attenuated with the increase of the presence of solid particles.

Ultrasound is thought to enhance the nucleation rate and rate of crystal growth in a saturated or super cooled medium by producing a large number of nucleation sites in the medium throughout the ultrasonic exposure, which may be due to cavitation bubbles serve as nuclei for crystal growth [63] and/or by the disruption of seeds or crystals already present within the medium thus increasing the number of nucleation sites. The acoustic cavitation caused by power ultrasound is the most important effect in food freezing which promotes nucleation, and also enhances heat and mass transfer due to the violent agitation created by acoustic micro streaming [116]. The influence of ultrasound on conventional cooling of food provides rapid and even seeding, thereby reducing the dwell time [1]. And also there are more number of seeds, smaller size of crystals thereby reducing the cell damage [92]. The ultrasound accelerates the cooling by improving the heat transfer [55, 56]. Suslick [93] concluded that power ultrasound can significantly increase the nucleus number in a concentrated sucrose solution.

The application of the low frequency high-energy power ultrasound (20–100 kHz range) is being applied in food processing and preservation due to its ability to control/modify nucleation and crystal growth [1, 2, 63]. It is also chemically noninvasive and operate in a non-contact mode [1, 18]. The power ultrasound has been applied to control freeze drying, freezing of oil-in-water emulsions and tempering of chocolate by Acton and Morris [1]. And many studies have been conducted, which showed the potential of using power ultrasound in accelerating the freezing rate and improving the quality of frozen foods like potatoes [56] and apples [18]. Price [81] has applied power ultrasound for production of molded frozen products such as sorbets and ice lollipops, which resulted in product of small ice crystals and uniform crystal size distribution and also improved the adhesion of lollipop to the wooden stick. Table 3.2 summarizes the applications of ultrasound technology in freezing of foods.

3.2.4   DEHYDROFREEZING

In the process of dehydrofreezing, the foods are dehydrated before freezing [100], where the reduction in amount of water in food decreases the number and size of ice crystals, and tissue damage during freezing. For better or improved texture after freezing and thawing at least 50% of the water must be removed by dehydration. The dehydration results in increased solute concentration, thereby decreasing the freezing point and increase the glass transition temperature, leading to more super cooling and better stability. Other advantages of the dehydrofreezing are better retention of pigment, vitamin and aroma, reduction in freezing time (less water to freeze) and reduction packaging and transport cost due to reduced weight. The dehydrofreezing technology is applied in fruits and vegetables. The dehydration process can be of different methods: air drying or osmotic dehydration (Osmo-dehydrofreezing).

TABLE 3.2   Ultrasound Applications in Freezing of Food

Food	Conditions	Application
Radish- Cylinder [109]	• 20 kHz • Time duration: 0, 3, 7, 10 and 15 s • Ultrasound intensity: 0.09, 0.17, 0.26 and 0.37 W/cm² • Onset temperature: –0.5, –1, –1.5 and –2°C	Initiation of Nucleation
Sucrose solution [14]	• 20 or 67 kHz • Ultrasonic output level: 1–10% • Duty cycle: 10–100% pulsed • Time duration: 1–20 min	
Super cooled water [37]	• 28 kHz, 0–100 W	
Agar gel sample [42]	• 25 kHz • Irradiation duration: 90, 180 and 270 s • Ultrasound intensity: 0.07, 0.25 and 0.42 W/cm²	Control size of ice crystals
Potato [36]	• 25 kHz • Actual ultrasound power: 0, 7.34, 15.85 and 28.89 W • Exposure time: 0, 1, 1.5, 2 and 2.5 min	Freezing rate acceleration
Frozen dough [56]	• 25 kHz • Electric power levels: 0, 175, 224, 288, 360 and 418 W • Duty cycle: 30 s on/30 s off	
Potato [92]	• 25 kHz • Actual ultrasound power: 0,7.34, 15.85 and 28.89 W • Exposure time: 2 min	Quality improvement
Broccoli [108]	• Ultrasound power: 0, 125, 150, 175 and 190 W • Duty cycle: 60 s on/60 s off	

The osmo-dehydrofreezing involves the immersion of food in a concentration solution of a solute. The partial dehydration by osmotic concentration of food has shown to decrease in enzymatic browning [17], reduction in structural collapse and drip loss during thawing [27]. For fruits, the popular solutes in use are sucrose, fructose, lactose, maltodextrin, and corn syrup. For vegetables, sodium chloride is commonly used. And oligofructose, trehalose and high-DE maltodextrin are also used [19]. Addition of sugars due to osmotic dehydration increases firmness of thawed/dehydrated tissues with reduction in subsequent ice formation in apple as shown by Tregunnol and Goff [101]. Moyano et al. [69] mentioned that the optimum conditions of 65° Brix at 20°C for 60 minutes osmotic dehydration of frozen papaya, resulted in higher overall acceptability of the product.

The osmotic dehydrofreezing has been suggested as a suitable process for production of reduced moisture content in fruits and vegetables with a natural flavor, color, texture and functional properties without preservatives [7, 27, 99].

3.2.5 ANTIFREEZE PROTEINS AND ICE-NUCLEATION PROTEINS

With the importance of controlling the growth of ice crystals in frozen foods, antifreeze protein and ice-nucleation protein are directly added to food. They interact with the food and influence the crystal size and crystal growth [35]. Both are opposite in function on ice crystals. The antifreeze protein can lower the freezing temperature and retard crystallization, whereas the ice-nucleation protein raise the temperature of the ice nucleation and reduce the degree of super cooling [23, 57].

3.2.5.1 Antifreeze Proteins

This approach is used in some fish and insects to prevent their blood from freezing at -1.9°C. They prevent their blood from freezing by producing antifreeze proteins, which binds to the surface of the ice crystals, and prevent their growth. The antifreeze protein is also said to be present in

insects, invertebrates, plants and bacteria [33]. The antifreeze is classified as two based on the presence or absence of carbohydrates: glycopro-
teins (AFGP) and non-glycoproteins (AFP) [35]. The main function of
antifreeze proteins is to reduce the freezing temperature and suppress the
growth of ice nuclei, resulting in inhibition of ice formation and growth
rate [35].

One of the potential applications in food is incorporation in dairy
foods such as ice-cream and de-icing agents to prevent the crystal growth
during storage, especially when temperature fluctuates [106]. Warren
et al. [106] patented the addition of antifreeze protein to commercially
available food products. Other application is seen in chilled and frozen
meat. Less drip loss and smaller crystal formation was observed when the
AFGP of 0.01 g/kg was injected 24 h before slaughter [24]. The application of antifreeze in food depends on the cost. The use of genetically
engineered AFP or synthetic AFP can help to overcome the problem of
cost [58].

3.2.5.2 Ice-Nucleation Protein

This was first taken from some amphibians and reptiles, where they generate ice nucleating protein as soon as the body temperature reaches –2
to –3°C and also produce cryoprotectants (glucose/glycerol) to lower the
freezing point of sensitive organs to delay freezing. Ice nucleation active
(INA$^+$-Produce INA substance) bacterial cells and their products such as
ice-nucleation protein (INP) are having great potential in food application. INP elevate the temperature of ice nucleation, reduce freezing time
and change the texture of frozen foods, resulting in less energy cost and
improve quality.

Studies were conducted with liquid (milk, juice), semi-solid (ice cream)
and solid (ground beef) [57]. The degree of super cooling was significantly
reduced when 70 g protein (ECIN – extracellular ice nucleation) in 10 mL
samples freezing at –6°C. But no apparent effect was noted on nucleation
temperatures for solid food (ground beef). INA is also used for improving
the freezing texture [3]. Studies were also conducted in egg white, hydro-
gels (proteins and polysaccharides), rice flour pasta, tofu, meat and poultry

protein. The application of INA bacterial cells and their products in food is safe, non-toxic and non-pathogenic.

3.3 THAWING

The success of the freezing application for enhancing the shelf life of food products depends on the optimized thawing conditions. Thawing is a reverse process of freezing but occurs slower than freezing. When center of the food is reached to 0°C, then we can say thawing process is complete. The thawing process is slower than freezing process and is a time consuming process (10 times slower) than the freezing process, and allows further damage to the food by accelerating the physiochemical changes and microbial growth. Therefore thawing of frozen foods is very important stage in terms of minimizing the amount of proteinaceous exudate (drip) loss from different food products on thawing resulting reduction in quality. For maintaining the food quality of frozen foods, rapid thawing at low temperatures and excessive dehydration of food is recommended [25, 41]. By generating the heat within the food, the thawing process can be made more rapid. During thawing, the frozen material is surrounded by continually expanding immobilized water layer resulting in resistance to heat flow. So, the thawing process for foods that are not fluid in nature is slower than the freezing process. The effects of thawing on food is more damaging than freezing due to much longer residence time in the most damaging temperature zone below the freezing point. The reason for the damage is due to the heat transfer through unfrozen portions is much slower than through frozen layer

The size and location of ice crystals formed during freezing has effect on exudation of moisture (drip loss) on thawing [12, 61]. Kalichevsky et al. [41] have shown that the volume of drip produced on thawing is closely related to the rate of freezing. Freezing and thawing prior to dehydration improves the rehydration rate due to generation of porosity through crystal formation. The most common method of thawing is apply heat to the surface of the material and allows to conduct through the food to center. Apart from the conventional method, a number of novel and innovative thawing technologies including high pressure, microwave, ohmic heating and ultrasound are being explored.

3.3.1 APPLICATION OF EMERGING TECHNOLOGIES: THAWING

3.3.1.1 High-Pressure Thawing

Application of high-pressure for thawing has advantages like: (i) preserve food quality; and (ii) reduction in thawing time [59, 114, 115]. The advantage of thawing food under high pressure is that the freezing point is dispersed, by which the temperature driving force (difference between air temperature and product temperature) can be increased while keeping the temperature low to avoid microbial growth. The reduction of thawing time in this process is directly related to this relative increase in the temperature difference or gap caused by the decrease of the melting point of ice until 210 MPa. The minimum equilibrium freezing point is about −22°C at 210 MPa. It has been reported that the use of pressures of 100–200 MPa may reduce the time required for thawing. The high-pressure application for thawing has both positive and negative effects on quality [40].

The thawing rate depends only on the conduction of heat, as pressure is transmitted uniformly through the sample [41]. The pressure level and treatment time, have effect on thawing rate and product quality [115]. But the size and initial temperature of the food did not have effect on thawing rate. So this can be followed for thawing of larger amount of product at higher pressure. The limitations for the application in food are: (i) high cost, (ii) high pressure freezing encounters, (iii) protein denaturation (animal tissues), and (iv) meat discoloration [41, 66].

The process of high pressure thawing is categorized as pressure assisted thawing and pressure induced thawing [50]. In Figure 3.3, DCBA corresponds to pressure assisted thawing and GFEBA to GFEBA. And IHFEBA is the process of thawing the frozen food to Ice III foods. This process involves the thawing of Ice III, which has higher densities than liquid water and Ice I.

3.3.1.2   Pressure Assisted Thawing

It corresponds to thawing under a constant pressure (DCBA in Figure 3.3). Frozen food is pressurized (DC) until it reaches thawing point (Figure 3.3). Heat is then applied under pressure to melt the ice. The pressure on the food is released when the food is completely thawed.

3.3.1.3   Pressure Induced Thawing

This process is reversion process of pressure shift freezing. This process corresponds to the phase transition initiated with a change and continued at a constant pressure (Figure 3.3: GFEBA). Table 3.3 indicates applications of thawing technology.

3.3.1.4   Ultrasound Thawing

Ultrasound has been explored in thawing of frozen food during the 50 years. It is an innovative technology in thawing. Studies have been conducted by various researchers using ultrasound at various frequencies and power levels to study the effect on thawing of frozen food quality [13, 47, 48, 52, 68, 116]. The ultrasound is associated with disadvantages such as

TABLE 3.3   High Pressure Thawing Applications

Food	Conditions	Observations
Beef [20]	50–200 MPa at 20°C for 30 min	• 50 MPa optimal • Higher pressure induced denaturation
Frozen meat [59]	–	• Reduction in thawing time.
Konjac glucomannan gel [96]	200–400 MPa	• Improvement of texture
Pollock whiting fillets [13]	150 MPa	• Reduction in drip loss
Strawberries [21]	600 MPa, 15 min, 50°C	• Improvement in sugar uptake • Reduction in vegetative microorganisms

poor penetration, localized heating and high power requirements resulting in wide application of this method [71, 116]. In 1999 Miles et al. [68] showed that overheating occurs at surface of food when application of high intensities and both high and low frequencies of ultrasound, which they overcame by adjusting the frequency (500 kHz) and intensity (0.5 W·cm^{-2}) for beef, pork and cod when thawed at 7.6 cm within about 2.5 h. They reported that the reason for overheating at the surface may be due to increase in attenuation with frequency and onset of cavitation at lower frequency. By using low frequency ultrasound. The thawing time can be shortened [48].

Kissam et al. [48] shortened the thawing time by 71% by raising the temperature of blocks of frozen cod from –29°C to –1°C, with no effect on product quality, and by using a combination of 1500 Hz and 60 W. Relaxation mechanism has shown that large amount of acoustic energy is absorbed by frozen foods when frequency is in the relaxation frequency range of ice crystals and also faster than conductive heating. It was also found that the thawing process required 71% less time by using acoustically assisted water immersion than using water immersion thawing when 1500 Hz acoustic energy at 60 W is applied. In meat and fish, studies indicated that acceptable thawing was achieved around 500 kHz, in confirmation with relaxation mechanism. Ultrasound thawing can shorten the thawing time, reduce drip loss and improve product quality.

3.3.1.5   Microwave Thawing

Application of Microwave for thawing of foods has been studied by many researchers [6, 79, 82, 104, 108]. The unique property of the microwaves is to penetrate and produce heat deep in the food, which has the potential of decreasing the thawing time. Microwave thawing is a faster method: the thawing time is reduced as much as a factor of seven than the convective thawing at ambient temperatures [98, 103]. In 1993, Tong et al. [98] designed a microwave oven with variable continuous power and feedback temperature controller to maintain a desired temperature gradient, which resulted in reduction of thawing time as much as a factor of seven compared to convective thawing. The advantages of the microwave thawing are decrease in thawing time, less floor space for processing and reduces drip loss, texture damage, microbial problems and chemical deterioration as well as discoloration compared to conventional thawing [65, 95, 103, 107]. The microwave thawing application is limited because of the localized overheating (run-away heating- some parts may cook while others remain frozen during thawing) resulting in non -uniform thawing. The preferential absorption is the major cause for runaway heating, resulting excess water loss and thermally chemical deterioration [103]. The non-uniformity of heating is due to uneven power distribution and increasing preferential absorption of microwaves in liquid regions. The thawing rate of the food is dependent on the size, shape, thickness and dielectric property of the food, and magnitude and frequency of microwave radiation [77, 85, 94]. The increase in the electric field intensity input increases the heating rate as well as thawing rate [85].

The non-uniform heating in microwave thawing is overcome by using power cycling or lower power levels in a continuous manner by Chamchong and Datta [11]. The water in the food is dissolved with various components, when frozen and thawed over a range of temperatures results in heterogeneous region, coexisting solid and liquid phase known as mushy zone [5]. Tahir and Farid [94] showed that the thawing time is one fifth less than that necessary in conventional thawing in microwave cyclic thawing of frozen meat using effective heat capacity method. They also observed that the thawing started from the surface and progressed slowly down to the bottom as penetration depth was small. Xin et al. [108] used different power levels ranging from 300 to 700 W, in frozen Hami-melon, and observed that the melon treated at 500 W retained better quality and highest sensory score compared to high pressure thawing at 10–250 MPa.

The microwave thawing has been studied by many researchers by developing different models. Pangrle et al. [77] studied the thawing process of frozen cylinders by coupling electromagnetic and thermal model using a plane wave as opposed to resonant cavity and developed a one-dimensional model for thawing of cylindrical samples. Later two-dimensional models for thawing were developed and predictions were compared with experimental data [5]. Basak and Ayyappa [5] studied with fixed grid based effective heat capacity method coupled with Maxwell's equations of microwave thawing of 2D frozen cylinders. Ratanadecho et al. [85] conducted the first analysis of a two-dimensional model based on Maxwell's equations coupled with the arbitrary moving front equation.

Microwave technology has been used at atmospheric pressure for thawing for many years. By reducing the pressure, the product temperature can be kept low resulting in better thawing and quality. Under vacuum conditions, the temperature rise can be mediated. In 1984 James thawed blocks of meat at 1 kPa pressure inside the chamber and microwaves at 915 MHz, stored at 20°C. These blocks were thawed in less than 2 h and the maximum surface temperature reached was in the range of 14.9°C to 26.7°C. He also mentioned that the high-energy requirements and capital cost limits the industrial application. Ito et al. [39] designed an equipment to expose the frozen foods to alternating cycles of vacuum application and release, with microwave treatment imitated partway through the vacuum release phase of each cycle and this is a patented technology.

3.3.1.6   Ohmic Heating

The ohmic heating is being applied to thaw frozen foods. The principle of ohmic heating is to increase in temperature of the food by generation of heat in food when electric current passes through food with high electrical resistance [28]. Ohmic heating has advantages like all the energy enters the food as heat and it has no limitation in depth of penetration. Ohmic heating is a volumetric heating method. Volumetric heating methods do not require large amount of water and thawing is more rapid [16]. The increase in voltage gradient in ohmic heating decreases the thawing time without any change in thawing loss [34]. With the problems in dielectric and microwave thawing like cooking of the surface before the center is thawed; Ohmic heating is an alternative for thawing of frozen foods. The electrical conductivity of frozen foods is twice lower than thawed foods, resulting in cooking of the thawed portion of the block while the rest is still frozen [32]. The factors affecting the ohmic heating application for thawing are electrical resistance or electrical conductivity of the food. Also frequency of the current is very important factor for ohmic thawing effectiveness. The wider application of ohmic heating is hindered by the hot spots in the food during and also the shape limitation, no complete contact of odd shaped products treated. Ohmic heat application effectiveness also depends on the physical state of the food product. It is difficult to conduct electricity in solid bodies [91].

Many studies have been conducted in application of Ohmic heating for thawing of frozen foods. Roberts et al. [86] have developed an ohmic thawing unit, for thawing of shrimp blocks without heating problems. Similarly Ohtsuki [71, 72] has patented an ohmic thawing process for frozen food stuffs treated with high voltage electrostatic field where the frozen foods are positioned with negative electrons and can be used for thawing rapidly in a temperature range −3 to 3°C. Naveh et al. [70] and Wang et al. [105] reported the method of liquid contact thawing for frozen meat by ohmic heating. Naveh et al. [70] used an electrode solution to overcome the shape limitation of foods to be thawed. Wang et al. [105] also showed that the use of brine as carrying fluid resulted in faster thawing process, when brine concentration increased and largest surface of food is perpendicular to electric field. The drip, cell damage and softening in frozen carrots was prevented by ohmic heating [29]. Similar results of reduced drip loss and improved water holding capacity were obtained by Yun Lee and Park [112] in frozen chucks of meat treated with lower voltages in combination of conventional water immersion thawing at 60–210 V (a.c.) and frequencies of 60 Hz–60 kHz. Yan Mingduo et.al. [110] observed that the ohmic heating treated pork at 0.4% saline solution and 100V speeded up the thawing process and maintained the meat quality. Application of ohmic heating for the thawing of frozen food has a high potential to maintain the higher quality.

3.4 CONCLUSIONS

The application of different emerging technologies in freezing and thawing, the quality of the food is improved or retained. Many studies have shown positive effects of the freezing and thawing. Only disadvantage observed is the initial cost of the equipment and the additional cost of the treated food. Research can be further conducted at industry levels to popularize the emerging technologies application in freezing and thawing. And also studies can be conducted in other methods like RF assisted freezing, Microwave assisted freezing and electric freezing (alternating current and direct current electric freezing).

3.5 SUMMARY

Freezing and thawing has played an important role in the food preservation since olden times. However the problems like non-uniform crystal development, destruction of food material, food quality loss, drip loss, etc., observed in conventional freezing and thawing methods, have given rise to application of new emerging techniques. With the emergence of novel technologies such as high-pressure, ultrasound etc., the freezing and thawing process have significantly improved leading to better retention of the quality during preservation. This chapter discusses application of high pressure (freezing and thawing), ultrasound (freezing and thawing), osmo-dehydrofreezing, antifreeze proteins and ice-nucleation proteins, ohmic heat (thawing) and microwave thawing in food.

KEYWORDS

- **Acoustic**
- **Antifreeze proteins**
- **Cavitation**
- **Clausis–Clapeyron equation**
- **Cryocomminution**
- **Crystallization**
- **Dehydrofreezing**
- **Drip loss**
- **Electrical conductivity**
- **Freeze concentration**
- **Freeze drying**
- **Freezing**

- **Freezing rate**
- **High pressure assisted freezing/thawing**
- **High pressure induced freezing/thawing**
- **High pressure shift freezing**
- **Homogenization**
- **Ice crystals**
- **Ice nucleation proteins**
- **Le Chatelier's principle**
- **Microwave**
- **Nucleation**
- **Ohmic heating**
- **Osmo-dehydrofreezing**
- **Quality**
- **Quick freezing**
- **Sharp freezing**
- **Tenderization**
- **Texturizing**
- **Thawing**
- **Thawing rate**
- **Ultrasound-high frequency low energy**
- **Ultrasound-low frequency high energy**

REFERENCES

1. Acton, E., & Morris, G. J. (1992). Method and apparatus for the control of solidification in liquids. In: *WO Pat. WO*, 99/20420.
2. Acton, E., & Morris, G. (1993). Process to control the freezing of foodstuffs. In: *WO Patent WO*, 1993/014,652.
3. Arai, S., & Watanabe, M. (1986). Freeze texturing of food materials by ice-nucleation with the bacterium Erwiniaananas. *Journal of Biological Chemistry*, 50(1), 169–175.
4. Awad, T. S., Moharram, H. A., Shaltout, Q. E., Asker, D., & Youssef, M. M. (2012). Applications of ultrasound in analysis, processing and quality control of food: a review. *Food Res. Int.*, 48, 410–427.

5. Basak, T., & Ayappa, K. G. (1997). Analysis of microwave thawing of slabs with effective heat capacity method. *AIChE Journal,* 43(7), 1662–1674.
6. Bialod, D. (1980). Electromagnetic Waves in Industrial High- frequency Ovens. Model Simulation of a Periodic Structure Containing a Heterogeneous Charge. Electricite de France Bulletin de la Direction des
7. Biswal, R. N., Bozorgmehr, K., Tompkins, F. D., & Liu, X. (1991). Osmotic concentration of green beans prior to freezing. *Journal of Food Science,* 56(4), 1008–1012.
8. Bozkurt, H., & Icier, F. (2012). Ohmic thawing of frozen beef cuts. *J. Food Process Eng,* 35, 16–36.
9. Brody, A. L., & Antenevich, J. N. (1959). Ultrasonic defrosting of frozen foods. *Food Technology,* 13, 109–110.
10. Burke, M. J., George, M. F., & Bryant, R. G. (1975). Water in plant tissues and frost hardiness. In: *Water Relations of Foods* (Food Science and Technology Monographs) by Duckworth, R. B. ed. New York: Academic Press, pp. 111–135.
11. Chamchong, M., & Datta, A. K. (1999). Thawing of foods in microwave I: effect of power levels and power cycling. *Journal of Microwave Power and Electromagnetic Energy,* 34(1), 9–21.
12. Chevalier, D., Sentissi, M., Havet, M., Ghoul, M., & Le Bail, A. (2000). Comparison between air blast freezing and pressure shift freezing of lobsters. *J. Food Sci.,* 65, 329–333.
13. Chourot, J. M. (1997). Contribution aletude de la decongelation par haute pression. These de doctorat Universite de Nantes France. 151 p.
14. Chow, R., Blindt, R. Chivers, R., & Povey, M. (2003). The sono crystallization of ice in sucrose solutions: primary and secondary nucleation. *Ultrasonics,* 41, 595–604.
15. Chow, R., Blindt, R., Chivers, R., & Povey, M. (2005). A study on the primary and secondary nucleation of ice by power ultrasound. *Ultrasonics,* 43, 227–230.
16. Clements, R. A. (2006). Development of an Ohmic Thawing Apparatus for Accurate Measurement of Electrical Resistance PhD Thesis University of Florida.
17. Convay, J., Castaigne, F., Picaroift, G., & Vovan, X. (1983). Mass transfer consideration in the osmotic dehydration of apples. *Canadian Institute of Food Science and Technology Journal,* 16, 25–29.
18. Delgado, A. E., Zheng, L., & Sun, D. W. (2009). Influence of ultrasound on freezing rate of immersion-frozen apples. *Food and Bioprocess Technology,* 2(3), 263–270.
19. Dermesonlouoglou, E., & Taoukis, P. (2006). Osmodehydrofreezing of sensitive fruit and vegetables: Effect on quality characteristics and shelf life. IUFoST 2006 Nantes.
20. Deuchi, T., & Hayashi, R. (1992). High pressure treatments at subzero temperature: application to preservation rapid freezing and rapid thawing of foods. *High pressure and Biotechnology* Vol. 224. Paris: John Libbey Eurotext, 353–355.
21. Eshtiaghi, M., & Knorr, D. (1996). High hydrostatic pressure thawing for the processing of fruit preparations from frozen strawberries. *Food Technology,* 10(2), 143–148.
22. Etudes, et Recherches. Serie B: Reseaux Electriques Materiels Electriques. Electr Fr Bull Dir EtudSer B.
23. Feeney, R. E., & Yeh, Y. (1993). Antifreeze proteins: properties mechanism of action and possible applications. *Food Technology,* January, pp. 82–88.

24. Feeney, R. E., & Yeh, Y. (1998). Antifreeze proteins: current status and possible food uses. Trends in *Food Science and Technology*, 9, 102–106.

25. Fennema, O. R., Powrie, W. D., & Marth, E. H. (1973). *Low Temperature Preservation of Foods and Living Matter.* Muncel Dunker Inc., New York, 1–598.

26. Fernandez-Martin, F., Otero, L., Solas, M. T., & Sanz, P. (2000). Protein denaturation and structural damage during high pressure shift freezing of porcine and bovine muscle. *J. Food Sci.,* 65, 1002–1008.

27. Forni, E., Torreggiani, K., Crivelli, G., Maestrelle, A., Bertolo, G., & Santelli, F. (1990). Influence of osmosis time on the quality of dehydrofrozen kiwi fruit. *Acta Horticulture,* 282, 425–434.

28. Fu, W. R., & Hsieh, C. C. (1999). Simulation and verification of 2-dimensional ohmic heating in static system. *J Food Sci.,* 64(6), 946–949.

29. Fuchigami, M., Hyakuumoto, N., Miyazaki, K., Nomura, T., & Sasaki, J. (1994). Texture and histological structure of carrots frozen at a programmed rate and thawed in an electrostatic field. *Journal of Food Science,* 59(6), 1162–1167.

30. Fuchigami, M., Kato, N., & Teramoto, A. (1997). High pressure freezing effects on textural quality of carrots. *J. Food Sci.,* 62, 804–808.

31. Fuchigami, M., Kato, N., & Teramoto, A. (1998). High pressure freezing effects on textural quality of Chinese cabbage. *J. Food Sci.,* 63, 122–125.

32. Goullieux, A., & Pain, J. P. *(*2005*).* Ohmic heating. *In: Emerging Technologies for Food Processing.* Sun, D. W. *(Ed.).* Italy: *Elsevier Academic Press,* 469–505.

33. Griffith, M., & Ewart, K. V. (1995). Antifreeze proteins and their potential use in frozen foods. *Biotechnology Advance,* 13(3), 373– 402.

34. Hayriye, B., & Filiz, I. C. (2012). Ohmic thawing of frozen beef cuts. *Journal of Food Process Engineering,* 35, 16–36.

35. Hew, C. L., & Yang, D. S. C. (1992). Protein interaction with ice. *European Journal of Biochemistry,* 203, 33–42.

36. Hu, S. Q., Liu, G., Li, L., Li, Z. X., & Hou, Y. (2013). An improvement in the immersion freezing process for frozen dough via ultrasound irradiation. *J. Food Eng.,* 114, 22–28.

37. Inada, T., Zhang, X., Yabe, A., & Kozawa, Y. (2001). Active control of phase change from super cooled water to ice by ultrasonic vibration 1. Control of freezing temperature. *Int. J. Heat Mass Transfer,* 44, 4523–4531.

38. Islam, M. N., Zhang, M., Adhikari, B., Cheng, X. F., & Xu, B. G. (2014). The effect of ultrasound assisted immersion freezing on selected physicochemical properties of mushrooms. *Int. J. Refrig.,* 42, 121–133.

39. Ito, N., Sugiyama, Y., & Asahara, T. (2003). Vacuum microwave thawing method and vacuum microwave thawing machine. Patent Japan. Kokai Tokkyo Koho JP 2003061635.

40. Kalichevsky, M. T., Knorr, D., & Lillford, P. (1993). Potential food applications of high-pressure effects on ice–water transitions. *Trends Food Science and Technology,* 6, 253–271.

41. Kalichevsky, M. T., Knorr, D., & Lillford, P. (1995). Potential applications of high-pressure effects on ice-water transitions. *Trends in Food Science and Technology,* 6, 53–259.

42. Kanda, Y., Aoki, Y. M., & Kosugi, T. (1992). Freezing of tofu (soybean curd) by pressure-shift: freezing and its structure. *J. Japan Soc. Food Sci. Technol.*, 39, 608–614.
43. Kiani, H., Zhang, Z., & Sun, D. W. (2013). Effect of ultrasound irradiation on ice crystal size distribution in frozen agar gel samples. *Innovative Food Sci. Emerg.*, 18, 126–131.
44. Kiani, H., Sun, D. W., Delgado, A., & Zhang, Z. (2012). Investigation of the effect of power ultrasound on the nucleation of water during freezing of agar gel samples in tubing vials. *Ultrason. Sonochem.*, 19, 576–581.
45. Kim, J. Y., Hong, G. P., Park, S. H., Lee, S., & Min, S. G. (2006). Effects of ohmic thawing on the physicochemical properties of frozen pork. *Food Sci. Biotechnol.*, 15, 374–379.
46. Kim, T. H., Choi, J. H., Choi, Y. S., Kim, H. Y., Kim, S. Y., Kim, H. M., & Kim, C. J. (2011). Physicochemical properties of thawed chicken breast as affected by microwave power levels. *Food Sci. Biotechnol.*, 20, 971–977.
47. Kissam, A. D. (1985). Acoustic thawing of frozen food. In US Patent (40504498 B2). USA.
48. Kissam, A. D., Nelson, R. W., Ngao, J., & Hunter, P. (1982). Water thawing of fish using low frequency acoustics. *J. Food Sci.*, 47(1), 71–75.
49. Kissam, A. D., Nelson, R. W., Ngao, J., & Hunter, P. (1981). Water-thawing of fish using low frequency acoustics. *J. Food Sci.*, 47, 71–75.
50. Knorr, D., Schlueter, O., & Heinz, V. (1998). Impact of high hydrostatic pressure on phase transitions on foods. *Food Technol.*, 52(9), 42–45.
51. Koch, H., Seyderhelm, I., Wille, P., Kalishevsky, M. T., & Knorr, D. (1996). Pressure-shift freezing and its influence on texture color microstructure and rehydration behavior of potato cubes. *Nahrung*, 40, 125–131.
52. Kolbe, E. (2003). Frozen food thawing. Encyclopedia of agricultural food and biological engineering, 416.
53. Le Bail, A., Chevalier, D., Mussa, D. M., & Ghoul, G. (2002). High pressure freezing and thawing of foods: a review. *Int. J. Refrig.*, 25, 504–513.
54. Levy, J., Dumay, E., Kolodziejczyk, E., & Cheftel, J. C. (1999). Freezing kinetics of a model oil-in-water emulsion under high pressure or by pressure release. Impact on ice crystals and oil droplets. *Lebensm.-Wiss. Technol.*, 32, 396–405.
55. Li, B., & Sun, D. W. (2002). Novel methods for rapid freezing and thawing of foods—A review. *Journal of Food Engineering*, 54(3), 175–182.
56. Li, B., & Sun, D.W. (2002). Effect of power ultrasound on freezing rate during immersion freezing of potatoes. *Journal of Food Engineering*, 55(3), 277–282.
57. Li, J., & Lee, T. C. (1998). Bacterial extracellular ice nucleator effects on freezing of foods. *Journal of Food Science*, 63(3), 375–381.
58. Liu, S., Wang, W., Von Moos, V., Jackman, J., Mealing, G., Monette, R., & Ben, R.N. (2007). In Vitro Studies of Antifreeze Glycoprotein (AFGP) and a C-Linked AFGP Analogue. *Biomacromolecules*, 8, 1456–1462 .
59. Makita, T. (1992). Application of high pressure and thermo-physical properties of water to biotechnology. *Fluid Phase Equilibrium*, 76, 87–95.
60. Mallett, C. P. (1993). *Frozen Food Technology*. Chapman and Hall London UK.

61. Martino, M. N., Otero, L., Sanz, P. D., & Zaritzky, N. E. (1998). Size and location of ice crystals in pork frozen by high-pressure-assisted freezing as compared to classical methods. *Meat Science, 50*(3), 303–313.

62. Mason, J., Chemat, F., & Vinatoru, M. (2011). The extraction of natural products using ultrasound or microwaves. *Current Organic Chemistry, 15*(2), 237–247.

63. Mason, T. J., Paniwnyk, L., & Lorimer, J. P. (1996). The uses of ultrasound in food technology. *Ultrasonics Sonochemistry, 3*(3), S253–S260.

64. Meisel, N. (1972). Tempering of meat by microwaves. *Microwave Energy Applications Newsletters, 5*(3), 3–7.

65. Meisel, N. (1973). Microwave applications to food processing and food systems in Europe. *Journal of Microwave Power, 8*(2), 143–146.

66. Mertens, B., & Deplace, G. (1993). Engineering aspects of high pressure technology in the food industry. *Food Technology, 47*(6), 164–169.

67. Mertens Rosenberg, U., & Bogl, W. (1987). Microwave thawing drying and baking in the food industry. *Food Technology,* June, 85–91.

68. Miles, C., Morley, M., & Rendell, M. (1999). High power ultrasonic thawing of frozen foods. *Journal of Food Engineering, 39*(2), 151–159.

69. Moyano, P. C., Vega, R. E., Bunger, A., Garreton, J., & Osorio, F. A. (2002). Effect of combined processes of osmotic dehydration and freezing on papaya preservation. *Food Science and Technology International, 8*(5), 295–301.

70. Naveh, D., Kopelman, I. J., & Mizrahi, S. (1983). Electroconductive thawing by liquid contact. *Journal of Food Technology, 18,* 171–176.

71. Norton, T., Delgado, A., Hogan, E., Grace, P., & Sun, D. W. (2009). Simulation of high pressure freezing processes by enthalpy method. *J. Food Eng., 91,* 260–268.

72. Ohtsuki, T. (1991). Process for thawing foodstuffs. US Patent, 5034236.

73. Ohtsuki, T. (1993). Process for thawing foodstuffs. European Patent, 0409430.

74. Otero, L., & Sanz, P. D. (2000). High-Pressure Shift Freezing. Part 1. Amount of Ice Instantaneously formed in the Process. *Biotechnol. Prog., 16,* 1030–1036

75. Otero, L., Martino, M., Zaritzky, N., Solas, M., & Sanz, P. D. (2000). Preservation of microstructure in peach and mango during high pressure- shift freezing. *Journal of Food Science, 65,* 466–470.

76. Otero, L., Solas, M. T., Sanz, P. D., Elvira, C. D., & Carasco, J. A. (1998). Contrasting effects of high-pressure-assisted freezing and conventional air-freezing on eggplant microstructure. *Z Lebens Unters Forsch, 206,* 338–342.

77. Pangrle, B. J., Ayappa, K. G., Davis, H. T., Davis, E. A., & Gordon, J. (1991). Microwave thawing of cylinders. *AlChE Journal, 37*(12), 1789–1800.

78. Pangrle, B. J., Ayappa, K. G., Sutanto, E., Davis, H. T., & Davis, E. A. (1991). Microwave thawing of semi-infinite slabs. *Chem. Eng. Comm., 112,* 39.

79. Phan, P. A. (1977). Microwave thawing of peaches a comparative study of various thawing processes. *Journal of Microwave Power, 12*(4), 261–271

80. Pingret, D., Fabiano-Tixier, A. S., & Chemat, F. (2013). Degradation during application of ultrasound in food processing: a review. *Food Control, 31,* 593–606.

81. Price, G. (1992). Ultrasonically assisted polymer synthesis. In: *Current Trends in Sonochemistry,* G. Price (Ed.), 87.

82. Priou, A., Fournet-Fayas, C., Deficis, A., & Gimonet, E. (1978). Microwave thawing of larger pieces of beef. Proceedings of 8th European Microwave Conference Paris France, 589–593

83. Rastogi, N. K. (2011). Opportunities and challenges in application of ultrasound in food processing. *Crit. Rev. Food Sci.,* 51, 705–722.

84. Ratanadecho, P., Aoki, K., & Akahori, M. (2002). Characteristics of microwave melting of frozen packed bed using a rectangular wave guide. *IEEE Trans. Microwave Theory Tech.,* 50(6), 1487–1494.

85. Rattanadecho, P. (2004). Theoretical and experimental investigation of microwave thawing of frozen layer using a microwave oven (effects of layered configurations and layer thickness). *International Journal of Heat and Mass Transfer,* 47, 937–945.

86. Roberts, J. S., Balaban, M.O., Zimmerman, R., & Luzuriaga, D. (1998). Design and testing of a prototype ohmic thawing unit. *Computers and Electronics in Agriculture,* 19, 211–222.

87. Saclier, M., Peczalski, R., & Andrieu, J. (2010). Effect of ultrasonically induced nucleation on ice crystals' size and shape during freezing in vials. *Chem. Eng. Sci.,* 65, 3064–3071.

88. Sanz, P. D., Otero, L., Elvira, C., & Carrasco, J. A. (1997). Freezing processes in high-pressure domains. *International Journal of Refrigeration,* 20(5), 301–307.

89. Schlüter, O., & Knorr, D. (2002). Impact of the metastable state of water on the design of high pressure supported freezing and thawing processes. 2002 ASAE Annual Meeting, 026197.

90. Schluter, O., George, S., Heinz, V., & Knorr, D. (1998). Phase transitions in model foods induced by pressure-assisted freezing and pressure-assisted thawing In Proceedings of the IIR International Conference, 1998, 23–26 September; Sofia Bulgarie. Paris: International Institute of refrigeration.

91. *Seyhun, N. (2008).* Modeling of tempering of frozen potato puree by microwave infrared assisted microwave and ohmic heating methods. PhD dissertation Middle East Technical University.

92. Sun, D. W., & Li, B. (2003). Microstructural change of potato tissues frozen by ultrasound-assisted immersion freezing. *Journal of Food Engineering,* 57(4), 337–345.

93. Suslick, K.S. (1988). Chemical biological and physical effects. In: *Ultrasound* (Suslick, K. S., ed.). New York: VCH, 123–163.

94. Taher, B. J., & Farid, M. M. (2001). Cyclic microwave thawing of frozen meat: experimental and theoretical investigation. *Chemical Engineering and Processing,* 40(4), 379–389.

95. Taoukis, P., Davis, E. A., Davis, H. T., Gordon, J., & Takmon, Y. (1987). Mathematical modeling of microwave thawing by the modified isotherm migration method. *Journal of Food Science,* 52(2), 455–463.

96. Teramoto, A., & Fuchigami, M. (2000). Changes in temperature texture and structure of konnyaku (konjacglucomannan gel) during high-pressure- freezing. *Journal of Food Science,* 65(3), 491–497.

97. Tironi, V., Lamballerie, M., & Le-Bail, A. (2010). Quality changes during the frozen storage of sea bass (Dicentrarchuslabrax) muscle after pressure shift freezing and pressure assisted thawing. *Innovative Food Sci. Emerg.,* 11, 565– 573.

98. Tong, C. H., & Lund, D. B. (1993). Microwave heating of baked dough products with simultaneous heat and moisture transfer. *J. Food Engineering,* 19, 319–339.

99. Torreggiani, D., Forni, E., Erba, M. L., & Longoni, F. (1995). Functional properties of pepper osmodehydrated in hydrolyzed cheese whey permeate with or without sorbitol. *Food Research International,* 28(2), 161–166.

100. Torreggiani, D., & Bertolo, G. (2001). Osmotic pre-treatments in fruit processing: chemical physical and structural effects *J. Food Eng.,* 49, 247–253.

101. Tregunnol, N. B., & Goff, H. D. (1996). Osmodehydrofreezing of apples: structural and textural effects. *Food Research International,* 29(5–6), 471–479.

102. Urrutia, B. G., Schlüter, O., & Knorr, D. (2004). High-pressure low-temperature thawing. *Suggested Definitions and Terminology,* 5 (4), 413–427.

103. Virtanen, A. J., Goedeken, D. L., & Tong, C. (1997). Microwave Assisted Thawing of Model Frozen Foods Using Feed-back Temperature Control and Surface Cooling. *Journal of Food Science,* 62(1), 150–154.

104. Wang, H., Luo, Y., Shi, C., & Shen, H. (2015). Effect of Different Thawing Methods and Multiple Freeze-Thaw Cycles on the Quality of Common Carp (*Cyprinuscarpio*). *Journal of Aquatic Food Product Technology,* 4(2), 153–162.

105. Wang, W. C., Chen, J. I., & Hua, H. H. (2002). Study of liquid-contact by ohmic heating. In *2002 IFT Annual Meeting Book of Abstracts* Paper 91F-4. Chicago: Institute of Food Technologists.

106. Warren, C. J., Mueller, C. M., & Mckown, R. L. (1992). Ice crystal growth suppression polypeptides and methods of preparation. US Patent, 5118792.

107. Wen, X., Hu, R., Zhao, J.H., Peng, Y., & Ni, Y. Y. (2015). Evaluation of the effects of different thawing methods on texture color and ascorbic acid retention of frozen hami melon (*Cucumismelo var. saccharinus*). *International Journal of Food Science & Technology,* 50(5), 1116–1122

108. Xin, Y., Zhang, M., & Adhikari, B. (2014). Ultrasound assisted immersion freezing of broccoli (Brassica oleracea L. var. botrytis L.). *Ultrason. Sonochem.,* 21, 1728–1735.

109. Xu, B. G., Zhang, M., Bhandari, B., & Cheng, X. F. (2014). Influence of power ultrasound on ice nucleation of radish cylinders during ultrasound-assisted immersion freezing. *Int. J. Refrig.,* 46, 1–8.

110. Yang, M., Zhang, C., & Li, G. (2009). Ohmic heating pork thawing and fabrication of fast food intestinal. *Transactions of the Chinese Society of Agricultural Engineering 2009,* S1.

111. Yin, Y., & Walker, C. E. (1995). A quality comparison of breads baked by conventional versus non-conventional ovens: a review. *Journal of the Science of Food and Agriculture,* 67, 283–291.

112. Yun, C. G., Lee, D. H., & Park, J. Y. (1998). Ohmic thawing of a frozen meat chunk. *Journal of Food Science and Technology (Korean),* 30(4), 842–847.

113. Zareifard, M. R., Ramaswamy, H. S., Trigui, M., & Marcotte, M. (2003). Ohmic heating behavior and electrical conductivity of two-phase food systems. *Innovative Food Science and Emerging Technologies,* 4, 45–55.

114. Zhao, Y. Y., Fores, R. A., & Olson, D. G. (1996). The action of high hydrostatic pressure on the thawing of frozen meat. *Annual Meeting of Institute of Food Technologist* New Orleans LA (June), 22–26.

115. Zhao, Y. Y., Fores, R. A., & Olson, D. G. (1998). High hydrostatic pressure effects on rapid thawing of frozen beef. *Journal of Food Science,* 63(2), 272–275.
116. Zheng, L., & Sun, D. W. (2006). Innovative applications of power ultrasound during food freezing processes—a review. *Trend Food Sci. Technol.,* 17, 16–23.
117. Zhu, S., Ramaswamy, H. S., & Simpson, B. K. (2004). Effect of high-pressure versus conventional thawing on color drip loss and texture of Atlantic salmon frozen by different methods. *LWT – Food Sci. Technol.,* 37, 291–299.

CHAPTER 4

PRINCIPLES OF NOVEL FREEZING AND THAWING TECHNOLOGIES FOR FOOD APPLICATIONS

DEEPIKA GOSWAMI,[1] JAGBIR REHAL,[2] HRADESH RAJPUT,[3] and HARSHAD M. MANDGE[4]

[1]Scientist, Food Grains and Oilseeds Processing Division, ICAR-CIPHET, Ludhiana – 141004, Punjab, India. Phone: +91-9592317693; E-mail: deepikagoswami@rediffmail.com

[2]Assistant Fruit and Vegetable Technologist, Department of Food Science and Technology, Punjab Agricultural University, Ludhiana – 141004, Punjab, India, E-mail: kaur_jagbir@yahoo.com

[3]Senior Research Fellow, Department of Food Science and Technology, Punjab Agricultural University, Ludhiana – 141004, Punjab, India, E-mail: hrdesh802@gmail.com

[4]Assistant Professor (PHT), College of Horticulture, Banda University of Agriculture and Technology, Banda – 210001, Uttar Pradesh, India, E-mail: mandgeharshad@gmail.com

CONTENTS

4.1 INTRODUCTION

Food is an essential part of living being and differs in its storability from non-perishable to highly perishable one. The present changing lifestyle and socio-economic status of the society puts forth the need of food products that can be stored for long period. However, the product storability is a challenge because of the various factors such as microbial growth, enzymatic activity, pest infestation, inappropriate temperature, presence of light, oxygen etc. There are various principles being practiced at present for food preservation. Preservation by cold (under low temperatures), one of the oldest methods, is among them. This principle includes freezing and refrigeration of food products. As a preservation method, freezing takes over where refrigeration and cold storage leave off. Properly frozen products maintain more of their original nutrients, color, flavor and texture and hence have the ability to satisfy the consumers demand for products closest to fresh foods. In recent years the rapid increase in sales of frozen foods may closely be associated with the increased ownership of domestic freezers and microwave ovens.

The various methods of freezing, effects of freezing on food quality, novel-freezing techniques, thawing methods are discussed in this chapter.

4.2 FREEZING

Freezing is one of the easiest, quickest, most versatile and convenient methods of preserving foods and has been used for thousands of years because of high quality of foods [27] even after long-term preservation. The frozen food industry had a humble beginning in the early part of the twentieth century, when it was restricted mainly to freezing of fruits, vegetables, meats and fish. Now-a-days a number of food products such as bakery goods, ice creams, desserts, juices etc. are also part of this industry. In the freezing technique, preservation is achieved by the following:

a. **Low temperature:** The optimum temperature range for most bacteria, yeasts and moulds is 16–38°C. As the temperature is lowered down, the microbial activity gets reduced. The enzymatic activity also gets reduced by lowering the temperature of the product.

b. **Reduced water activity:** During the freezing process, there is change in the state of water from liquid to solid. The water immobilization and increased concentration of the dissolved solutes in the unfrozen water lowers the water activity (a_w) of the food.

c. **Pre-treatment by blanching:** Freezing of fruits and vegetables requires pre-treatment by blanching that itself is a preservation technique.

4.2.1 COMMERCIALLY FROZEN FOODS: EXAMPLES

a. Fruits such as strawberries, oranges, raspberries, blackcurrants either as a whole or in the form of puree or juice concentrates.

b. Vegetables such as peas, green beans, sweetcorn, spinach, sprouts and potatoes.

c. Fish fillets and seafoods such as cod, plaice, shrimps and crab meat including fish fingers, fish cakes or prepared dishes with an accompanying sauce.

d. Meats from beef, lamb, poultry as carcasses, boxed joints or cubes, and the meat products like sausages, reformed steaks etc.

e. Baked goods.

f. Prepared foods such as desserts, ice cream, complete meals and cook–freeze dishes.

4.2.2 FOODS NOT SUITABLE FOR FREEZING

There are certain foods that are not suitable for freezing purposes as this process brings undesirable changes in their texture, color, flavor and nutritional value. Some examples of such foods are given in Table 4.1.

4.2.3 THE FREEZING CURVE

Freezing is a unit operation in which the temperature of a food is reduced below its freezing point and in this process a proportion of the water undergoes a change in state to form ice crystals. During freezing a

TABLE 4.1   Foods for Which Freezing is Not Suitable

Foods	Normal use	Conditions after thawing
Cabbage, celery, cucumbers, lettuce, parsley, radishes	Raw salad	Lose crispness, limp and waterlogged, development of off flavor, odor and color.
Cheese or crumb toppings	On casseroles	Soggy
Cooked macaroni, spaghetti, rice	Side dish, in casseroles	Mushy, warmed over-flavor
Cream or custard fillings	Pies, baked goods	Gets separated, become watery and lumpy
Egg whites, cooked sandwiches, gravy, desserts	Salads and creamed foods	Tough, rubbery and spongy
Fried foods (except commercial frozen foods)	Snacks	Lose crispness, soggy
Fruit jelly	Sandwiches	Bread may get soaked
Gelatin	In salads or desserts	Weeping
Icings made from egg whites	cakes, cookies	Frothy, weepy
Mayonnaise, salad dressings	On sandwiches	Gets separated during freezing
Milk sauces	In casseroles, gravies	Curdling or separation
Raw potatoes	Food uses	Darkening, mealy texture
Sour cream	Topping in salads	Gets separated, watery

food product do not cool uniformly. The point that cools most slowly is called as the thermal center of a food. When we monitor the temperature of this thermal point during freezing, a characteristic curve is obtained (Figure 4.1) that has six components.

The six components of the freezing curve for a food in Figure 4.1 are given below:

AS   The food is cooled to below its freezing point θ_f, which with the exception of pure water, is always below 0°C. For example, fruits that contain 87–95% moisture have a freezing point of 0.9 to –2.7°C. At point S, there occurs supercooling: a phenomenon in which although the temperature reaches below the freezing point,

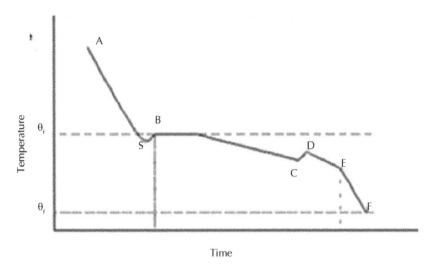

FIGURE 4.1   Freezing curve.

the water remains in liquid form. This supercooling may be up to as much as 10°C below the freezing point.

SB   **Nucleation or agitation** initiates formation of ice crystals and causes the release of latent heat of crystallization. It results into rapid temperature rise to the freezing point, which is just below 0°C because of the dissolved solids in water phase.

BC   With the progress in freezing, the **solute concentration** in the unfrozen liquid increases that exert a gradual freezing point depression on the remaining solution. Hence, the temperature falls slightly. It is during this stage that the major part of the ice is formed.

CD   **Supersaturation** of one of the solutes crystallizes it out. The latent heat of crystallization is released and the temperature rises to the eutectic temperature for that solute.

DE   **Crystallization** of water and solutes continues. The total time t_f taken (the *freezing plateau*) is determined by the rate at which heat is removed.

EF   The temperature of the ice–water mixture gets further decreased to the temperature of the freezer. All of the water is not frozen in foods and the proportion of water frozen depends on the type and composition of the food and the temperature of storage. For example at a storage temperature of –20°C, the percentage of water frozen is 88% in lamb, 91% in fish and 93% in egg albumin.

4.2.4 FREEZING METHODS

In commercial freezing, there are three common freezing methods, which can further be subdivided into various methods (Table 4.2):

1.   Freezing in air
     - Still-air "sharp" freezing.
     - Air blast freezing.
     - Fluidized-bed freezing.
     - Spiral freezing.

TABLE 4.2   Comparison of Freezing Methods

Method of freezing	Typical film heat transfer coefficients ($W\ m^{-2}\ K^{-1}$)	Typical freezing time (min.): For a specified food to $-18°C$	Food
Still air	6–9	180–4320	Meat carcass
Blast (5 m s^{-1})	25–30	15–20	Unpackaged peas
Blast (3 m s^{-1})	18	–	–
Spiral belt	25	12–19	Hamburgers, fish fingers
Fluidized bed	90–140	3–4	Unpackaged peas
		15	Fish fingers
Plate	100	75	25 kg blocks of fish
		25	1 kg carton vegetables
Scraped surface	–	0.3–0.5	Ice cream (layer approximately 1 mm thick)
Immersion (Freon)	500	10–15	170 g card cans of orange juice
		0.5	Peas
		4–5	Beef burgers, fish fingers
Cryogenic (liquid nitrogen)	1500	1.5	454 g of bread
		0.9	454 g of cake
		2–5	Hamburgers, seafood
		0.5–6	Fruits and vegetables

Adapted from Refs. [4, 5, 11, 14, 24].

2. Freezing by indirect contact with refrigerant
    - Single plate.
    - Double contact plate.
    - Pressure plate.
    - Slush freezer.
3. Freezing by direct immersion in refrigerating medium
    - Heat exchange fluid.
    - Compressed gas.
    - Refrigerant spray.

Selection of the freezing methods for a particular food depends on the following factors:

- Required rate of freezing
- Size and shape of the foods.
- Packaging requirements of the food.
- Type of operation—batch type or continuous type.
- Scale of production.
- Range of products to be frozen.
- Capital and operational costs.

4.2.4.1 Air Freezing

In the air freezing method, cold air at varying temperature and velocity is used for freezing of packaged or unpackaged non-fluid foods. The air temperatures range from –18° to –40°C and the degree of velocity increases from sharp freezing to air blast freezing. In the fluidized bed freezing, the air velocity is used to subdivide and move the particles of the product being frozen.

4.2.4.1.1 Still Air Sharp Freezing

Sharp freezing is the oldest and least expensive air freezing method under which the product is simply placed in an insulated cold room usually maintained in the temperature range of –23° to –30°C. Although the air circulates within the room by convection, usually little or no provision is made for forced connection. Sometimes gentle movement of air is caused by circulating fans in the room. However, the velocity of air is not as much as in case of air blast freezing. As the still air is a poor conductor of heat, the foods placed even in low temperatures are frozen comparatively slow. Hence, many hours or even days can be required for the completion of the freezing process. In sharp freezing, the freezing time is generally 3–72 h or more based on the conditions and the size of product. Transferring of flavors and temperature from warm product to yet to be frozen product may hamper the product quality. Moreover, dehydration due to slow freezing rate and temperature fluctuations may be excessive in sharp freezing.

4.2.4.1.2 Air Blast Freezing

Air blast freezers are operated at comparatively higher air velocity and lower temperature range than still air sharp freezers. Typically, temperatures of –30° to –45°C with forced air velocities of 10–15 m/s are employed in air blast freezing. The forced air circulation significantly reduces the freezing time for the product. In a low temperature room, the product to be frozen is placed on freezing coils held on a tray either loose or in packages and cold air is blown over the product. Air blast freezers are of both batch type and continuous type. The cold air flow may be adjusted so as to pass over or under. Foods of variety of sizes and shapes can be accommodated in this method. There are following limitations:

- Excessive dehydration of unpackaged foods may occur under uncontrolled conditions which necessitates frequent defrosting of equipment.
- Undesirable bulging of packaged foods which are not confined between rigid surfaces during freezing.
- Drying out of food at its surface resulting into 'freezer burn.'

4.2.4.1.3 Fluidized Bed Freezing

Fluidized bed freezing is a modification of air-blast freezing, where cold air is blowed up through the wire mesh belt that supports and conveys the product. The cold air with velocity exceeding the velocity of free fall of the particles causes fluidization of the product by partially lifting or suspending the particles in a manner that resembles a boiling liquid. The vibration thus imparted to the product accelerates the freezing rate. Fluidization performs the following purposes:

- Subdivision of the product.
- Promotes intimate contact of each particle with the cold air.
- Keep clusters from freezing together.

Solid food particles ranging in size from peas to strawberries can be fluidized by forming a bed of particles 2–12 cm. Products that are relatively small and uniform in size such as peas, limas, cut green beans, strawberries, whole kernel corn, brussel sprouts, etc., give best results in fluidized bed freezing. Fluidized bed freezing is advantageous over conventional air-blast freezing as follows:

- Efficient heat transfer and rapid freezing rate.
- Lesser product dehydration and hence lesser frequent defrosting of equipments.
- Short freezing time that prevents loss of moisture.

The major disadvantage of this freezing method is its inability to fluidize large or nonuniform products at a reasonable air velocities.

4.2.4.1.4 Spiral Freezers

The spiral freezers are characterized by a self-stacking, self-enclosing stainless steel belt that can be bent laterally. The belt is designed to provide compactness, reliability and improved air flow. The capacity of the freezers can be varied by adjusting the belt width and number of tiers in the belt stack. The advantage of the spiral freezers lies in the fact that infeeds and outfeeds can be located to suit most line layouts. These are most suitable for products that require careful handling and a long freezing time (generally 10 minutes to 2 hours).

4.2.4.2 Plate Freezing

Plate freezing is an indirect contact freezing method. Here the food products are placed in contact with a metal surface, which are chilled either by cold brine or vaporizing refrigerants but separate the food from the refrigerant. Hence, the product (in package also) is in direct contact with the cold wall but in indirect contact with the refrigerant. Indirect contact of the food with the refrigerant permits for use of those refrigerants that might otherwise adversely affect the food or its package. These freezers may be of batch type, semi automatic, or automatic type. Efficiency of plate freezers depends on the extent of contact between the plates and the food. This necessitates for the food packages to be well filled or slightly overfilled to make good pressure contact with the plates. Solid compact foods freeze more rapidly than those products having air spaces in between. For liquid food products and purees, the freezing system is in the form of a tubular scraped-surface heat exchanger where a refrigerant is pumped in place of the steam. Freezing is never carried out to completion. Rather, the product is frozen to a slush condition, packaged and then further frozen in an air-blast or immersion type freezer.

Plate freezers consist of a series of flat hollow refrigerated metal plates and the packaged food products may rest on, slide against, or be pressed between these plates. The plates are mounted parallel to each other and may be either horizontal or vertical. The product to be frozen is in the form of parallel-sided blocks and, during the freezing process, heat flow is perpendicular to the faces of the plates. Plate freezing is an economical method and is advantageous as follows:

- Minimum product dehydration.
- Lesser need for defrosting of the equipment.
- No package bulging.
- However, some limitations are also associated with plate freezing systems as follows:
- Requirement of food packages with uniform thickness.
- Comparatively slower freezing rate as than other methods.

4.2.4.3 Immersion Freezing

In the immersion freezing, there is a direct contact of the food product, either packaged or unpackaged, with the refrigerant either by immersion in or by spraying with a cold liquid. The packaged food is passed through a bath of the refrigerant solutions on a submerged mesh conveyor. There does not occur any change in the state of the refrigerant (cold liquid) throughout the freezing process. The aqueous solutions of the substances like propylene glycol, glycerol, sodium chloride, calcium chloride, and mixtures of salt and sugar are used as freezing (cold liquid). The immersion freezing refrigerants are broadly classified into: low-freezing point liquids and cryogenic liquids.

4.2.4.3.1 Low-Freezing Point Liquids

These refrigerants are chilled by indirect contact with another refrigerant. Examples are solutions of sugars, sodium chloride and glycerol.

4.2.4.3.2 Cryogenic Liquids

These refrigerants exhibit the cooling effect by their own evaporation, for example, compressed liquefied nitrogen. The concentration of the solution is such that it remains liquid and effective at −18°C or lower temperature.

The major advantages of immersion freezing method are as follows:

- Rapid freezing due to the intimate contact between the food and the refrigerant that minimizes the resistance to heat transfer.
- Suitability for rapid freezing of irregularly shaped food pieces also.
- Prevention of oxidation of sensitive foods as there is minimum contact of air with the food during freezing.
- Product quality unattainable by any other freezing methods can be obtained by using cryogenic liquids.
- Easier adaptation to continuous operations.

However limitations on the refrigerants that may be used, especially for contact with non-packaged food, do exist with this method also. These refrigerants when used for non-packaged foods must be: Non-toxic, pure, clean, free from foreign tastes and bleaching agents, colorless and odorless. While using for contact with packaged foods, these refrigerants must be non-toxic and non-corrosive to the packaging material.

4.2.4.3.3 Immersion Freezing with Cryogenic Liquids (Cryogenic Freezing)

Cryogenic liquids are liquefied gases of extremely low boiling point. For example liquid nitrogen having a boiling point of −196°C is a cryogenic liquid. Cryogenic freezing is different from liquid-immersion freezing in the fact that removal of heat occurs during a change of state by the cryogenic liquid. Liquid nitrogen and liquid carbon dioxide (boiling point is −79°C) are the most common food grade cryogenic liquids. Table 4.3 elaborates the properties of these food cryogens.

Use of liquid nitrogen in immersion freezing has some advantages as below:

- Slow boiling point of liquid nitrogen provides a great driving force for heat transfer.
- Minimize the resistance to the heat transfer by intimately contacting all the portions of irregularly shaped foods.
- No need of another refrigerant to cool this medium.
- Non-toxic and inert to the food constituents.
- Prevents oxidation of foods during freezing by displacing air from the food.
- The speed of freezing gives product with a quality that is not attainable by other freezing methods.
- Minimum dehydration loss from the product

However, the higher cost of liquid nitrogen is a limiting factor in liquid nitrogen freezing.

TABLE 4.3 Properties of Food Cryogens [9]

Property	Liquid nitrogen	Carbon dioxide
Density (kg m^{-3})	784	464
Specific heat (kJ kg^{-1} K^{-1})	1.04	2.26
Latent heat (kJ kg^{-1})	358	352
Total usable refrigeration effect (kJ kg^{-1})	690	565
Boiling point (°C)	−196	−78.5 (sublimation)
Thermal conductivity (Wm^{-1} K^{-1})	0.29	0.19
Consumption per 100 kg of product frozen (kg)	100–300	120–375

4.2.5 ADVANCES IN FREEZING TECHNIQUES

Some chemical and physical aids to freezing and thawing have been developed currently aiming towards energy saving and an improved quality product. These are as follows:

- High pressure freezing.
- Ultrasound assisted freezing.
- Dehydrofreezing and osmodehydrofreezing.
- Immersion freezing in ice slurry.
- Antifreeze protein and ice nucleation protein.

4.2.5.1 High Pressure Freezing

Water changes its state from liquid to solid on freezing. The solid state called as 'ice' has a density lower than the water and due to this reason when water is frozen under atmospheric pressure its volume increases. The volume increase can be about 9% on freezing at 0°C and about 13% at –20°C [12]. This expansion in volume has detrimental effect on the tissues of the food.

According to Le Chatelier's principle, an increase in pressure will cause a decrease in the freezing point. Hence in high-pressure application, the food can be cooled much lower to its initial freezing point without ice formation. After giving time enough for the food to equilibrate throughout to a low temperature, the pressure is rapidly released. The faster is the release of pressure, the lower will be the nucleation pressure. The rapid pressure release results into a large degree of supercooling that leads to simultaneous and uniform nucleation, and the formation of a large number of small crystals [12]. There occurs formation of instantaneous and homogenous ice throughout the whole volume of the product [18, 25]. The ice crystals thus formed have densities higher than water itself. Hence during phase transition, there is no expansion in volume of ice and no tissue damages of food. Due to smaller ice crystals formation, the damage to cells is minimized and hence there is significant improvement in product quality [3, 18].

The release of pressure in this freezing method results into supercooling and hence, this technology is especially useful for freezing of foods having large dimensions where thermal gradients are pronounced and there is possibility of damage by freeze-cracking while applying other freezing methods, including cryogenic freezing also [18, 25].

4.2.5.2 Dehydrofreezing

Dehydrofreezing is a freezing technique where a food is first dehydrated to a desirable moisture (typically 70% moisture removal) and then is frozen [30, 31]. The dehydration of the product can be done either by air drying or by osmotic dehydration (osmo-dehydrofreezing). The osmodehydration of the food product is achieved by immersion in a concentrated solution of some solute. For fruits, solution of sucrose, glucose, fructose, lactose, maltodextrin and corn syrup may be used; and for vegetables sodium chloride solution is used.

Reduction in moisture content prior to freezing leads to formation of lesser number of ice crystals of smaller sizes. Hence, this technique is especially useful for the food items such as fresh fruits and vegetables having higher moisture content and susceptibility to tissue damage is low due to large ice crystal formation during freezing. An increased freezing rate can also reduce the large ice crystal formation but a higher level of moisture in such foods makes the tissue damage inevitable. The moisture removal also results into increased solute concentration that leads to decrease in the freezing point, more supercooling and better stability [1, 30]. The freezing time and refrigeration load is also reduced because there remains less water to freeze. Retention of pigments, vitamins and aroma is also improved. Hence, osmodehydrofreezing is a freezing technique by which not only a product quality comparable to conventional products is obtained, but the cost of packaging, transportation and storage is also reduced [1].

4.2.5.3 Ultrasound Assisted Freezing

Freezing of food products can also be improved by combining with ultrasound technique. This technique has aids in freezing by exerting the effects as given below:

- **Agitation:** The heat and mass transfer is enhanced by the agitation and freezing near surfaces of food occurs in faster rate.
- **Nucleation:** Ultrasound triggers nucleation leading to formation of more and smaller crystals.

However, it also has certain adverse effects on the freezing process such as:

- **Heating:** Reduces the freezing rate.
- **Cavitation:** Ultrasound causes formation of gas bubbles near food surfaces (cavitation) that reduces the freezing.

The heating effect can be reduced by applying the ultrasound intermittently and at the right power level. A power level of 15.85 W applied for 2 minutes may give the best reduction in freezing time. Ultrasound assisted freezing did not show disruptions and separation of cells in foods as seen in foods frozen without ultrasound. This may be attributed to the triggering of intracellular nucleation by ultrasound, which stops cell loss of water and shrinkage [15].

4.2.5.4 Immersion Freezing in Ice Slurry

This technique is a variant of immersion freezing. In conventional immersion freezing the product is usually wrapped to prevent absorption of refrigerant. However, in some processed foods the absorption may be an advantage like in desserts where the sugar-ethanol solution based ice slurries are used for immersion freezing. This technique offers the following advantages:

- short freezing time due to high heat transfer rate from ice slurry;
- high quality due to small crystal size;
- absorption of food additives (antioxidants, flavorings, aromas and micronutrients);
- improved quality and shelf life of the product.

4.2.5.5 Use of Antifreeze Proteins (AFP) and Ice Nucleation Proteins (INP)

The primary concern in freezing process and frozen food products is controlling the growth of ice crystals. The present approach for this is use of antifreeze proteins and ice nucleation proteins- two functionally distinct classes of proteins [10]. It has been observed that animals can survive subfreezing body temperatures by generating these two proteins for maintaining their body fluids in a supercooled state (*freeze avoidance*) or for controlling the freezing process in their body (*freeze tolerance*), respectively.

4.2.5.5.1   Antifreeze Proteins

Some fish and insects can prevent their blood from freezing (freezing point of blood is about –0.8°C) even at around –1.9°C the freezing point of seawater. At this temperature, there are ice crystals floating around too that might cause nucleation. But the antifreeze proteins produced by the fish and insects prevent the growth of these ice crystals and hence make there survival possible. The antifreeze proteins find their potential application in ice creams to prevent crystal growth during storage, especially when temperature fluctuates [36] in meat or intravenous injection before slaughter to obtain frozen meat with smaller ice crystals and less drip. The higher cost involved in use of antifreeze proteins can be overcome by using genetically engineered antifreeze proteins or by using synthetic antifreeze proteins [16].

4.2.5.5.2   Ice Nucleation Proteins

Some amphibians and reptiles control the freezing by producing ice nucleation proteins. These proteins initiate ice nucleation as soon as body temperature reaches –2° to –3°C. However, a cryoprotectant is also produced to lower the freezing point of the most sensitive organs, thereby the freezing is delayed and minimized.

4.2.6   EFFECT OF FREEZING ON FOOD QUALITY

Freezing is a quick and convenient way to preserve fruits and vegetables while maintaining high quality. It is a process of reducing the product temperature that slows down the quality deteriorating processes such as oxidation of fat, growth of microorganisms, enzyme activity and the loss of surface moisture (dehydration).

4.2.6.1   Chemical Changes During Freezing

The chemical changes occurring in fresh fruits and vegetables continue even after harvest, leading to their spoilage and quality deterioration. These changes lead to loss of color, loss of nutrients, flavor changes, and color changes in the product. Freezing of these perishable food products, immediately after harvest at their peak time, is one way for their preservation. The chemical changes in fruits and vegetables may be attributed to the presence of certain enzymes that catalyze the chemical reactions. Hence, inactivation of these enzymes is a must to prevent or lower down these chemical changes. Blanching, or exposure of the food product to boiling water or steam for a brief period of time, is such a process for enzyme inactivation in vegetables. This process also reduces the number of microorganisms on the surface of the product.

4.2.6.2   Textural Changes During Freezing

The textural quality of frozen food is highly associated to freezing and thawing procedures. To have good quality frozen product, rapid freezing of the product and storing at a constant sub-freezing temperature [28] is essential. The freezing process causes softening of tissue, dehydration and shrinkage in some products. These are some factors that affect the food quality.

4.2.6.2.1   Ice Crystal Damage

In fruits and vegetables the rigid cell walls give them the support structure, and texture. These cell walls contain the water and other chemical substances. The freezing process causes formation and expansion of ice crystals that might lead to the cell walls rupture. The rupturing of cell walls results into a frozen product that, upon thawing, will be much softer than it was when raw. High starch vegetables such as peas, corn, etc., have less noticeable textural changes in freezing. For products that are to be consumed in raw form such as celery, lettuce, and tomatoes etc. this textural change is noticeable. Hence these products are usually not frozen. When frozen, these are served and consumed in the partially thawed state so as to make the effect of freezing on the fruit tissue less noticeable. The textural changes due to freezing are not as apparent in products which are cooked before eating because cooking also softens cell walls.

4.2.6.2.2 *Dehydration and Shrinkage*

Among the constituents of living cell water, some oils and fats crystallize out during freezing. In cellular foods such as meat that have water present both inside and outside the cells, normal freezing will not cause ice crystal formation inside the cell due to the phenomenon of supercooling. However, conversion of water, present outside the cell, into ice crystals makes the remaining extracellular liquid more concentrated than the intracellular liquid. The osmotic pressure thus exerted causes movement of water from cells to the outside through the cell walls and hence leads to product dehydration and shrinkage.

4.2.6.2.2 *Starch Retrogradation*

In many frozen starch-based food products such as pasta, noodles and steamed rice, the moisture content is much lower (50–60% w/w) [33]. At such low moisture content or high starch concentration, newly-prepared rice-based products are soft, pliable and elastic; but they become harder and/or undergo syneresis during freezing and long-term storage due to starch retrogradation [37].

4.2.6.2.3 *Concentration Effects*

As the water freezes, the remaining solution becomes more and more concentrated in solutes. This concentration effect cause several kinds of damages as follows:

- **Precipitation of solutes out of the solution**: For e.g. precipitation of excessive level of lactose out from ice cream, giving a sandy and gritty texture to the product.
- **Salting out effect**: Solutes that do not precipitate but remain in concentrated solution may cause protein denaturation.
- **Protein coagulation:** Acidic solutes, remaining in the concentrated solution, may drop the pH below the isoelectric point of protein leading to its coagulation.
- **Imbalance of a colloidal suspension:** Colloidal suspension has a balance of anions and cations. Precipitation of any of these during freezing may disturb the balance.
- **Removal of gases from solution:** Gas containing solution when frozen causes concentration followed by supersaturation of the gases. Ultimately it leads to the gas removal from the solution.
- **Dehydration** of adjacent tissues at micro-environmental level.

4.2.6.3 Rate of Freezing and Product Quality

The rate of freezing and the formation of small ice crystals in freezing are critical to minimize tissue damage and drip loss in thawing. The rate of freezing has direct effect on product quality. Fast freezing gives a product of higher quality and hence is preferred over slow freezing. Fast freezing is advantageous in following ways:

- A faster freezing rate does not give time to water for diffusing through the cell walls, and hence the cells have significant supercooling before its water is lost. There occurs a large number of nucleation leading to formation of large number of small intracellular ice crystals. These intracellular crystals ensure that the cell is not distorted or dehydrated, and so quality may be improved. On the contrary, a slower freezing rate results into formation of a few and larger ice crystals.
- Fast freezing minimizes the concentration effects by decreasing the contact time of concentrated solutes with food tissues, colloids and individual constituents while converting from the unfrozen to fully frozen state.

4.2.6.4   Factors Affecting the Product Quality During Freezing and Storage

The condition of the food at the time of freezing determines the final quality of the frozen food. Frozen food can be no better than the food was before it was frozen. Freezing does not sterilize foods as canning does. It simply retards the growth of microorganisms and slows down chemical changes that affect quality or cause food spoilage. Hence, frozen or low temperature storage is a must for frozen foods.

4.2.6.4.1   Moisture Loss

Exposure of a frozen product to air or another gaseous medium leads to an inevitable loss of water vapor. This moisture loss can be indicated by the frost accumulating on the coil surfaces. Fast cooling is the measure to reduce dehydration. Packaging in proper packaging material and in a proper way protects the frozen products from loss of moisture, color, flavor and texture.

4.2.6.4.2   Freezer Burn

Loss of moisture may also be when ice crystals evaporate from an area at the surface resulting into 'freezer burn' - a dry, grainy, brownish spot that becomes tough in frozen storage. This surface freeze-dried spot is although not harmful to the product but is organoleptically unacceptable and is very likely to develop off flavors.

4.2.6.4.3   Enzymes

Enzymes are the important catalysts of various chemical reactions in fruits and vegetables even after their harvest. However, their continuous activity may lead to quality deterioration in terms of color, flavor and nutrient contents. Hence, these need to be controlled or inactivated prior to the freezing process. Freezing, heating and chemical compounds can control enzyme actions. Freezing slows enzyme activity so that many frozen foods, such as meats and many fruits, will keep satisfactorily with little or no further treatment.

4.2.6.4.4 Air

Oxygen present in the air may cause flavor and color changes if the food is improperly packaged.

4.2.6.4.5 Microorganisms' Growth

Freezing process does not actually sterilize or reduce the microorganisms in a product. It simply arrests the growth of microorganisms which are already present in the product. While blanching prior to freezing destroys some microorganisms and there is a gradual decline in the number of these microorganisms during freezer storage, sufficient populations are still present to multiply in numbers and cause spoilage of the product when it thaws. Hence, any temperature fluctuation leading to thawing of the product and/or any accidental microbial contamination increases the chances of microbial spoilage of the product.

4.2.6.4.6 Ice Crystals

Large ice crystals associated with slow freezing tend to rupture the cells, causing an undesirable texture change. Hence a fast freezing leading to formation of minute ice crystals is preferred.

4.2.6.4.7 Freezer Temperature

The storage life of frozen foods is shortened as the temperature rises. Fluctuating temperatures result in growth in the size of ice crystals, further damaging cells and creating a mushier product. Changes in temperature can also cause water to migrate from the product.

4.2.7. PACKAGING OF FROZEN PRODUCTS

Packaging of a food product performs various functions like containment, retention of nutrients, protection from light, air, moisture, informative function, etc. For the frozen products packaging material should have the following characteristics:
- Moisture and vapor-proof.
- Food grade material.
- Durable and leak-proof.
- No brittleness and cracks at frozen temperature.
- Odorless, tasteless and grease-proof.
- Protect foods from off flavors and odors.
- Easy to fill, seal and use.
- Easy to label and store.
- Reasonable cost.

4.3 THAWING

Most of the frozen products need to be thawed before further use. Thawing is usually done by taking the frozen product out and keeping at ambient temperature for some time. However during thawing, foods are subjected to damage by chemical and physical changes and microorganisms also. Therefore, optimum thawing procedures are of concern to food technologists [6, 12]. Quick thawing at low temperature to avoid notably rise in temperature and excessive dehydration of food is desirable to assure food quality. With time some advancements have been developed in the thawing processes such as by using high pressure, microwave, ohmic and acoustic thawing.

4.3.1 HIGH PRESSURE THAWING

In the high-pressure thawing technique, the frozen food is pressurized until it reaches its thawing point. Heat is then applied under pressure to melt the ice and upon complete thawing of the food the pressure is released. The advantage of thawing under pressure is that the freezing point is depressed therefore the temperature driving force (different between air temperature and product temperature) can be increased several times, while keeping the temperature low to avoid microbial growth. However, high pressure has adverse effect on animal tissues due to protein denaturation whereas plant tissue seems little affected.

High-pressure thawing has been researched upon by some [17, 38, 39] to find that this technique preserved food quality and reduced the necessary thawing time. High pressure thawing reduced the thawing time to one-third of the time necessary at atmospheric pressure while maintaining the organoleptic qualities comparable to those of conventionally thawed products [17]. It was effective in texture improvement in frozen tofu as compared to that in tofu, thawed at atmospheric-pressure. Zhao and others [39] reported a minor undetectable drip loss and no negative effects on color, penetration force or cooking loss of beef thawed using high pressure.

In this technique thawing rate is affected by pressure level and treatment whereas product characteristics, such as size and initial temperature, did not affect thawing rate, indicating that it is advantageous to thaw a larger amount of product at high pressure. Some of the limitations with this technique are high cost, pressure-induced protein denaturation and meat discoloration [12, 20].

4.3.2 MICROWAVE THAWING

Microwaves are electromagnetic waves of radiant energy with wavelength in the range of 0.025–0.75 m that corresponds to frequencies of about 20,000–400 MHz. These can penetrate and produce heat deep within food materials [34] and hence find applications in accelerated thawing. The advantages of microwave thawing are shorter thawing time, less space requirement, reduction in drip loss, microbial spoilage and chemical deterioration [19, 29, 35, 32]. However, localized overheating is the major limitation of this technique which may be attributed to the preferential absorption of microwaves by liquid water.

The thawing rates of frozen samples in microwave thawing depend on material properties and dimensions and the magnitude and frequency of the electromagnetic radiation [26]. Factors such as thermal properties varying with temperature, irregular shapes and heterogeneity of the food make the thawing process more complicated [32].

4.3.3   OHMIC THAWING

When an alternating electric current is passed through a conducting food with high electrical resistance, heat is generated instantly inside the food, thus increasing the temperature of the food item [7]. This is termed as ohmic heating or electro-heating. Ohmic heating is advantageous over microwave heating in that nearly all of the energy enters the food as heat and it has no limitation of penetration depth. It has potential application in thawing of frozen foods by introduction of a frozen food (with negative electrons) into a high voltage electrostatic field [22, 23]. This technique prevents the drip, cell damage and softening in products such as carrot [8].

Studies on application of this technique are still in its infancy stage. Hence, more work needs to be carried out to better explore and commercialize this thawing technique.

4.3.4   ACOUSTIC THAWING

Acoustic thawing was investigated by Brody and Antenevich [2] and they observed a poor penetration, localized heating and high power requirement that hindered the development of this method. Kissam and others [13] used the relaxation mechanism for acoustic thawing and observed that application of a frequency, in the relaxation frequency range of ice crystals in the food, resulted into absorption of more acoustic energy by frozen foods. Acoustic thawing is a potential technology for the food industry if proper frequencies and acoustic power are chosen [21].

4.4   SUMMARY

Freezing is one of the easiest, quickest, most versatile and convenient methods of preserving foods that gives a product with high quality similar to as a fresh product. Freezing preserves the food by lowering the temperature, reducing the water activity and microbial destruction (in blanching). Some foods cannot be frozen because of some deleterious effect of freezing on the texture, color and integrity of the products.

There are several freezing methods such as using air freezers (still air, blast, fluidized bed and spiral type), freezing by indirect contact with the refrigerant (plate freezing) and by direct immersion in refrigerant (heat exchange, spraying of refrigerant, cryogenic freezing, etc.). Over the period, several advances have been developed in the freezing process. High pressure freezing, acoustic freezing, dehydrofreezing, use of anti-freeze proteins and ice nucleation proteins are to name a few. The rate of freezing is a determining factor for the product quality as small ice crystals that do not cause detrimental effect can be formed in the fast rate freezing process only. Dehydrofreezing involves dehydration of the product followed by the freezing process. There are certain techniques that can accelerate the freezing process. Some further studies and efforts are still required to minimize the cost of operation during freezing.

KEYWORDS

- Acoustic thawing
- Air freezing
- Concentration effects
- Cryogenic freezing
- Dehydrofreezing
- Freezer burn
- Freezing
- Freezing curve
- Freezing rate
- High pressure thawing
- Ice crystal damage
- Immersion freezing
- Microwave thawing
- Nucleation
- Ohmic thawing
- Plate freezing

- **Salting in**
- **Salting out**
- **Solute concentration**
- **Solute crystallization**
- **Super cooling**
- **Supersaturation**
- **Thawing**

REFERENCES

1. Biswal, R. N., Bozorgmehr, K., Tompkins, F. D., & Liu, X. (1991). Osmotic concentration of green beans prior to freezing. *Journal of Food Science*, 56, 1008–1011.
2. Brody, A. L., & Antenevich, J. N. (1959). Ultrasonic defrosting of frozen foods. *Food Technology*, 13, 109–110.
3. Chevalier, D., Sentissi, M., Havet, M., & Le Bail, A. (2000). Comparison of air-blast and pressure shift freezing on Norway lobster quality. *Journal of Food Science*, 65, 329–333.
4. Desrosier, W., & Desrosier, N. (1978). *Technology of Food Preservation*. 4th edn. AVI, Westport, Connecticut, pp. 110–151.
5. Earle, R. L. (1983). *Unit Operations in Food Processing*. 2nd edn. Oxford University Press, Oxford, pp. 78–84.
6. Fennema, O. R., Powrie, W. D., & Marth, E. H. (1973). *Low Temperature Preservation of Foods and Living Matter*. New York, USA: Marcel Dekker, p. 598.
7. Fu, W. R., & Hsieh, C. C. (1999). Simulation and verification of two-dimensional ohmic heating in static system. *Journal of Food Science*, 64, 946–949.
8. Fuchigami, M., Hyakuumoto, N., Miyazaki, K., Nomura, T., & Sasaki, J. (1994). Texture and histological structure of carrots frozen at a programmed rate and thawed in an electrostatic field. *Journal of Food Science*, 59, 1162–1167.
9. Graham, J. (1984). *Planning and Engineering Data, 3, Fish Freezing*. FAO Fisheries Circular No 771. FAO, Rome.
10. Hew, C. L., & Yang, D. S. C. (1992). Protein interaction with ice. *European Journal of Biochemistry*, 203, 33–42.
11. Holdsworth, S. D. (1987). Physical and engineering aspects of food freezing. In: *Developments in Food Preservation, Vol. 4*. S. Thorne (ed.). Elsevier Applied Science, Barking, Essex, pp. 153–204.
12. Kalichevsky, M. T., Knorr, D., & Lillford, P. J. (1995). Potential food applications of high-pressure effects on ice-water transitions. *Trends in Food Science and Technology*, 6, 253–258.

13. Kissam, A. D., Nelson, R. W., Ngao, J., & Hunter, P. (1981). Waterthawing of fish using low frequency acoustics. *Journal of Food Science*, 47, 71–75.
14. Leeson, R. (1987). *Applications of Liquid Nitrogen in Individual Quick Freezing and Chilling*. BOC (UK) Ltd, London SW19 3UF.
15. Li, B., & Sun, D.W. (2002). Effect of power ultrasound on freezing rate during immersion freezing, *Journal of Food Engineering*, 55. 277–282.
16. Liu, S., Wang, W., von Moos, E., Jackman, J., Mealing, G., Monette, R., & Ben, R.N. (2007). In Vitro Studies of Antifreeze Glycoprotein (AFGP) and a C-Linked AFGP Analogue. *Biomacromolecules*, 8, 1456–1462.
17. Makita, T. (1992). Application of high pressure and thermophysical properties of water to biotechnology. *Fluid Phase Equilibrium*, 76, 87–95.
18. Martino, M. N., Otero, L., Sanz, P. D., & Zaritzky, N. E. (1998). Size and location of ice crystals in pork frozen by high-pressure-assisted freezing as compared to classical methods. *Meat Science*, 50, 303–313.
19. Meisel, N. (1973). Microwave applications to food processing and food systems in Europe. *Journal of Microwave Power*, 8, 143–146.
20. Mertens, B., & Deplace, G. (1993). Engineering aspects of high-pressure technology in the food industry. *Food Technology*, 47, 164–169.
21. Miles, C. A., Morley, M. J., & Rendell, M. (1999). High power ultrasonic thawing of frozen foods. *Journal of Food Engineering*, 39: 151–159.
22. Ohtsuki, T. (1991). Process for thawing foodstuffs. US Patent 5034236.
23. Ohtsuki, T. (1993). Process for thawing foodstuffs. European Patent 0409430.
24. Olsson, P., & Bengtsson, N. (1972). *Time–Temperature Conditions in the Freezer Chain*. Report No. 30 SIK. Swedish Food Institute, Gothenburg.
25. Otero, L., Martino, M., Zaritzky, N., Solas, M., & Sanz, P. D. (2000). Preservation of microstructure in peach and mango during high-pressure- shift freezing. *Journal of Food Science*, 65, 466–470.
26. Pangrle, B. J., Ayappa, K. G., Davis, H. T., Davis, E. A., & Gordon, J. (1991). Microwave thawing of cylinders. *AlChE Journal*, 37, 1789–1800.
27. Persson, P. O., & Londahl, G. (1993). Freezing technology. In: *Frozen Food Technology*, C. P. Mallett (Ed.), Glasgow, UK: Blackie Academic & Professional, pages 20–58.
28. Petzold, G., & Aguilera, J. M. (2009). Ice morphology: fundamentals and technological applications in foods. *Food Biophysics*, 4, 378–396.
29. Rosenberg, U., & Bogl, W. (1987). Microwave thawing, drying, and baking in the food industry. *Food Technology*, pp. 85–91.
30. Robbers, M., Singh, R. P., & Cunha, L. M. (1997). Osmotic-convective dehydrofreezing process for drying kiwifruit. *Journal of Food Science*, 62, 1039–1042, 1047.
31. Spiazzi, E. A., Raggio, Z. I., Bignone, K. A., & Mascheroni, R. H. (1998). Experiments on dehydrofreezing of fruits and vegetables: mass transfer and quality factors. *Advances in the Refrigeration Systems*, Food Technologies and Cold Chain, IIF/IIR, 6, 401–408.
32. Taoukis, P., Davis, E. A., Davis, H. T., Gordon, J., & Takmon, Y. (1987). Mathematical modeling of microwave thawing by the modified isotherm migration method. *Journal of Food Science*, 52, 455–463.

33. Tran, T., Thitipraphunkul, K., Piyachomkwan, K., & Sriroth, K. (2008). Effect of starch modifications and hydrocolloids on freezable water in cassava starch systems. *Starch/Stärke,* 60, 61–69.
34. Tong, C. H., Lentz, R. R., & Lund, D. B. (1993). A microwave oven with variable continuous power and a feedback temperature controller. *Biotechnology Progress,* 9, 488–496.
35. Virtanen, A. J., Goedeken, D. L., & Tong, C. H. (1997). Microwave assisted thawing of model frozen foods using feedback temperature control and surface cooling. *Journal of Food Science,* 62, 150–154.
36. Warren, C. J., Mueller, C. M., & Mckown, R. L. (1992). *Ice Crystal Growth Suppression Polypeptides and Methods of Preparation.* US Patent 5,118,792.
37. Yu, S., Ma, Y., & Sun, D.-W. (2010). Effects of freezing rates on starch retrogradation and textural properties of cooked rice during storage. *LWT–Food Science & Technology,* 43, 1138–1143.
38. Zhao, Y. Y., Fores, R. A., & Olson, D. G. (1996). The action of high hydrostatic pressure on the thawing of frozen meat. *Annual Meeting of Institute of Food Technologist,* New Orleans, LA (June), 22–26.
39. Zhao, Y. Y., Fores, R. A., & Olson, D. G. (1998). High hydrostatic pressure effects on rapid thawing of frozen beef. *Journal of Food Science,* 63, 272–275.

CHAPTER 5

OVERVIEW OF APPLICATIONS OF DRYERS FOR FOODS: SOLAR, NOVEL, INFRARED AND INDUSTRIAL

ARCHANA MAHAPATRA[1] and P. PUNAM TRIPATHY[2]

[1]*Research Scholar, Agricultural and Food Engineering Department, Indian Institute of Technology, Kharagpur, Kharagpur–721302, West Bengal, India*

[2]*Assistant Professor, Agricultural and Food Engineering Department, Indian Institute of Technology, Kharagpur, Kharagpur – 721302, West Bengal, India, E-mail: punam@agfe.iitkgp.ernet.in*

CONTENTS

5.1   INTRODUCTION

Global food production has increased substantially in recent years due to advancements in agricultural sector, such as development of improved disease-resistant staple crop varieties, use of chemical fertilizers and pesticides, more irrigated crops and improved farm machineries. The global production of different agricultural commodities is shown in Figure 5.1 [40].

The per capita food supply has increased from 2200 kcal/day in 1960 to more than 2800 kcal/day in 2009 [40]. Food supply varies from region to region, such as Europe has the highest supply at 3370 kcal/person/day whereas emerging economies like Western Asia, Northern Africa, Latin America and Eastern Asia have per capita supply of about 3000 kcal/day. Sub-Saharan Africa and Southern Asia still have less than 2500 kcal/person/day food supply. Although enough food is produced to feed the world population, still 11.3% of the population remained undernourished in 2012–2014 [41]. The global food wastage of edible parts of the food produced is approximately 1.3 billion tones as shown in Figure 5.2, which is roughly one third of the total food produced [39]. Food loss occurs at early stages of food chain which is due to financial, managerial and technical constraints in harvesting techniques, inadequate storage and cooling facilities. Quantitatively, 30% cereals; 40–50% root crops, fruits and vegetables; 20% oilseeds, meat and dairy products; and 35% fish are lost or wasted annually (Figure 5.3).

Drying is one of the most accessible post-harvest operations employed to reduce the moisture content of food products to a safe moisture limit so that micro-organisms cannot grow and the shelf life of the products can be enhanced for a longer period. Thermal drying, which is the most widely used method for drying agricultural commodities in industrial sector, requires

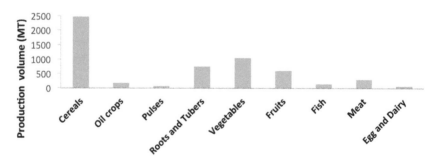

FIGURE 5.1   Worldwide production volume of different agricultural produce as on 2010 [40].

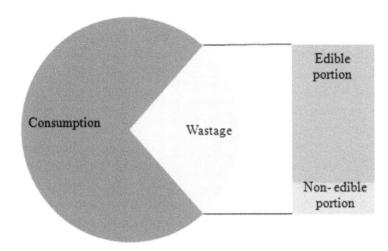

FIGURE 5.2   Global food consumption-wastage scenario.

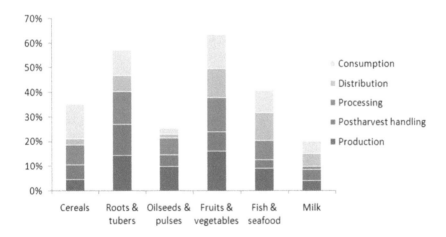

FIGURE 5.3   Global food losses and waste trend.

energy harnessing from various sources like electricity, fossil fuel, natural gas, forest residues like wood, bark etc. The rapid depletion of natural fuel resources and the increasing cost of fossil fuel make thermal drying expensive. In addition, the use of fossil fuel results in serious environmental pollution due to the increasing concentration of carbon dioxide in the atmosphere leading to global warming. Further, in developing countries the availability of grid-connected electricity is very scarce or intermittently available.

In thermal dryers, the drying air temperature increases beyond the optimum temperature for drying of agricultural products, thus altering the final product quality. Hence, nowadays much emphasis has been given on the use of renewable sources of energy for drying applications. The application of solar energy for drying of agricultural commodities has utmost importance since it can easily provide the low temperature heating required for drying of food stuffs [37, 100].

Research and developments in drying technology have accelerated over the past few years. There is an increasing market demand for sustainable and novel drying methods resulting in high output, reduced energy consumption and carbon footprint with improved food safety and security for better quality products. Different types of agricultural commodities such as grains, seeds, fruits, vegetables and herbs are being dried using various types of dryers out of which the solar, novel, infrared and industrial dryers are the promising techniques for better product quality and growing consumer acceptability.

This chapter discusses the principle, design and applications of the dryers for drying of various agricultural commodities.

5.2 DRYING METHODS

5.2.1 SOLAR DRYING

Solar energy for drying of agricultural products is being used since ages and it ranges from open sun drying to advanced solar dryers. In traditional open sun drying, the food products are directly exposed to solar radiation as a result of which the products are easily contaminated by dirt, dust and environmental pollution, degradation due to rain, storm etc. causing huge loss of the quality and quantity of the product. In order to overcome the drawbacks of open sun drying, solar dryers are most promising alternatives for getting better quality end products. The solar energy drying systems are used for drying various grains such as rough rice [12], corn seed [91]; fruits such as apple, banana, copra, coconut, figs, grape, mango, pineapple, strawberry; vegetables such as cassava, cauliflower, chilli, green peas, onion, tomato; spices like turmeric; and tea leaves, cocoa [45], apricot [129], potato [132], mint [2], eggplant [8], pistachio nuts [81], coffee beans [86].

5.2.2 NOVEL DRYING TECHNIQUES

The food industry is largely in demand for ingredient development and providing novel products to the society. In recent times, many advances have taken place for technology development associated with industrial drying of food products that includes various methods of pre-treatment and quality analysis. More than 85% of industrial dryers are the conventional type and they utilize hot air or combustion gases as the heat transfer medium [140]. The end products obtained from conventional drying methods suffer from poor quality with a huge loss in nutritional, functional and sensorial attributes. In order to get best quality end products, in the recent years there have been significant technological advancements in food drying in terms of different pre-treatments, techniques, equipments and quality [28]. These advancements address the growing need to find improved drying techniques to preserve the quality of the dried product at proper utilization of energy.

Hence, the novel drying methods emerge as energy efficient technology for enhancing the product quality; reduce energy consumption and environmental emissions, higher output and flexibility, production of new products, and ease in combining different operations in the same unit with better process control. Some of the novel drying techniques are: refractance window drying, superheated steam drying, high electric field drying, microwave and radio frequency drying, heat pump drying, modified atmosphere drying, impinging stream drying, contact sorption drying. These advanced drying techniques are either developed by combination of two or more existing drying methodologies or by applying a recently developed concept as shown in Figure 5.4.

5.2.3 INFRARED DRYING

Infrared drying is based on the principle of transfer of infrared (IR) energy from the heating source to the food surface, resulting in less drying time. During the drying of food products, the IR radiation energy in the wavelength range of 0.75 to 1.4 μm is transferred from the heating element to the product surface without heating the surrounding air [71]. The radiation impinges on the exposed food material, penetrates it and then is

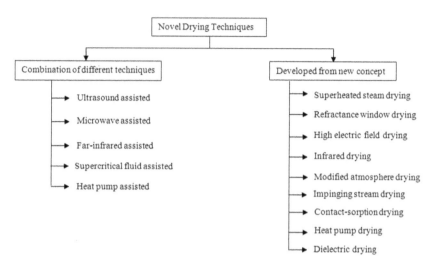

FIGURE 5.4   Classification of novel drying systems.

converted to sensible heat. IR drying has been investigated as a potential drying method for various type of foodstuffs including fruits, vegetables, grains and seeds, e.g., paddy [5]; pear, carrot, sweet corn [97]; banana [88]; cashew kernel [53]; sweet potato [32]; barley [1]. The IR drying has gained popularity because of its superior thermal efficiency, high heat transfer coefficients and fast drying rate compared to conventional drying methods.

5.2.4   INDUSTRIAL DRYING

Food materials need to have suitable moisture content for further processing and for satisfactory packaging. The cost of transportation also depends on the moisture content of the product, and a balance must be maintained between the cost of transportation and the cost of drying. Hence, adequate consideration should be given to the methods employed in saving the energy in different dryers. In addition, the type of products and the necessary precautions needed during drying at industrial scale should be properly taken care. Although the fundamental principle of heat and mass transfer in drying is common for all kind of products, yet the wide variation in physical and chemical composition of the products make some of the drying methods most suitable for a specific product, which may not be suitable for another product. Some unique features of drying that creates a challenge to the industry are the wide variation in product size and porosity, drying time, operating temperature and pressure, production capacity etc.

Selection of a drying method depends on several factors: form of raw material and its properties, desired physical form and characteristics of the product, necessary operating conditions, availability of dryer, cost of operation, final product quality, energy consumption, and quality of dried products. Some of the most commonly used industrial dryers include tray dryer [113], tunnel dryer [50], conveyor dryer [110], rotary dryer [62], drum dryer [59], fluidized bed dryer [114], freeze dryer [107], spray dryer [24], pneumatic dryer [142], spouted bed dryer [80] and osmotic dryer [26].

5.3 FUNDAMENTALS OF DRYING PROCESS

Drying is a transient heat and mass transfer process where heat transfer to the product occurs from the drying agent (hot air, water, steam etc.) either by conduction, convection or radiation or a combination of these; and moisture migration occurs from inside of the product to the surrounding. These transport parameters are affected by external parameters, like temperature of drying medium, humidity of air, type of product, rate and direction of air flow etc. and internal parameters e.g. moisture content of the product and food product temperature. The initial stage of drying is mainly affected by the surrounding external parameters whereas the internal parameters control the drying rate after the critical moisture content is reached. In this section, fundamental aspects of drying process as well as the effect of process parameters on the drying rate for different food products are discussed.

5.3.1 DRYING KINETICS OF FOOD PRODUCTS

A typical drying rate curve during the drying of food materials is shown in Figure 5.5. As can be seen, the food surface is at a lower temperature in the beginning than the drying medium and thus evaporation rate is higher

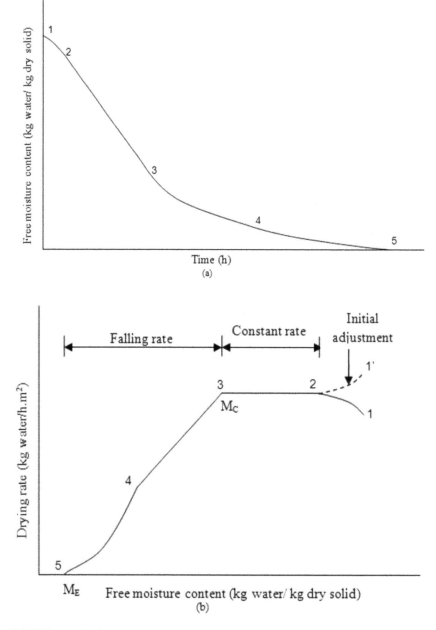

FIGURE 5.5 Drying rate curve (a) Free moisture content ~ time and (b) Drying rate ~ free moisture content.

until the surface comes to wet bulb temperature (line 1–2). If the surface is at higher temperature, it will reduce to reach wet bulb temperature at the surface (line 1–2). After the initial unsteady-state adjustment period, the drying process begins to start. Line 2–3 shows the constant rate-drying period. During this period, the solid surface is very wet and acts as a free water surface. The surface temperature of product remains close to wet bulb temperature of drying air. The constant rate of drying continues until the critical moisture content of the product is reached (point 3). The drying rate in this period depends upon the temperature difference between drying air and product surface at constant velocity and relative humidity.

The moisture movement within the product is the controlling factor in falling rate drying and it is often divided into two stages:

    a.   Unsaturated surface drying or first falling rate (line 3–4) in which the wet surface area decreases gradually till the surface becomes completely dry; and

    b.   Drying or second falling rate (line 4–5) in which rate of moisture diffusion inside the product is slow. In some materials, e.g., non-hygroscopic products, the second falling-rate period is usually absent, whereas hygroscopic materials have two or more periods.

5.3.2 EQUILIBRIUM MOISTURE CONTENT

The knowledge of the state of thermodynamic equilibrium between the surrounding air and food material is a basic prerequisite for drying. In drying modeling, it is very important to consider the moisture content of the material that comes into equilibrium with the drying air. When a wet solid is brought into contact with continuous stream of air at a constant humidity and temperature, the solid will either loose or gain moisture depending on the moisture gradient. The process will continue until the vapor pressure of moisture in the solid becomes same as that of air and both the solid and air are said to be in equilibrium. The corresponding moisture content of the solid is known as equilibrium moisture content (EMC) at the specific condition. EMC is useful to determine whether a product will gain moisture or loose moisture at a certain temperature and humidity condition during drying and storage. A plot of relative humidity and EMC is known as

equilibrium moisture curve or isotherm. The equilibrium moisture of agricultural food products can be achieved either by adsorption or by desorp-
tion, as expressed by the respective isotherms given in Figure 5.6.

5.3.3 FACTORS AFFECTING DRYING RATE OF FOOD PRODUCTS

The major process parameters affecting the drying rate are moisture content, airflow rate, thickness of the product and drying air temperature.

5.3.3.1 Moisture Content

Free moisture escapes from surface, as the vapor pressure is higher than
that of atmosphere. As drying continues, the free water is removed from
the product surface and vapor pressure decreases gradually resulting in a

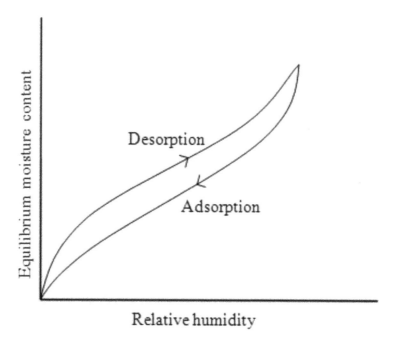

FIGURE 5.6   Hysteresis loop for food products.

decrease in drying rate. The drying kinetics of red pepper was studied by Akpinar et al. [3] and they observed that the rate of moisture migration to the product surface and subsequent evaporation from surface to air was decreased with decrease in the moisture content, resulting in a reduced drying rate. Similarly, drying rate was decreased continuously with drying time due to the reduction of moisture content in apricots [130].

5.3.3.2   Air-Flow Rate

Increasing the airflow rate at a constant drying air temperature shortens the drying time to reach the equilibrium temperature inside the product resulting in enhanced drying rate. Fikiin et al. [44] have studied that with the increase in airflow rate from 1 to 10 m/s, the drying rate for apricots was increased leading to a higher heat transfer coefficient. The drying time for carrots was decreased by 25% as the air flow rate increased from 0.5 m/s to 1 m/s resulting in increase in the drying rate [31].

5.3.3.3   Thickness of the Product

For longer drying periods, the rate of diffusion is the controlling factor for drying rate and the drying rate varies with thickness of the food product. It was observed that with the increase in the thickness of carrot slices, the drying time also increased due to the increased diffusion path [31]. In case of eggplant drying, it was noticed that with the increase in slice thickness from 0.635 cm to 2.54 cm, drying time was increased by 294%, implying a higher mean drying rate at lower slice thicknesses. Thinly sliced products dry out faster due to the reduced distance of moisture movement and increased exposed surface area [38].

5.3.3.4   Drying Air Temperature

At higher drying air temperature, the effective drying time is less resulting in an increase in the drying rate. At high temperature, the heat transfer between drying air and the food product increases significantly favoring the evaporation of the water leading to increased drying rate [3].

5.4 DESCRIPTION OF DIFFERENT DRYING TECHNOLOGIES

5.4.1 SOLAR ENERGY DRYING SYSTEMS

Solar energy drying system consists of a flat plate collector, drying chamber and other auxiliary units. The flat plate collector is a heat exchanger which converts solar radiation to heat energy. It transforms energy in the low and medium temperature range up to 100°C using both beam as well as diffuse solar radiation. A flat plate collector consists of an absorber plate painted black in order to absorb maximum solar radiation, may be flat, grooved or corrugated with fins or ducts attached. It is insulated properly to minimize heat losses from sides and back of the absorber plate. The collector also consists of the glazing made up of one or more layer of glass or radiation transmitting material. Figure 5.7 explains the different energy losses in a flat plate solar collector.

Solar radiation falling on the glass cover is partly absorbed, partly reflected and the rest is transmitted through the glazing and falls on to the absorber plate. The absorber plate absorbs the radiation and a fraction of the absorbed energy is reradiated back in the form of long wavelength radiation. The energy losses that take place in a flat plate solar collector are top loss, bottom loss and edge loss. Top loss relates to the heat energy lost

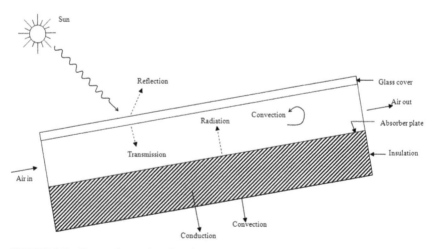

FIGURE 5.7 Energy losses in a flat plate solar collector.

from the top surface of the collector and takes place mainly by radiation and convection whereas bottom loss is the heat energy lost from the bottom surface of the collector enclosure which occurs mainly by conduction. Edge loss is the loss of heat energy from the sides of the collector which is also mainly by conduction and convection to the surrounding.

The performance of the flat plate solar collector is represented by an energy balance considering the useful energy, thermal losses and optical losses from incident radiation [33]. Collection efficiency, defined as the ratio of useful energy gain to the incident solar energy over a specified time period, is one of the major indices of collector performance. Mathematically it can be expressed by Hottel-Whillier-Bliss equation [33]:

$$\eta = \frac{Q_u}{IA_c} \qquad (1)$$

The useful solar energy gain is given by:

$$Q_u = \dot{m} C_p \left(T_{out} - T_{in} \right) \qquad (2)$$

$$Q_u = A_c F_R \left[S - U_L \left(T_{in} - T_{amb} \right) \right] \qquad (3)$$

where, S is the total solar radiation considering optical losses and is given as:

$$S = I\tau\alpha \qquad (4)$$

$$Q_u = A_c F_R \left[I\tau\alpha - U_L \left(T_{in} - T_{amb} \right) \right] \qquad (5)$$

Now, combining Eqs. (1) and (5), we get:

$$\eta = F_R \left(\tau\alpha \right) - F_R U_L \frac{\left(T_{in} - T_{amb} \right)}{I} \qquad (6)$$

The collector heat removal factor, F_R, is the quantity that relates the actual useful energy gains of the collector to that if it were at the fluid inlet temperature and given as:

$$F_R = \frac{\dot{m} C_p}{A_c U_L} \left[1 - \exp\left(-\frac{A_c U_L F'}{\dot{m} C_p} \right) \right] \qquad (7)$$

The thermal energy lost from collector by conduction, convection and radiation through glass cover, bottom and edges of the collector is accounted in the overall heat loss coefficient U_L.

$$U_L = U_t + U_b + U_e \tag{8}$$

where, U_t, U_b and U_e represent the top, bottom and edge loss coefficients and can be expressed by the following equations:

$$U_t = \left[\frac{1}{h_{c,p-g} + h_{r,p-g}} + \frac{1}{h_w + h_{r,p-amb}} \right] \tag{9}$$

$$U_b = \frac{K_{in}}{L_{in}} \tag{10}$$

$$U_e = U_b \frac{A_e}{A_c} \tag{11}$$

These heat transfer parameters are very important in analyzing the thermal performance of solar drying systems.

Figure 5.8 illustrates a detailed classification of solar energy drying systems depending on the design of dryer components and mode of solar heat utilization. Solar dryers can be broadly classified as: passive, active and hybrid dryers. In the passive or natural convection dryers, the solar heated air circulates inside the drying chamber by buoyancy forces whereas external fan or blowers are used to circulate drying air in case of active or forced convection dryers. These dryers are further classified as direct, indirect and mixed mode types based on the incidence of solar radiation on the product.

5.4.1.1 Direct Solar Dryer

A simple solar cabinet dryer (Figure 5.9) consists of a rectangular box with transparent glass on the top in order to transmit solar radiation into the dryer. The interior surfaces of the dryer are properly insulated and blackened for maximum absorption of radiation. The product is spread as a thin

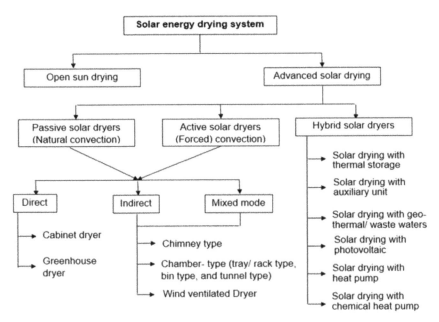

FIGURE 5.8   Classification of solar energy drying system.

layer on the perforated tray present inside the drying chamber. Ambient air flows into the cabinet through an inlet opening provided in the bottom of the chamber and passes over the products and moist air escapes from the rear side outlet opening. The whole process is facilitated by natural circulation of drying air. Direct solar drying has been investigated for various fruits: ber, sapota; vegetables: spinach, okra, tomato, ginger, onion, peas, cabbage, sweet potato, bitter gourd, sugar beet [127]; grapes, figs [17], pineapple [51].

5.4.1.2   Indirect Solar Dryer

An indirect solar dryer mainly consists of a solar air heater and a drying chamber connected in series with each other (Figure 5.9). Both the collector and dryer assembly are made of galvanized sheet metal and painted black in order to absorb maximum solar radiation. The collector top is covered with either glass or polycarbonate sheet to transmit solar radiation into the drying system whereas the top of the drying chamber is opaque so

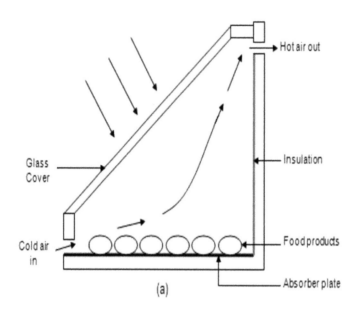

(a)

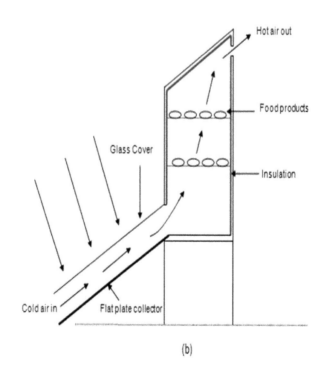

(b)

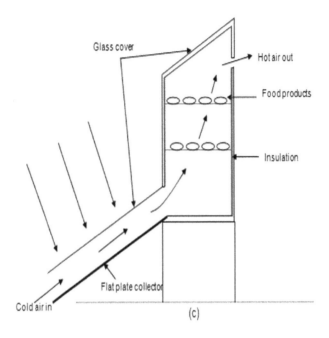

FIGURE 5.9   Schematic diagram of (a) direct (b) indirect and (c) mixed mode solar dryer.

that the food products are not directly exposed to solar radiation. Ambient air flows through the collector inlet and gets heated as it passes through the collector length.

The hot air then enters the drying chamber where food products are spread in thin layers in one or more perforated trays. Moisture is evaporated from the product and the moist air exits through the outlet opening.

Air circulation in the drying chamber is facilitated by using natural convection draft, for example, chimney or forced convection, for example, fans and blowers. Sometimes reflectors are provided on both sides of the collector in order to enhance more of the incidence radiation into the collector and to increase the overall drying efficiency. In these dryers, the solar radiation does not fall on the product directly; as a result the dried product is of superior quality retaining the nutritional content without any discoloration. Indirect solar drying of fruits, vegetable, herbs and spices have been investigated by various researchers, such as green peas [56], thymus and mint [36], bitter gourd [135], cumin seeds [141], corn seed [91], orange and lemon [78]; non-parboiled paddy [118].

5.4.1.3   Mixed-Mode Solar Dryer

A mixed-mode solar dryer uses the principle of both direct and indirect solar dryer for drying of food products (Figure 5.9). The design and working principle is very much similar to that of indirect type solar dryer. The top of the drying chamber is transparent here instead of opaque as in indirect type. Hence in these types of dryers, the solar radiation is incident both on the collector as well as on the drying chamber due to which the product dries out quickly resulting in higher drying rate as compared to both direct as well as indirect type dryers. Application of mixed-mode solar drying has been studied in detail for various food products such as: rough rice [16], beans [47], potato [131], apple, apricot and carrot [73], red pepper and grapes [35].

5.4.1.4   Solar Greenhouse Dryer

Greenhouse is composed of transparent walls and roofs so that the inside temperature is higher than the ambient temperature. The dryers are made of structural frame covered with clear polyethylene material, glass or, rigid panel. The solar radiation incident over the surface is transmitted to the dryer, a part of it is absorbed by the product and part of the long wavelength radiation is reflected back to atmosphere. As the long wavelength radiation cannot escape through the chamber, hence the temperature of the chamber further increases. Air ventilation is facilitated either by chimney on top, or actively by using fans. Depending upon the structure, these dryers can be dome shaped or roof even type. The dome shaped structure absorbs maximum solar radiation whereas proper mixing of air is achieved in the roof even type dryer. Greenhouse drying is used for drying of grain, fruits, vegetables, medicinal plants as well as cash crops. Apart from drying, greenhouse is also used for crop production, soil solarization, poultry and aquaculture farming. Greenhouse dryers are suitable for low temperature drying of cereal grains, fruits, vegetables like banana [54], onion [65], tomato [57], and amaranth grains [109].

5.4.1.5   Hybrid Solar Drying System

The availability of solar energy depends upon the variable weather condition as well as on the location. Hence, a steady supply of solar energy is not possible, which acts as a major constraint in its extensive use. In order to facilitate uninterrupted drying, hybrid solar drying is a promising option, which employs the use of additional energy source as a back-up in some supplementary units to facilitate drying during off sunshine hours and places where adequate solar energy is not available. The supplementary unit includes use of thermal storage medium (e.g., rock-bed, sand, phase change material); electricity, biomass burner, LPG gas burner, diesel; geothermal energy and wastewater; photovoltaic (PV) units, heat pump; and dehumidification system, etc.

5.4.1.5.1   Solar Dryer With Heat Storage Unit

The solar thermal energy can be stored as sensible heat, latent heat or chemical form [11]. For storage using sensible heat, temperature of the storage medium is raised and the sensible heat content increases depending upon the specific heat, temperature change and mass of the material. Water acts as an excellent medium for heat storage due to its high heat capacity and is inexpensive. Rock bed and pebble bed are also used as sensible heat storage medium. Heat storage by latent heat occurs by change of the state of material from solid to liquid or liquid to gas and vice versa at constant temperature. Solid- liquid phase change materials are commonly used for providing the necessary heat during drying. As the material absorbs heat, it changes to liquid phase and heat is stored which is later released to the drying medium and the material eventually solidifies. Latent heat storage is more efficient than the sensible heat storage type since it stores heat at almost constant temperature. Since the latent heat capacity of a material is much higher than its sensible heat capacity, latent heat storage stores more heat than the later one. Some of the materials used as latent heat storage medium are paraffin, low melting metals, hydrates of salts. Heat storage using chemical energy occurs by reversible chemical reactions involving breaking and joining of bonds by absorption and release of thermal heat.

Some applications of heat storage medium in solar drying are presented in Table 5.1.

5.4.1.5.2   Solar Dryer With Auxiliary Unit

Apart from solar energy, secondary heat sources like electricity, biomass, petroleum, etc. supplement to solar heating during adverse weather conditions. Hybrid solar dryer consists of basic units like collector, blower, drying chamber etc. and an additional heating unit like electric heater; biomass burner and gas-to-gas heat exchanger or LPG (liquid propane gas) combustion heater. The sources of biomass include woody plant, grass, food crops, and agricultural, animal and other organic wastes. The gas-to-gas heat exchanger ensures clean, smoke, shot and ash free gas supply to the drying chamber in order to avoid contamination of food. Solar drying of some agricultural products with supplementary heat sources are presented in Table 5.2.

TABLE 5.1   Solar Drying of Agricultural Products With Various Heat Storage Medium

Product to be dried	Heat storage medium	Type of heat storage	Reference
Cocoa beans	• Molecular sieve- 13X $[(Na_{86}[(AlO_2)_{86}\cdot(SiO_2)_{106}]\cdot264H_2O)]$ • $CaCl_2$	Desiccant storage	[30]
Coconut	• Concrete • Sand • Rock-bed	Sensible heat	[7]
Garlic clove	Mixture of propylene glycol 60% and water 40% v/v	Latent heat	[120]
Mushroom	Paraffin wax	Latent heat	[106]
Seeded grape	Calcium chloride hegzahidrat	Latent heat	[20]
Tomato, onion, pepper, okra and spinach	Rock pebbles	Sensible heat	[9]

TABLE 5.2   Solar Drying of Agricultural Products With Auxiliary Units

Product to be dried	Auxiliary heating source	Reference
Peas and beans	Electrical heater	[66]
Chilli	Biomass burner	[74]
Pepper berry	Biomass burner	[108]
Pineapple	Biomass burner	[18]
Saladette tomatoes	LPG burner	[77]
Banana	LPG burner	[123]
Seed maize	LPG burner	[126]

5.4.1.5.3   Solar Dryer Integrated With PV Panel

Integration of PV cells in solar drying system makes them independent of any other energy source. PV cells are used in solar drying system either as a part of the collector unit to improve efficiency or to operate the fan or blowers to force air inside the drying chamber. The drying air is heated in the collector and circulated to the drying chamber by solar powered blower/fans. Solar dryers integrated with PV panel have been studied for various vegetables, fruits and herbs, such as chilli [13], saffron [83], jackfruit leather [25], apple, fenugreek and mint [84], white oyster mushroom [85].

5.4.1.5.4. Solar Dryer With Heat Pump

Heat pumps are ideal for low temperature drying requirements, matching to solar drying. They are efficient, recycle energy from exhaust air and control the drying air temperature. Heat pump in combination with solar drying system has been successfully used for grains, fruits, vegetables and herbs: round grained rice [75], copra [82], tomato, strawberry, mint, parsley [117].

5.4.2   NOVEL DRYING TECHNOLOGIES

The majority of industrial drying operations are convective type using either hot air or combustion gases as heat transfer medium for drying of food materials. This type of drying requires high energy inputs and the exhaust air is released to the surrounding ambient air. Due to the increasing interest both from the public and government to reduce environmental degradation, it is essential to improve the drying processes to reduce energy consumption and greenhouse gas (GHG) emissions, still providing a high quality dried product. In order to achieve these goals, there is a growing concern to identify novel drying technologies to preserve the food quality at better utilization of energy. The main aim of novel drying methods is to exploit different physical phenomena to enhance already existing commercial drying techniques, as in the case of ultrasound or microwave/dielectric drying, or utilizing newly revealed phenomena as in the case of superheated steam drying or refractance window drying.

5.4.2.1 Ultrasound Assisted Drying Technology

The application of ultrasound waves to porous food material induces the formation of micro-channels on its surface due to the deformation of the porous solid material when exposed to ultrasound waves. The high power ultrasound waves are characterized by low frequencies (20–100 kHz) at high intensities (typically 10–1,000 W/cm) [128]. Drying of foods using high power ultrasound waves is more significant at low temperatures, thereby decreasing the probability of deterioration of food. In recent years, ultrasound power is used in hot air drying for lowering the resistance to mass transfer [46], to assist in osmotic dehydration in order to lower the drying time [43], and in infrared drying to retain the quality of the end product [34]. In recent years, ultrasound assisted drying has been used for various fruits and vegetables such as: banana, genipap, jambo, melon, papaya, pineapple, pinha, sapota [43], apple [111], potato [94], red bell pepper [116], broccoli [139], and carrot [69].

5.4.2.2 Microwave and Dielectric Drying System

Microwave and dielectric or radio frequency (RF) drying systems use non-destructive technique while drying the food products. Drying is done at ambient temperature leading to uniform temperature distribution inside the food product by eliminating chances of case-hardening and solvent migration. These dryers are highly efficient with less drying time. Dielectric and microwaves are high frequency electromagnetic waves. The electromagnetic waves penetrate into a large depth of the food material causing volumetric heating of water molecules present in the foods, offering a higher energy conversion rates and therefore shorter process time. Dielectric heating is done between 1–100 MHz and microwave heating between 300 MHz and 300 GHz [115]. Dipolar molecules like water which have non-uniform distribution of positive and negative charges on the atoms tend to align themselves under the influence of electric field. As the electric field changes from positive to negative, the dipole also tends to align accordingly. The cycle continues several times per second and so does the alignment of the dipole, resulting in friction and development of heat within the product. This phenomenon predominates in microwave drying. The heat produced due to dipolar rotation is given as [134]:

$$P = 1.41 f \left(\frac{E}{d}\right)^2 \varepsilon' \tan \delta \times 10^{12} \tag{12}$$

where, P is power, f is the frequency of electromagnetic field, E is the voltage, d is the distance between electrodes, ε' is the dielectric constant of the material, and $\tan \delta$ is the loss tangent = ratio of dielectric loss (ε'') and dielectric constant (ε').

The degree of heating of a product depends upon its dielectric properties which are further governed by certain parameters like product moisture content, temperature, frequency of electromagnetic wave etc. High moisture food has high dielectric constant and high loss factor. As a result, they absorb the microwave and RF. At lower frequency of wave, the penetration depth increases. Penetration depth of microwave is given as [42]

$$d_0 = \frac{\lambda_0}{2\pi \sqrt{\varepsilon' \tan \delta}} \tag{13}$$

where, d_0 is the depth of penetration and λ_0 is the wavelength.

Microwave-assisted drying has proven to be a successful technique in overcoming the issues of low industrial yields and inferior kernel quality, as encountered in conventional air drying processes for enhancing the rehydration capacity of carrots [124], to dry macadamia nuts [121], to overcome the quality losses in spinach due to over-heating of the product [93]. Similarly, the radio frequency dryers find its application in drying of tea leaves [119] and RF combined hot air drying of in-shell Macadamia nuts [138].

5.4.2.3 Refractance Window Dryer

It is a novel drying system used to dry liquid food products to powder, flakes, and sheets with value addition. The drying method was developed by MCD Technologies, Inc. (Tacoma, Washington). The product does not come in direct contact with the drying medium, avoiding cross contamination. The schematic sketch of a typical refractance window dryer is shown in Figure 5.10. The wet food material is spread as a thin film on a plastic conveyor belt and the thermal energy is transferred from the hot water to the food materials.

In this dryer, heat transfer to the product occurs by conduction and radiation. Initially a large deviation of the refractive index was observed between the water and air when no product is placed on the belt. As a result the radiation from water to air through the plastic is reflected back to water without any loss of energy. Once the product is spread on the belt, the refractive index of water and the product becomes closer since food products contains a huge water fraction. Accordingly, thermal radiation is transmitted to the product with very less reflection from the plastic-water interface. The radiation transmits through the plastic material, which acts as a window to the wet product with mere loss of energy, hence the process is known as refractive window drying. As the drying proceeds,

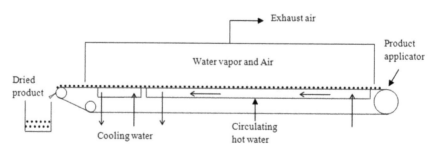

FIGURE 5.10 Schematic sketch of refractance window dryer.

moisture content in the product decreases increasing the refractive index and the transmission of thermal radiation decreases closing the radiation window. After this, conduction from the hot water to product is the major heat transfer process in drying. The drying process is rapid as the belt is in contact with the hot water and finally the dry product is collected using a scrapping knife.

Several investigators have suggested that the final product quality in terms of color retention, vitamins and antioxidants are superior as compared to other drying methods. RW dried mango powder was found to be smooth and flaky with uniform thickness compared to freeze dried powder [21]. Tomato powder with suitable characteristics such as better solubility, short dispersion time and color retention can be produced by this dryer [22]. It is used to dry various commercial products such as: herbal extracts and nutritional supplements, dehydrated fruits and vegetables powder, scrambled egg mixes and avocado powders [90].

5.4.2.4   Superheated Steam Drying (SSD)

In superheated steam (SS) drying, superheated steam at atmospheric pressure is an alternative drying medium for drying of food materials which can withstand temperatures above 100°C. When saturated steam is circulated over the food materials, it results in boiling and steam evolution from center of the food. Once moisture is removed from the food product, then the superheated steam returns to saturated conditions. This drying leads to a longer constant rate period and lower critical moisture content compared to conventional drying methods. Any direct or indirect dryer can be operated as an SSD, with slight modifications. Application of superheated steam has been studied on pilot scale and commercial scale of spray dryer, flash dryer, fluidized bed dryer, impinging jet and impinging stream dryer, conveyor dryer, agitated bed and packed bed dryers at various pressure conditions.

A typical superheated steam-drying unit is shown in Figure 5.11. It mainly consists of a steam generator, steam regulator, pipelines, drying chamber and auxiliary heater. Saturated steam at high pressure and temperature is generated in the steam generator and passes through the regulator, where the pressure drops.

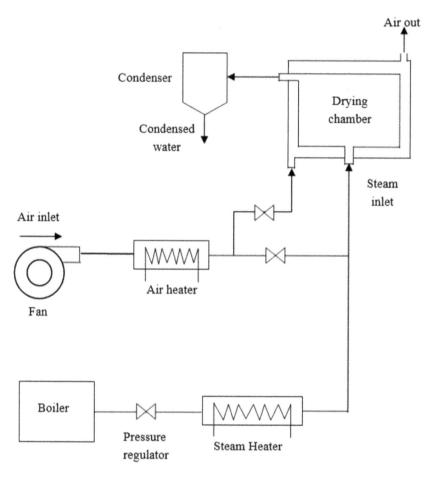

FIGURE 5.11   Schematic sketch of superheated steam dryer (SSD).

Due to pressure drop, saturated steam becomes superheated steam and flows to the drying chamber by pipelines. The temperature of steam is maintained by electrical heaters and heating tapes in the pipe, minimizing heat loss and condensation of steam. Since the exhaust from dryer is also steam, it can be further utilized in other food processing operations like blanching, pasteurization, sterilization etc. Various studies have been done on application of SSD to food products. Deventer et al. [29] carried out drying experiments to study the effect of superheated steam on vegetables (carrot, potato, cauliflower, celeriac, asparagus and leek), herbs (oregano), cacao nuts, wheat, flour and spices (paprika powder, onion powder). Superheated steam drying of sugar-beet pulp, potatoes, Asian noodles and spent grains have been investigated by Pronyk et al. [102]. Cenkowski et al. [23] have observed that the use of superheated steam at 150°C was efficient in drying distillers' spent grain. They reported that superheated steam is a better option for obtaining an end product without any adverse effect on its protein and phenolic content. This novel drying method is also suitable for heat sensitive products such as basil [14] and pepper seeds [70].

5.4.2.5   Pneumatic or Flash Drying

Pneumatic or flash dryers are convective type dryers widely used in the food industry to dry particulate solids. The particles are transported in a hot gas stream, which provides the heat of vaporization as well as carries away the evaporated moisture. The contact surface between the product and drying medium causes high heat and mass transfer resulting in increased drying rate and reduced drying time. Mostly heat sensitive products are processed with superheated steam as the heating medium in flash drying systems.

A typical flash drying system (Figure 5.12) consists of a heater, feeding unit, drying duct, air-product separator, exhaust fan and product collecting unit. Hot air/superheated steam is mixed with wet product in the feeding unit using some mixing device. The stream then flows upwards in the drying duct at velocity higher than the terminal velocity of the product. The thermal contact time between the food particles and conveying medium is very short (0.5–10 sec) [19]. More of surface moisture is removed as compared to internal moisture in this process. After drying, the solid is separated from the air stream using cyclone separator, fabric filter, electrostatic precipitator etc. The product can be recirculated if required to remove further moisture.

Food products mostly dried in a flash dryer include salt, bread crumb, corn starch, corn gluten, casein, gravy powder, soup powder, vegetable protein, spent tea, wheat starch, soybean protein, meat residue, and flour, etc. [19].

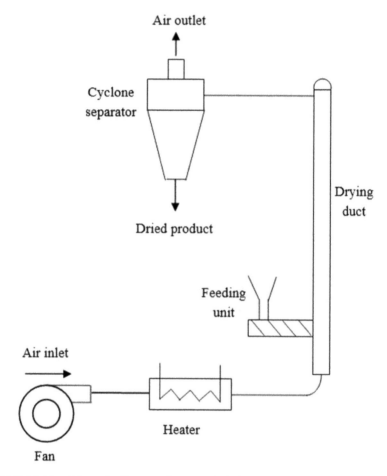

FIGURE 5.12   Schematic diagram of pneumatic drying system.

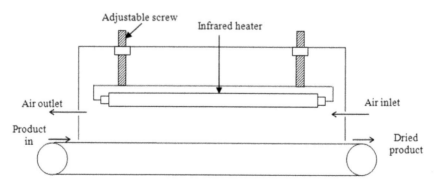

FIGURE 5.13   Schematic diagram of infrared dryer.

5.4.3 INFRARED DRYING

Infrared radiation is a part of electromagnetic spectrum within the wavelength range of 0.75–1000 µm. It can be further classified into near infrared (0.75–1.4 µm), mid-infrared (1.4–3 µm) and far infrared (3–1000 µm) zones [96]. An infrared dryer consists mainly of an emitter or heat source and reflector. The emitter may be gas-heated type or electrically heated type.

When the radiator is heated, it emits infrared radiation, which is absorbed, transmitted or reflected by the product. The absorbed radiation penetrates the product, causes vibration and rotation of the molecules and generates heat, eventually drying the product (Figure 5.13). A better process control, uniform heating of the products leading to higher heat transfer rates can be achieved in this dryer. This technique is used to dry low moisture foods like: flour, grains, pasta, vegetables [48]; cocoa, tea, bread crumb [122]; fruits [58]. It can be used in combination with other drying techniques such as hot air, vacuum, microwave, vibration to increase the drying rate. Drying of green peas in infrared assisted vibratory dryer resulted in reduced drying time and specific energy consumption by maintaining the original product quality [15]. The far-infrared drying of red ginseng offered faster drying rate, higher saponin content, lower color change, and less energy consumption than hot-air drying [89].

5.4.4 INDUSTRIAL DRYERS

With the progress of time, the drying technologies have become more diverse and complex. Further, there is always an industrial need to meet higher production rates, higher energy costs, quality criteria and the stringent environmental regulations. It is therefore very necessary for dryer designers to take care of the detail specifications along with the type and nature of the food to be handled by the dryer. A structural approach for suitable dryer selection is based on the following parameters:

  a. Specifications of the key process parameters.
  b. Throughput of the dryer and mode of production (batch/continuous).
  c. Physico-chemical properties of the wet feed as well as desired product specifications.

d. Drying kinetics and sorption isotherms of moist solids.
e. Quality parameters of dried products.
f. Economic evaluation of the drying system.
g. Pilot scale trials and safety aspects.

The dryers are classified into numerous types depending on the mode of operation, type of heat input, drying medium, drying temperature, residence time in the dryer. Each type of dryer has its specific characteristics that make it suitable or unsuitable for a specific application. Some types of dryers are very expensive while others are efficient in their mode of operation. So, it is very necessary to be aware of the wide variety of industrial dryers available in the market as well as their advantages and disadvantages. In the current section some of the industrial dryers commonly used for drying of fruits, vegetables, grains and seeds are discussed.

5.4.4.1 Tray Drying

Tray dryers are batch type dryer also known as shelf, cabinet and compartment type dryer, used to dry solid food products and pastes. A typical tray dryer (Figure 5.14) consists of a stack of trays placed in a large insulated

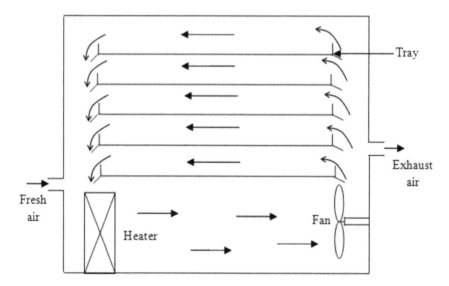

FIGURE 5.14   Schematic diagram of cabinet dryer.

drying chamber in which heated air is circulated with a fan located inside or outside of the drying cabinet. The mode of heat transfer to the product surface is by conduction from heated trays or by radiation from heated surfaces. For drying of granular materials, tray can be replaced with screens and hot air passes through the bed, resulting in higher heat transfer and reduced drying time. Tray dryer is mostly used to dry vegetables and semi perishable products, dye and pharmaceutical products. Since it is highly labor intensive for loading and unloading the product, the operational cost of the dryer is higher as compared to the dryer cost. The drying time is also very long (10–60 hours). A number of studies have been carried out by several researchers on tray drying of agricultural products such as: tomato [113], papaya [72] and cassava starch [6].

5.4.4.2 Conveyor Dryers

Conveyor dryers are used to dry granular materials. Particulate solids of 25–150 mm thick layer are slowly conveyed through the moving screen conveyor into the drying chamber. The drying chamber is usually long and divided to a number of sections as shown in Figure 5.15. Each section has a separate fan and heating coil. At the inlet, hot air passes from bottom of the screen to top, whereas near the discharge end, hot air passes from top of the product to bottom.

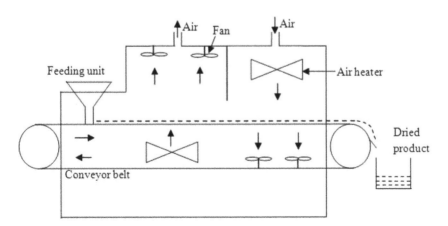

FIGURE 5.15   Schematic diagram of a conveyor dryer.

Various types of products can be dried by this dryer such as: breakfast cereal, baked snacks, nuts, animal feed, charcoal, rubbers [101]. Lutfy et al. [79] have developed a nonlinear autoregressive with exogenous input (NARX) network model to handle the complexity and nonlinearity of the conveyor belt dryer during drying of paddy grains.

5.4.4.3 Rotary Dryer

Rotary dryer consists of a slowly rotating cylindrical shell that is typically inclined to the horizontal at 1–5° towards the outlet. It is used to dry granular free flowing solids. The wet solid enters the shell at one end and the flights present throughout the inside wall of the cylinder helps in lifting the solids and the dried products are withdrawn at the lower end (Figure 5.16). The drying medium flows axially through the drum either concurrently or counter currently with the food materials.

Various researchers have evaluated the performance of rotary dryer for drying agricultural products such as: grains, beans, nuts, vegetables, herbs, animal feeds, agricultural wastes, and by-products [27]. They have reviewed on the effects of product, drying air, design and operational parameters of rotary dryer on performance and dried product quality (rice, maize, coffee beans, groundnut, chili, red beet, carrot, peppermint, alfalfa).

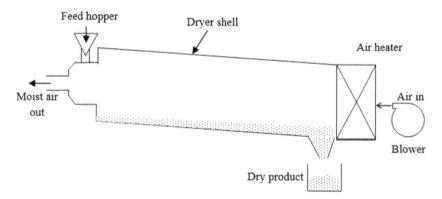

FIGURE 5.16   Schematic sketch of a counter current rotary dryer.

5.4.4.4 Freeze Dryer

Heat sensitive solid food products having high flavor content, chemicals, and pharmaceutical products are effectively dried by freeze drying. The drying occurs below the triple point of liquid water by sublimation of the frozen moisture into vapor. The vapor is then removed from the drying chamber by vacuum pumps. Freeze drying results in the highest product quality in terms of original flavor retention as compared to other drying process. The frozen layer offers structural rigidity at the sublimation front preventing collapse of the solid after drying. Hence, the dried material becomes porous without any shrinkage retaining the original shape and rehydrates completely. The process is used to dry products such as: coffee, soups, seafood, fruits and vegetables, mushroom, and dairy products.

The tray freeze dryer is the most common type in industrial scale where heat for sublimation is supplied by conduction through the bottom of the tray. The vacuum is created and maintained by refrigerated condenser, which is further supported by vacuum pump. The vacuum pressure is maintained below 25 Pa and the condenser operates at 40°C. Similarly, the tunnel freeze dryer, which is a semi-continuous type dryer, is used industrially for drying of agricultural foodstuff (Figure 5.17). It consists of a large vacuum cabinet into which tray carrying trolleys are loaded

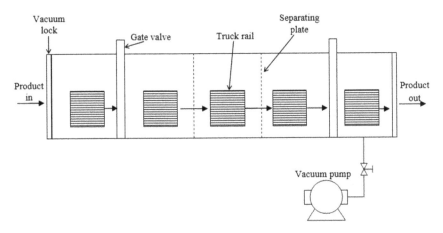

FIGURE 5.17   Schematic diagram of a tunnel freeze dryer.

at intervals through vacuum lock provided at the entry end of the tunnel whereas the dried products are unloaded through another vacuum lock at the exit end of the drying chamber. Heating plates are provided in the tunnel to facilitate necessary heat transfer to the product. As drying proceeds, heat requirement reduces and accordingly temperature and vacuum level is maintained in each section in the tunnel by separate heating and vacuum units. Tunnel freeze dryers are beneficial in terms of production volume.

Several authors have modified the existing design of freeze dryer in order to get faster drying rate along with better retention of aroma and other quality characteristics for some agricultural food products such as: apple [87], mushrooms [136], and coffee [55].

5.4.4.5   Fluidized Bed Dryer

Fluidized bed dryers are used for drying wet particulates and granular materials that can be fluidized easily in the food, chemical, pharmaceutical and agricultural industries. A typical fluidized bed dryer (Figure 5.18) consists of a fluidizing bed column, distributor plate, air heater and blower. Wet particle rests upon the distributor plate. Hot air from the air heater is passed to the bed of particle through the distributor plate. At low air velocity, the bed remains static. As the velocity of air increases, the pressure drop across the bed decreases and at a certain velocity the bed is supported by the air stream. This is called the minimum fluidization velocity. Beyond minimum fluidization velocity, the pressure drop across the bed remains almost constant with any further increase in velocity. Drying usually occurs at superficial velocity, 2–4 times higher than the minimum fluidization velocity.

The minimum fluidization velocity is higher for high moisture food, as they offer high cohesive forces. After fluidization, the air stream is subjected to a cyclone separator where fine particles are separated from the exhaust air. Fluidized bed drying offers various advantages: it has high thermal efficiency, good mixing of air and product, higher moisture evaporation rate, uniform drying of products, low maintenance and easy control. Various agricultural commodities are successfully dried in fluidized bed dryer such as: potato [10], capsicum [67], apple [63], soybean [105], rice and wheat seeds [60], corns and unshelled pistachios [92].

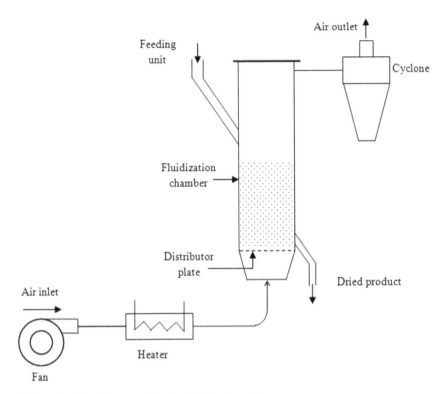

FIGURE 5.18   Schematic sketch of fluidized bed dryer.

5.4.4.6   Spouted Bed Dryer

Spouted bed dryers (Figure 5.19) are used for drying larger particles having particle size greater than 5 mm, which exhibits a slugging characteristic in fluidized beds. Although the working principle of spouted bed
is similar to that of fluidized bed, only a part of the bed containing the
particulate solid is subjected to jet of air stream at a time. The agitated
particles fall out of the spout on the annular region of product. The spout-
ing is regular and cyclic resulting in uniform drying. The process is more
energy efficient than fluidized bed drying. It is used to dry heat sensitive
granular products, foods, pharmaceuticals and plastics. Many researchers
have studied the performance of spouted bed drying for grains like corn,
wheat [95]; peas [112], sweet potato [76], carrot [137], guava [4].

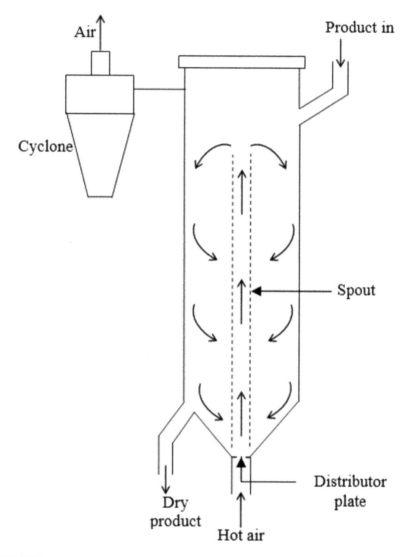

FIGURE 5.19 Schematic diagram of a spouted bed dryer.

5.4.4.7 Drum Dryer

The use of drum dryers in the food industry have numerous applications in drying of fruit and vegetable pulp, mashed potatoes, milk product, baby

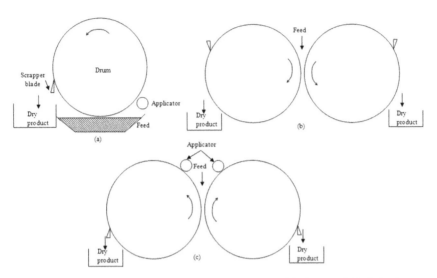

FIGURE 5.20   Drum dryers with (a) Single drum (b) Double drum and (c) Twin drum.

foods and breakfast cereal. This is an energy efficient drying method and is mainly effective for drying high viscous liquid foods or pureed food products. Various configurations of the drum dryer are available based on the number of drums, type of feeding mechanism and surrounding air pressure. The schematic sketch of single and double drum dryer is shown in Figure 5.20. A typical drum dryer is made up of one or two horizontally mounted slowly rotating metal roll(s) or cylinder(s) heated internally by steam, hot water or other high temperature heat transfer liquid. The steam at a temperature up to 200°C is used to heat the inner surface of the drum. The cylinders are made of high-grade cast iron or stainless steel, a supporting frame, a product feeding system, a scraper (commonly known as doctor blade), and some auxiliary units. The diameter of typical drums ranges from 0.5 m to 6 m and the length from 1 m to 6 m. The liquid feed or puree material is applied as thin layer onto the outer surface of revolving drums. The residence time of the product on the drum ranges from a few seconds to several seconds in order to reach the safe moisture content, afterwards the dried products are ground into flakes or powder.

Drum drying technique has been studied for various products, e.g., jackfruit powder [103], apple puree [68], orange and grapefruit [99], potato flakes [61], pre-gelatinized maize starch [64], rice starch and flour [125].

5.4.4.8 Spray Dryer

Spray drying operates on the principle of removal of moisture from the food products by rapid evaporation of liquid droplets under exposure to high temperature. This method is very suitable for heat sensitive materials such as foods and pharmaceuticals. A typical spray dryer consists of a cylindrical drying chamber with a short conical bottom and some auxiliary units like atomizing unit, air heater, and fan to draw atmospheric air, pump, and cyclone separator (Figure 5.21).

The atomizer or spray nozzle is used to disperse the liquid or slurry into a controlled drop size spray inside the drying chamber and the efficient contact between the spray and drying medium results in moisture evaporation from the food products. The heated drying air can be circulated as a

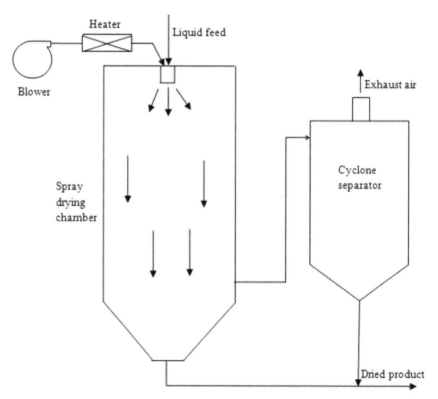

FIGURE 5.21   Schematic diagram of spray drying process.

co-current or counter-current flow to the atomizer direction inside the drying chamber. These operating conditions are included in the design of spray drying process to increase the product recovery and to get proper quality specifications of the end product. The major characteristics of spray-dried powder have lower moisture content, lower hygroscopicity, lower degree of caking, and higher rehydration abilities.

Spray drying can handle a wide variety of pumpable food materials e.g. heat sensitive, non- heat sensitive biological products. It is widely used in food industries to produce fruits and vegetable powders like: water melon [104], pomegranate [133], tomato [49], milk powder, whey [52], and instant tea [98].

5.5 CONCLUSIONS

In this chapter, an attempt has been made to provide a concise overview of different drying technologies adopted for drying of food products. Solar drying is a promising technique to overcome the energy crisis generated due to depletion of fossil fuels without polluting the environment. Novel dryers are assumed to preserve the functional, nutritional and sensory attributes of the product along with maximum utilization of energy. Industrial dryers like tray, tunnel, freeze, drum, rotary and spray dryers are extensively used for large scale drying of food commodities. There is a better control of drying temperature, pressure and production rate in these dryers, making them industry-friendly. Over increasing stringent consumer demands on quality of dried products, reduction of specific energy consumption and negative environmental impacts of fossil fuel combustion have provided impetus for renewable and novel drying technologies. It is therefore recommended that the newly emerging drying technologies should be evaluated closely before selecting the proper dryer for a suitable application. Finally, the user is encouraged to make a preliminary selection of one or more possible systems and appropriate selection of operating conditions in order to achieve better dryer performance. Since drying is an ever demanding and an important operation in the food industry, it deserves a multi-disciplinary and further exhaustive study with close industry-academia research collaboration.

5.6   SUMMARY

The increasing market demand for dehydrated food products and the consumer's concern to receive the highest quality specifications emphasized the need for a better understanding of the drying process. Again, the ever-increasing population along with rising cost of fossil fuel energy has promoted the possibility of using sustainable drying technologies. In this chapter, an overview of different types of dryers used for drying of grains, seeds, fruits and vegetables are reported. The key criterion for selection of suitable dryer for specific type of food product along with their classification is presented. The fundamentals of drying processes and the effect of different process parameters on the drying kinetics of food products have also been discussed. The detailed overview of different drying technologies would substantially help the policy makers, research institutions, standardizing agencies and food industries in establishing stable and sturdy drying system, which eventually assist the end users in appropriate selection.

KEYWORDS

- Contact sorption
- Conveyor dryer
- Dielectric dryer
- Drum dryer
- Drying kinetics
- Fluidized bed dryer
- Freeze dryer
- Industrial dryer
- Infrared dryer
- Microwave dryer
- Novel dryer
- Pneumatic dryer

- **PV panel**
- **Refractance window**
- **Rotary dryer**
- **Solar dryer**
- **Solar collector**
- **Spouted bed dryer**
- **Spray dryer**
- **Superheated steam dryer**
- **Tray dryer**
- **Ultrasound dryer**

REFERENCES

1. Afzal, T. M., & Abe, T. (2000). Simulation of moisture changes in barley during far infrared radiation drying. *Computers and Electronics in Agriculture*, 26(2), 137–145.
2. Akpinar, E. K. (2010). Drying of mint leaves in a solar dryer and under open sun: Modeling, performance analyses. *Energy Conversion and Management*, 51(12), 2407–2418.
3. Akpinar, E. K., Bicer, Y., & Yildiz, C. (2003). Thin layer drying of red pepper. *Journal of Food Engineering*, 59(1), 99–104.
4. Alsina, O. L. S. de, Almeida, M. M. de, Silva, J. M. da and Monteiro, L. F. (2014). Drying of Fruits Pieces in Fixed and Spouted Bed. *Transport Phenomena and Drying of Solids and Particulate Materials*, Advanced Structured Materials, 48, 141–159. Springer International Publishing.
5. Amaratunga, K. S. P., Pan, Z., Zheng, X., & Thompson, J. F. (2005). Comparison of drying characteristics and quality of rough rice dried with infrared and heated air. *In Proceedings of ASAE Annual International Meeting*, Tampa, Florida.
6. Aviara, N. A., Onuoha, L. N., Falola, O. E., & Igbeka, J. C. (2014). Energy and energy analyses of native cassava starch drying in a tray dryer. *Energy*, 73, 809–817.
7. Ayyappan, S., Mayilsamy, K., & Sreenarayanan, V. V. (2015). Performance improvement studies in a solar greenhouse drier using sensible heat storage materials. *Heat and Mass Transfer*, 1–9. doi: 10.1007/s00231-015-1568-5.
8. Azimi, A., Tavakoli, T., Beheshti, H. K., & Rahimi, A. (2012). Experimental study on eggplant drying by an indirect solar dryer and open sun drying. *Iranica Journal of Energy and Environment*, 3(4), 347–353.
9. Babagana, G., Silas, K., & Mustafa, B. G. (2012). Design and construction of forced/natural convection solar vegetable dryer with heat storage. *ARPN Journal of Engineering and Applied Sciences*, 7(10), 1213–1217.

10. Bakal, S. B., Sharma, G. P., Sonawane, S. P., & Verma, R. C. (2012). Kinetics of potato drying using fluidized bed dryer. *Journal of Food Science and Technology*, 49(5), 608–613.

11. Bal, L. M., Satya, S., & Naik, S. N. (2010). Solar dryer with thermal energy storage systems for drying agricultural food products: a review. *Renewable and Sustainable Energy Reviews*, 14(8), 2298–2314.

12. Bala, B. K., & Woods, J. L. (1994). Simulation of the indirect natural convection solar drying of rough rice. *Solar Energy*, 53(3), 259–266.

13. Banout, J., Ehl, P., Havlik, J., Lojka, B., Polesny, Z., & Verner, V. (2011). Design and performance evaluation of a double-pass solar drier for drying of red chilli (*Capsicum annum* L.). *Solar Energy*, 85(3), 506–515.

14. Barbieri, S., Elustondo, M., & Urbicain, M. (2004). Retention of aroma compounds in basil dried with low pressure superheated steam. *Journal of Food Engineering*, 65(1), 109–115.

15. Barzegar, M., Zare, D., & Stroshine, R. L. (2015). An integrated energy and quality approach to optimization of green peas drying in a hot air infrared-assisted vibratory bed dryer. *Journal of Food Engineering*, 166, 302–315.

16. Basunia, M. A., & Abe, T. (2001). Thin-layer solar drying characteristics of rough rice under natural convection. *Journal of Food Engineering*, 47(4), 295–301.

17. Belessiotis, V., & Delyannis, E. (2011). Solar drying. *Solar Energy*, 85(8), 1665–1691.

18. Bena, B., & Fuller, R. J. (2002). Natural convection solar dryer with biomass back-up heater. *Solar Energy*, 72(1), 75–83.

19. Borde, I., & Levy, A. (2006). Pneumatic and flash drying. *Handbook of Industrial Drying*, 397–410.

20. Çakmak, G., & Yıldız, C. (2011). The drying kinetics of seeded grape in solar dryer with PCM-based solar integrated collector. *Food and Bioproducts Processing*, 89(2), 103–108.

21. Caparino, O. A., Sablani, S. S., Tang, J., Syamaladevi, R. M., & Nindo, C. I. (2013). Water sorption, glass transition and microstructures of refractance window–and freeze-dried mango (Philippine "carabao" var.) powder. *Drying Technology*, 31(16), 1969–1978.

22. Castoldi, M., Zotarelli, M. F., Durigon, A., Carciofi, B. A. M., & Laurindo, J. B. (2014). Production of tomato powder by refractance window drying. *Drying Technology*, 33(12), 1463–1473.

23. Cenkowski, S., Sosa-Morales, M. E., & Flores-Alvarez, M. D. C. (2012). Protein content and antioxidant activity of distillers' spent grain dried at 150°C with superheated steam and hot air. *Drying Technology*, 30(11–12), 1292–1296.

24. Chegini, G. R., Khazaei, J., Ghobadian, B., & Goudarzi, A. M. (2008). Prediction of process and product parameters in an orange juice spray dryer using artificial neural networks. *Journal of Food Engineering*, 84(4), 534–543.

25. Chowdhury, M. M. I., Bala, B. K., & Haque, M. A. (2011). Energy and energy analysis of the solar drying of jackfruit leather. *Biosystems Engineering*, 110(2), 222–229.

26. da Silva, W. P., do Amaral, D. S., Duarte, M. E. M., Mata, M. E., DeSilva, C. M., Pinheiro, R. M., & Pessoa, T. (2013). Description of the osmotic dehydration and convective drying of coconut (*Cocos nucifera* L.) Pieces: A three-dimensional approach. *Journal of Food Engineering*, 115(1), 121–131.

27. Delele, M. A., Weigler, F., & Mellmann, J. (2015). Advances in the application of a rotary dryer for drying of agricultural products: A review. *Drying Technology*, 33(5), 541–558.
28. Dev, S. R., & Raghavan, V. G. (2012). Advancements in drying techniques for food, fiber and fuel. *Drying Technology*, 30(11–12), 1147–1159.
29. Deventer, Van, H. C., & Heijmans, R. M. (2001). Drying with superheated steam. *Drying Technology*, 19(8), 2033–2045.
30. Dina, S. F., Ambarita, H., Napitupulu, F. H., & Kawai, H. (2015). Study on effectiveness of continuous solar dryer integrated with desiccant thermal storage for drying cocoa beans. *Case Studies in Thermal Engineering*, 5, 32–40.
31. Doymaz, I. (2004). Convective air drying characteristics of thin layer carrots. *Journal of Food Engineering*, 61(3), 359–364.
32. Doymaz, I. (2012). Infrared drying of sweet potato (*Ipomoea batatas* L.) slices. *Journal of Food Science and Technology*, 49(6), 760–766.
33. Duffie, J. A., & Beckman, W. A. (1980). *Solar Engineering of Thermal Processes, Volume 3*. New York: John Wiley and Sons.
34. Dujmić, F., Brnčić, M., Karlović, S., Bosiljkov, T., Ježek, D., Tripalo, B., & Mofardin, I. (2013). Ultrasoundassisted infrared drying of pear slices: Textural issues. *Journal of Food Process Engineering*, 36(3), 397–406.
35. El-Khadraoui, A., Kooli, S., Hamdi, I., & Farhat, A. (2015). Experimental investigation and economic evaluation of a new mixed-mode solar greenhouse dryer for drying of red pepper and grape. *Renewable Energy*, 77, 1–8.
36. El-Sebaii, A. A., & Shalaby, S. M. (2013). Experimental investigation of an indirect-mode forced convection solar dryer for drying thymus and mint. *Energy Conversion and Management*, 74, 109–116.
37. El-Sebaii, A. A., &Shalaby, S. M. (2012). Solar drying of agricultural products: A review. *Renewable and Sustainable Energy Reviews*, 16(1), 37–43.
38. Ertekin, C., & Yaldiz, O. (2004). Drying of eggplant and selection of a suitable thin layer drying model. *Journal of Food Engineering*, 63(3), 349–359.
39. FAO, (2011). Global Food Losses and Food Waste, Food and Agriculture Organization of the United Nations.
40. FAO, (2013). Food and Agriculture Organization of the United Nations, Rome.
41. FAO, (2014). The State of Food Insecurity in the World, Food and Agriculture Organization of the United Nations.
42. Fellows, P. J. (2009). Food processing Technology: Principles and Practice. *Elsevier*.
43. Fernandes, F. A., & Rodrigues, S. (2008). Application of ultrasound and ultrasound-assisted osmotic dehydration in drying of fruits. *Drying Technology*, 26(12), 1509–1516.
44. Fikiin, A. G., Fikiin, K. A., & Triphonov, S. D. (1999). Equivalent thermophysical properties and surface heat transfer coefficient of fruit layers in trays during cooling. *Journal of Food Engineering*, 40(1), 7–13.
45. Fudholi, A., Sopian, K., Ruslan, M. H., Alghoul, M. A., & Sulaiman, M. Y. (2010). Review of solar dryers for agricultural and marine products. *Renewable and Sustainable Energy Reviews*, 14(1), 1–30.
46. García-Pérez, J. V., Rosselló, C., Cárcel, J. A., De la Fuente, S., & Mulet, A. (2007). Effect of air temperature on convective drying assisted by high power ultrasound. *Defect and Diffusion Forum*, 258, 563–574.

47. Gatea, A. A. (2011). Performance evaluation of a mixed-mode solar dryer for evaporating moisture in beans. *Journal of Agricultural Biotechnology and Sustainable Development*, 3(4), 65–71.

48. Ginzburg, A. S., 1969. Application of Infrared Radiation in Food Processing. *Chemical and Process Engineering Series, Leonard Hill, London.*

49. Goula, A. M., Karapantsios, T. D., Achilias, D. S., & Adamopoulos, K. G. (2008). Water sorption isotherms and glass transition temperature of spray dried tomato pulp. *Journal of Food Engineering*, 85, 73–83.

50. Goyal, R. K., Kingsly, A. R. P., Manikantan, M. R., & Ilyas, S. M. (2007). Mathematical modeling of thin layer drying kinetics of plum in a tunnel dryer. Journal *of Food Engineering*, 79(1), 176–180.

51. Gudiño-Ayala, D., & Calderón-Topete, Á. (2014). Pineapple drying using a new solar hybrid dryer. *Energy Procedia*, 57, 1642–1650.

52. Hall, C. W., & Hedrick, T. I. (1971). *Drying of milk and milk products (2nd edition).* Westport, Connecticut, USA, AVI Publishing Co. Inc.

53. Hebbar, H. U., & Rastogi, N. K. (2001). Mass transfer during infrared drying of cashew kernel. *Journal of Food Engineering*, 47(1), 1–5.

54. Intawee, P., & Janjai, S. (2011). Performance evaluation of a large-scale polyethylene covered greenhouse solar dryer. *International Energy Journal*, 12(1), 39–52.

55. Ishwarya, S. P., & Anandharamakrishnan, C. (2015). Spray-Freeze-Drying approach for soluble coffee processing and its effect on quality characteristics. *Journal of Food Engineering*, 149, 171–180.

56. Jadhav, D. B., Visavale, G. L., Sutar, N., Annapure, U. S., & Thorat, B. N. (2010). Studies on solar cabinet drying of green peas (*Pisum sativum*). *Drying Technology*, 28(5), 600–607.

57. Janjai, S. (2012). A greenhouse type solar dryer for small-scale dried food industries: development and dissemination. *International Journal of Energy and Environment*, 3, 383–98.

58. Jaturonglumlert, S., & Kiatsiriroat, T. (2010). Heat and mass transfer in combined convective and far-infrared drying of fruit leather. *Journal of Food Engineering*, 100(2), 254–260.

59. Jittanit, W., Chantara-In, M., Deying, T., & Ratanavong, W. (2011). Production of tamarind powder by drum dryer using maltodextrin and Arabic gum as adjuncts. *Songklanakarin Journal of Science and Technology*, 33(1), 33–41.

60. Jittanit, W., Srzednicki, G., & Driscoll, R. H. (2013). Comparison between fluidized bed and spouted bed drying for seeds. *Drying Technology*, 31(1), 52–56.

61. Kakade, R. H., Das, H., & Ali, S. (2011). Performance evaluation of a double drum dryer for potato flake production. *Journal of Food Science and Technology*, 48(4), 432–439.

62. Kaleemullah, S., & Kailappan, R. (2005). Drying kinetics of red chillies in a rotary dryer. *Biosystems Engineering*, 92(1), 15–23.

63. Kaleta, A., Górnicki, K., Winiczenko, R., & Chojnacka, A. (2013). Evaluation of drying models of apple (var. Ligol) dried in a fluidized bed dryer. *Energy Conversion and Management*, 67, 179–185.

64. Kalogianni, E. P., Xynogalos, V. A., Karapantsios, T. D., & Kostoglou, M. (2002). Effect of feed concentration on the production of pregelatinized starch in a double drum dryer. *LWT-Food Science and Technology*, 35(8), 703–714.

65. Kassem, A. M., Habib, Y. A., Harb, S. K., & Kallil, K. S. (2011). Effect of architectural form of greenhouse solar dryer system on drying of onion flakes. *Egyptian Journal of Agricultural Research*, 89(2), 627–638.

66. Khalifa, A. J. N., Al-Dabagh, A. M., & Al-Mehemdi, W. M. (2012). An experimental study of vegetable solar drying systems with and without auxiliary heat. *ISRN Renewable Energy*, 2012.

67. Khan, M. A., Patel, K. K., Kumar, Y., & Gupta, P. (2014). Color kinetics of capsicum during microwave assisted fluidized bed drying. *Journal of Food and Nutritional Disorders*, 4(1). *doi: 10.4172/2324-9323.1000164*.

68. Kitson, J. A., & MacGregor, D. R. (1982). Technical note: Drying fruit purees on an improved pilot plant drum dryer, *Journal of Food Technology*, 17(2), 285–288.

69. Kowalski, S. J., Szadzińska, J., & Pawłowski, A. (2015). Ultrasonic-assisted osmotic dehydration of carrot followed by convective drying with continuous and intermittent heating. *Drying Technology*, 33(13), 1570–1580.

70. Kozanoglu, B., Flores, A., Guerrero-Beltrán, J. A., & Welti-Chanes, J. (2012). Drying of pepper seed particles in a superheated steam fluidized bed operating at reduced pressure. *Drying Technology*, 30(8), 884–890.

71. Krust, P. W., & Mcquistan, R. B. (1962). *Elements of Infrared Technology*. John Wiley and Sons: New York.

72. Kurozawa, L. E., Hubinger, M. D., & Park, K. J. (2012). Glass transition phenomenon on shrinkage of papaya during convective drying. *Journal of Food Engineering*, 108(1), 43–50.

73. Lamnatou, C., Papanicolaou, E., Belessiotis, V., & Kyriakis, N. (2012). Experimental investigation and thermodynamic performance analysis of a solar dryer using an evacuated-tube air collector. *Applied Energy*, 94, 232–243.

74. Leon, M. A., & Kumar, S. (2008). Design and performance evaluation of a solar-assisted biomass drying system with thermal storage. *Drying Technology*, 26(7), 936–947.

75. Li, H., Dai, Y., Dai, J., Wang, X., & Wei, L. (2010). A solar assisted heat pump drying system for grain in-store drying. *Frontiers of Energy and Power Engineering in China*, 4(3), 386–391.

76. Liu, P., Zhang, M., & Mujumdar, A. S. (2014). Purple-fleshed sweet potato cubes drying in a microwave-assisted spouted bed dryer. *Drying Technology*, 32(15), 1865–1871.

77. López-Vidaña, E. C., Méndez-Lagunas, L. L., & Rodríguez-Ramírez, J. (2013). Efficiency of a hybrid solar–gas dryer. *Solar Energy*, 93, 23–31.

78. Lotfalian, A., Ghazavi, M. A., & Hosseinzadeh, B. (2010). Comparing the performance of two type collectors on drying process of lemon and orange fruits through a passive and indirect solar dryer. *Journal of American Science*, 6(10), 248–251.

79. Lutfy, O. F., Selamat, H., & Mohd Noor, S. B. (2015). Intelligent modeling and control of a conveyor-belt grain dryer using a simplified type-2 neuro-fuzzy controller. *Drying Technology*, 33(10), 1210–1222.

80. Markowski, M., Sobieski, W., Konopka, I., Tańska, M., & Białobrzewski, I. (2007). Drying characteristics of barley grain dried in a spouted-bed and combined IR-convection dryers. *Drying Technology*, 25(10), 1621–1632.

81. Midilli, A., & Kucuk, H. (2003). Mathematical modeling of thin layer drying of pistachio by using solar energy. *Energy Conversion and Management*, 44(7), 1111–1122.

82. Mohanraj, M. (2014). Performance of a solar-ambient hybrid source heat pump drier for copra drying under hot-humid weather conditions. *Energy for Sustainable Development*, 23, 165–169.

83. Mortezapour, H., Ghobadian, B., Minaei, S., & Khoshtaghaza, M. H. (2012). Saffron drying with a heat pump–assisted hybrid photovoltaic–thermal solar dryer. *Drying Technology*, 30(6), 560–566.

84. Munir, A., Sultan, U., & Iqbal, M. (2013). Development and performance evaluation of a locally fabricated portable solar tunnel dryer for drying of fruits, vegetables and medicinal plants. *Pakistan Journal of Agricultural Sciences*, 50(3), 493–498.

85. Mustayen, A. G. M. B., Rahman, M. M., Mekhilef, S., & Saidur, R. (2015). Performance evaluation of a solar powered air dryer for white oyster mushroom drying. *International Journal of Green Energy*, 12(11), 1113–1121.

86. Mwithiga, G., & Kigo, S. N. (2006). Performance of a solar dryer with limited sun tracking capability. *Journal of Food Engineering*, 74(2), 247–252.

87. Nakagawa, K., & Ochiai, T. (2015). A mathematical model of multi-dimensional freeze-drying for food products. *Journal of Food Engineering*, 161, 55–67.

88. Nimmol, C., Devahastin, S., Swasdisevi, T., & Soponronnarit, S. (2007). Drying of banana slices using combined low-pressure superheated steam and far-infrared radiation. *Journal of Food Engineering*, 81(3), 624–633.

89. Ning, X., Lee, J., & Han, C. (2015). Drying characteristics and quality of red ginseng using far-infrared rays. *Journal of Ginseng Research*, doi: 10.1016/j.jgr.2015.04.001.

90. Ochoa-Martínez, C. I., Quintero, P. T., Ayala, A. A., & Ortiz, M. J. (2012). Drying characteristics of mango slices using the Refractance Window™ technique. *Journal of Food Engineering*, 109(1), 69–75.

91. Oko, C. O. C., & Nnamchi, S. N. (2013). Coupled heat and mass transfer in a solar grain dryer. *Drying Technology*, 31(1), 82–90.

92. Özahi, E., & Demir, H. (2015). Drying performance analysis of a batch type fluidized bed drying process for corn and unshelled pistachio nut regarding to energetic and exergetic efficiencies. *Measurement*, 60, 85–96.

93. Ozkan, I. A., Akbudak, B., & Akbudak, N. (2007). Microwave drying characteristics of spinach. *Journal of Food Engineering*, 78(2), 577–583.

94. Ozuna, C., Cárcel, J. A., GarcíaPérez, J. V., & Mulet, A. (2011). Improvement of water transport mechanisms during potato drying by applying ultrasound. *Journal of the Science of Food and Agriculture*, 91(14), 2511–2517.

95. Pallai, E., Szentmarjay, T., & Mujumdar, A. S. (2006). Spouted bed drying. *Handbook of Industrial Drying*, CRC Press.

96. Pan, Zhongli., & Atungulu, Griffiths G. (2010). *Infrared Heating for Food and Agricultural Processing*. CRC Press, pp. 1–300.

97. Pan, Z., Olson, D. A., Amaratunga, K. S. P., Olsen, C. W., Zhu, Y., & McHugh, T. H. (2005). Feasibility of using infrared heating for blanching and dehydration of fruits and vegetables. *Paper Presentation at Annual International Meeting of American Society of Agricultural & Biological Engineers* at Tampa, Florida, USA, p. 12.

98. Pandey, R. K., & Manimehalai, N. (2014). Production of instant tea powder by spray drying. *International Journal of Agriculture and Food Science Technology*, 5(3), 197–202.

99. Passy, N., & Mannheim, C.H. (1983). The dehydration, shelf-life and potential uses of citrus pulps. *Journal of Food Engineering*, 2(1), 19–34.

100. Pirasteh, G., Saidur, R., Rahman, S. M. A., & Rahim, N. A. (2014). A review on development of solar drying applications. *Renewable and Sustainable Energy Reviews*, 31, 133–148.

101. Poirier, D. (2006). Conveyor dryers. Handbook of Industrial Drying. CRC Press.

102. Pronyk, C., Cenkowski, S., & Muir, W. E. (2004). Drying foodstuffs with superheated steam. *Drying Technology*, 22(5), 899–916.

103. Pua, C. K., Hamid, N. S. A., Tan, C. P., Mirhosseini, H., Rahman, R. B. A., & Rusul, G. (2010). Optimization of drum drying processing parameters for production of jackfruit (*Artocarpus heterophyllus*) powder using response surface methodology. *LWT-Food Science and Technology*, 43(2), 343–349.

104. Quek, S. Y., Chok, N. K., & Swedlund, P. (2007). The physicochemical properties of spray-dried watermelon powders. *Chemical Engineering and Processing: Process Intensification*. 46, 386–392.

105. Ranjbaran, M., & Zare, D. (2013). Simulation of energetic- and exergetic performance of microwave-assisted fluidized bed drying of soybeans. *Energy*, 59, 484–493.

106. Reyes, A., Mahn, A., & Vásquez, F. (2014). Mushrooms dehydration in a hybrid-solar dryer, using a phase change material. *Energy Conversion and Management*, 83, 241–248.

107. Reyes, A., Mahn, A., & Huenulaf, P. (2011). Drying of apple slices in atmospheric and vacuum freeze dryer. *Drying Technology*, 29(9), 1076–1089.

108. Rigit, A. R. H., Jakhrani, A. Q., Kamboh, S. A., & Kie, P. L. T. (2013). Development of an indirect solar dryer with biomass backup burner for drying pepper berries. *World Applied Sciences Journal*, 22(9), 1241–1251.

109. Ronoh, E. K., Kanali, C. L., Mailutha, J. T., & Shitanda, D. (2010). Thin layer drying kinetics of amaranth (Amaranthus cruentus) grains in a natural convection solar tent dryer. *African Journal of Food, Agriculture, Nutrition and Development*, 10(3).

110. Sabarez, H. T., & Noomhorm, A. (1993). Performance testing of an experimental screw conveyor dryer for roasting cashew nuts. *Postharvest Biology and Technology*, 2(3), 171–178.

111. Sabarez, H. T., Gallego-Juarez, J. A., & Riera, E. (2012). Ultrasonic-assisted convective drying of apple slices. *Drying Technology*, 30(9), 989–997.

112. Sahin, S., Sumnu, G., & Tunaboyu, F. (2013). Usage of solar-assisted spouted bed drier in drying of pea. *Food and Bioproducts Processing*, 91(3), 271–278.

113. Santos-Sánchez, N. F., Valadez-Blanco, R., Gómez-Gómez, M. S., Pérez-Herrera, A., & Salas-Coronado, R. (2012). Effect of rotating tray drying on antioxidant components, color and rehydration ratio of tomato saladette slices. *LWT-Food Science and Technology*, 46(1), 298–304.

114. Sarker, M. S. H., Ibrahim, M. N., Aziz, N. A., & Punan, M. S. (2013). Drying kinetics, energy consumption, and quality of paddy (MAR-219) during drying by the industrial inclined bed dryer with or without the fluidized bed dryer. *Drying Technology*, 31(3), 286–294.

115. Schiffmann, R. F. (1995). Microwave and dielectric drying. *Handbook of Industrial Drying*, 1, 345–372.

116. Schössler, K., Jäger, H., & Knorr, D. (2012). Novel contact ultrasound system for the accelerated freeze-drying of vegetables. *Innovative Food Science and Emerging Technologies*, 16, 113–120.

117. Sevik, S. (2014). Experimental investigation of a new design solar-heat pump dryer under the different climatic conditions and drying behavior of selected products. *Solar Energy*, 105, 190–205.

118. Shanmugam, S., Kumar, P., & Veerappan, A. (2014). Thermal performance of a solar dryer with oscillating bed for drying of non-parboiled paddy grains. *Energy Sources, Part A: Recovery, Utilization and Environmental Effects*, 36(17), 1877–1885.

119. Shinde, A., Das, S., & Datta, A. K. (2013). Quality improvement of orthodox and CTC tea and performance enhancement by hybrid hot air–radio frequency (RF) dryer. *Journal of Food Engineering*, 116(2), 444–449.

120. Shringi, V., Kothari, S., & Panwar, N. L. (2014). Experimental investigation of drying of garlic clove in solar dryer using phase change material as energy storage. *Journal of Thermal Analysis and Calorimetry*, 118(1), 533–539.

121. Silva, F. A., Marsaioli, A., Maximo, G. J., Silva, M. A. A. P., & Goncalves, L. A. G. (2006). Microwave assisted drying of macadamia nuts. *Journal of Food Engineering*, 77(3), 550–558.

122. Skj□ldebrand, C. (2001). Infrared heating. *Thermal Technologies in Food Processing*, 208–228.

123. Smitabhindu, R., Janjai, S., & Chankong, V. (2008). Optimization of a solar-assisted drying system for drying bananas. *Renewable Energy*, 33(7), 1523–1531.

124. Sumnu, G., Turabi, E., & Oztop, M. (2005). Drying of carrots in microwave and halogen lamp–microwave combination ovens. *LWT-Food Science and Technology*, 38(5), 549–553.

125. Supprung, P., & Noomhorm, A. (2003). Optimization of drum drying parameters for low amylose rice (KDML105) starch and flour. *Drying Technology*, 21(9), 1781–1795.

126. Taşkın, O., & Korucu, T. (2015). Feasibility Analysis of Solar Wall Application in Seed Maize Drying. *Journal of Agriculture and Sustainability*, 7(2).

127. Thanvi, K. P. (1993). Studies of drying okra in an inclined solar drier. *In Proceedings of the National Solar Energy Convention, Roles of Renewable Energy* held at Gujarat Energy development Agency, Vadodara, Gujarat on December (pp. 11–13).

128. Tiwari, B. K., & Mason, T. J. (2012). Ultrasound processing of liquid foods. *In:* Cullen, P. J., Tiwari, B. K., Valdramidis, V. (eds.) *Novel Thermal and Non-Thermal Technologies for Fluid Foods.* ISBN: 978-0-12-381470-8, pp. 135–165.

129. Toğrul, İ. T., & Pehlivan, D. (2002). Mathematical modeling of solar drying of apricots in thin layers. *Journal of Food Engineering*, 55(3), 209–216.

130. Toğrul, İ. T., & Pehlivan, D. (2003). Modeling of drying kinetics of single apricot. *Journal of Food Engineering*, 58(1), 23–32.

131. Tripathy, P. P. (2015). Investigation into solar drying of potato: Effect of sample geometry on drying kinetics and CO_2 emissions mitigation. *Journal of Food Science and Technology*, 52(3), 1383–1393

132. Tripathy, P. P., & Kumar, S. (2009). Neural network approach for food temperature prediction during solar drying. *International Journal of Thermal Sciences*, 48(7), 1452–1459.

133. Vardin, H., & Yasar, M. (2012). Optimization of pomegranate (*Punica granatum* L.) juice spray-drying as affected by temperature and maltodextrin content. *International Journal of Food Science and Technology*, 47, 167–176.

134. Vega-Mercado, H., Góngora-Nieto, M. M., & Barbosa-Cánovas, G. V. (2001). Advances in dehydration of foods. *Journal of Food Engineering*, 49(4), 271–289.
135. Vivek, R., Balusamy, T., & Srinivasan, R. (2015). Experimental testing and fabrication of low cost solar dryer for bitter gourd. *International Journal of Research and Innovation in Engineering Technology*, 1(10), 22–26.
136. Wang, H. C., Zhang, M., & Adhikari, B. (2015). Drying of shiitake mushroom by combining freeze-drying and mid-infrared radiation. *Food and Bioproducts Processing*, 94, 507–517.
137. Wang, S., Yang, R., Han, Y., & Gu, Z. (2015). Effect of three spouted drying methods on the process and quality characteristics of carrot cubes. *Advanced Engineering and Technology II*, 301.
138. Wang, Y., Zhang, L., Johnson, J., Gao, M., Tang, J., Powers, J. R., & Wang, S. (2014). Developing hot air-assisted radio frequency drying for in-shell macadamia nuts. *Food and Bioprocess Technology*, 7(1), 278–288.
139. Xin, Y., Zhang, M., & Adhikari, B. (2013). Effect of trehalose and ultrasound-assisted osmotic dehydration on the state of water and glass transition temperature of broccoli (Brassica oleracea L. var. botrytis L.). *Journal of Food Engineering*, 119(3), 640–647.
140. Zarein, M., Samadi, S. H., & Ghobadian, B. (2015). Investigation of microwave dryer effect on energy efficiency during drying of apple slices. *Journal of the Saudi Society of Agricultural Sciences*, 14(1), 41–47.
141. Zomorodian, A., & Moradi, M. (2010). Mathematical modeling of forced convection thin layer solar drying for Cuminum cyminum. *Journal of Agricultural Science and Technology*, 12, 401–408.
142. Zotarelli, M. F., Porciuncula, B. D. A., & Laurindo, J. B. (2012). A convective multi-flash drying process for producing dehydrated crispy fruits. *Journal of Food Engineering*, 108(4), 523–531.

PART II

ULTRASONIC TREATMENT
OF FOODS

CHAPTER 6

ULTRASONIC ASSISTED DERIVATIZATION OF FATTY ACIDS FROM EDIBLE OILS AND DETERMINATION BY GC-MS

M. RAMOS, A. BELTRAN,I. P. ROMAN, M. L. MARTIN, A. CANALS, and N. GRANE

Analytical Chemistry, Nutrition and Food Sciences Department, University of Alicante, 03080, Alicante, Spain

CONTENTS

6.1 INTRODUCTION

Oils and fats are good sources of energy, containing essential fatty acids, fat soluble vitamins, nutrients and antioxidants which provide benefits to the human health, playing functional roles in the daily diet [33]. Thus, the fatty acids determination can be used for characterization and quality control of oily products. For this reason, reliable analytical methods are required for the purpose of regulations for quality control and product labeling. Analysis of fatty acid profile has been generally performed by the use of gas chromatography [25]. Since direct determination is not recommended due to the low volatility and poor stability of fatty acids, derivatization to fatty acid methyl esters (FAME) is preferred in order to improve physical-chemical properties for chromatographic analysis [4, 19, 28]. The transesterification process of vegetable oils comprises the reaction of triglycerides with an alcohol in the presence of a strong acid or base catalyst producing a mixture of fatty acid alkyl esters and glycerol [36]. Basic catalysis is faster than acid catalysis, however the latter is strongly affected by mixing of the reactants and/or efficient heating, requiring strict anhydrous conditions [38–40, 42].

Ultrasounds are based on the process of propagation of the compression waves with frequencies above the range of human hearing, and are used to accelerate chemical processes by producing cavitation phenomena [40]. As a result, this process leads to the formation of micro fine bubbles with a very short lifetime producing intense local heating and high pressures [23]. This phenomenon is additionally supported by reactions in two-phase systems by an increase in active surface area [22]. Different applications of sono-chemistry have been developed in all areas of chemistry and related chemical technologies, as analytical chemistry, for the extraction of metals, organometals and organic compounds in solid samples. Ultrasonic assisted method has been a useful tool to determine the physicochemical properties of many foods [5, 11, 12, 16, 27, 35] and extract main compounds of natural products [20, 21, 30, 31, 43]. In addition, it is widely used in industry for emulsification of immiscible liquids [32] and as an efficient alternative for biodiesel production assisted by ultrasounds [10, 14, 26, 29, 34, 41].

Little importance has been given to the type of ultrasound device in use and/or the variables which should be optimized to obtain the best results from this energy. However, sonication via a sonotrode provides the necessary mechanical energy for mixing and the required activation energy for initiating the transesterification reaction [37]. In this sense, the use of ultrasonic-assisted derivatization method can be an alternative to perform transesterification reaction, since sonication can help mixing the reactants and accelerating the process. Keeping special care to know the influence of ultrasound on each step for an appropriate optimization of ultrasound variables [8].

The aim of this study is to develop a fast, easy to handle and efficient method for the determination of the fatty acid profile of edible oils using an ultrasonic-assisted derivatization process via sonotrode in order to carry out the transesterification followed by gas chromatography - mass spectrometry (GC-MS) analysis. For this purpose, important parameters, which affect this process, were optimized step by step. The method was applied to obtain the FAME of almond oils, in order to discriminate between two different almond cultivars (Marcona from Spain and Butte from America) based on their major fatty acid profiles.

6.2   MATERIAL AND METHODS

6.2.1   REAGENTS AND MATERIALS

Sodium hydroxide (NaOH, > 98%), methanol (MeOH, > 99.9%), n-hexane (for GC residue analysis, > 96%), sodium chloride (NaCl, > 99.5%) and sulphuric acid (H_2SO_4, 96%) were purchased from Panreac (Barcelona, Spain). Fatty acid methyl esters (FAME) standards for GC-MS analysis from palmitic (C16:0), palmitoleic (C16:1), stearic, (C18:0), oleic (C18:1) linoleic (C18:2) and linolenic (C18:3) acids were purchased from Sigma Aldrich (Steingeim, Germany). Methyl tridecanoate (C13:0) was used as internal standard, due to this fatty acid is not present in almonds [13], and it was purchased from Sigma Aldrich (Steingeim, Germany).

The vegetable oil used for the development of the ultrasonic-assisted derivatization method was soybean oil purchased from a local retail shop. The method was further applied to six samples from two different almond cultivars: three Marcona and three American Butte almond samples. Marcona was selected as being a representative cultivar grown in Spain, and it can be used for authentication of nougats with original quality. American Butte was chosen because it is one of the most widely grown cultivars in the world [1] and it is sometimes used as a substitute for Marcona almond cultivar in the production of nougat. All samples were grown in the same crop year. Marcona almonds were collected from different Spanish geographical areas (Jaén, Ibi and Muro de Alcoy). American Butte samples were grown in California and obtained from a Spanish importer (Colefruse, San Juan, Alicante).

6.2.2 TRANSESTERIFICATION PROCEDURE

The transesterification reaction was carried out according to fast ultrasonic-assisted derivatization method in triplicate, via sonotrode, where the following parameters were optimized: NaOH as a catalyst, amplitude of ultrasonic processor, ultrasonic wave cycle and the reaction time. For the transesterification process a sonic probe was used with an UP 200S ultrasonic processor (Hielscher ultrasonic Gmblt) operating at 200 W and 24 kHz frequency. The amplitude and wave cycle for the reaction could be adjustable from 20 to 100% and from 0.1 to 1 cycle per ton, respectively. A titanium sonotrode (S7) with a diameter of 7 mm and 100 mm length was used to transmit the ultrasounds into the solution.

Fatty acids from 0.3 g of soybean oil were transformed into their respective FAME by adding 6 mL of NaOH in MeOH solution in a 35 mL tube to be subjected to ultrasound waves. The sonotrode was directly submerged up to 20 mm into the solution. In order to optimize the parameters affecting transesterification process, the concentration of NaOH in methanol was varied from 0.1 to 1.5 mol L^{-1} (M). Two different amplitude values (60 and 80%) and four cycle values (0.3, 0.5, 0.7 and 0.9) were used and were adjusted by using the controllers; and the reaction times were 0.3, 1.0, 1.2, 1.3, 2.0 and 2.3 min.

After the sonication, 6 mL of H_2SO_4 in MeOH (5% weight) was added to complete the esterification of fatty acids and to neutralize the NaOH. This step is strongly recommended because it is necessary to remove the acid excess before GC analysis, since this reagent tends to damage the GC column [2]. After that, the solution was treated with 6 mL of n-hexane to extract the FAME. Finally, 25 mL of a saturated chloride sodium solution were added in order to facilitate the separation of two layers. Finally, appropriate dilution and internal standard addition were performed and these were analyzed by GC-MS in triplicate.

The previously developed method by using the optimal conditions was applied with almond oils from different cultivars, which were extracted using a commercial fat extractor (Selecta, Barcelona, Spain) [3]. Three aliquots of each almond oil sample (three Marcona and three American Butte) were analyzed in order to determine their major fatty acid composition. Transesterification process of fatty acids was carried out by ultrasonic irradiation via sonotrode using the developed optimized ultrasonic-assisted method. Then, the major fatty acid profile was determined by GC-MS as the average of triplicate analysis of the most abundant fatty acids found in the almond oil samples (palmitic, stearic, palmitoleic, oleic and linoleic acid).

6.2.3 GAS CHROMATOGRAPHY-MASS SPECTROMETRY (GC-MS) ANALYSIS

GC-MS analysis was performed by using an Agilent 6890N GC system coupled to an Agilent 5973N mass selective detector operating in electronic impact (EI) ionization mode (70 eV). A SGE BPX70 (60.0 m x 0.25 mm I.D., 0.25 μm film thickness) column and a split/splitless injectors were used. The column temperature was programmed from 120 to 245°C (for 15 min.) @ 3°C.min^{-1}. The injector temperature was set at 250°C and helium was used as carrier gas at a flow-rate of 1.0 mL.min^{-1}. Calibration standards were prepared by adding different amounts of commercial standards in n-hexane to cover the different concentration range of FAME in the samples (from 10 to 1000 ppm). The internal standard (C13:0) was added to all samples and standard solutions. Identification of FAME was performed in full-scan mode (30–450 m/z) by the combination of Wiley 275.L mass spectra library and gas chromatographic retention times of standards (including methyl tridecanoate internal standard). Quantification of major fatty acids of different almond oils was performed by extracting the following ions: 74 m/z (palmitic and stearic acid); 55 m/z (palmitoleic and oleic acid) and 67 m/z (linoleic acid).

6.2.4 STATISTICAL ANALYSIS

Each experiment was triplicated. The data were reported as means ± SD. Statistical analysis of results was performed with SPSS commercial software (Version 15.0, Chicago, IL). A one-way analysis of variance (ANOVA) was carried out.

6.3 RESULTS AND DISCUSSION

6.3.1 EFFECT OF CATALYST CONCENTRATION (NAOH IN METHANOL)

In order to investigate the effect of NaOH as a catalyst in the transesterification process, it was varied from 0.1 to 1.5 M in methanol, because this solution was used due to it is a widely applied in alternative method of transesterification for biodiesel production [9]. Moreover, the use of ultrasound could help to decrease the amount of catalyst due to the increased chemical activity in the presence of cavitations [17].

The Figure 6.1 shows the effects of NaOH in methanol concentration on the transesterification process at 80% of ultrasonic amplitude and 0.9 cycles per second of ultrasonic wave cycle. The final point of reaction was considered as the time when only one phase was observed. As it can be seen, the amount of major fatty acids was increased with increase in concentration of NaOH in methanol upto 1 M. However, a slight decrease was observed above this concentration. This behavior can be attributed to side reactions (as saponification) occurring simultaneously with transesterification and reducing the conversion of triglycerides to the desired ester [15]. This behavior was also observed by Kumar et al. [17] in a study on ethanolysis of coconut oils with potassium hydroxide [17]. For these reasons, 1 M was selected as the optimal NaOH concentration for the next steps of the optimization process.

6.3.2 EFFECTS OF ULTRASONIC AMPLITUDE

The optimal concentration of catalyst was maintained to study the effect of ultrasonic amplitude. The Figure 6.2 shows that the maximum amount of FAME was achieved at 80% of ultrasonic amplitude ($p < 0.05$). This can be attributed to better efficacy of ultrasound, to ensure sufficient mixing and emulsification of two immiscible reaction layers, at higher levels of

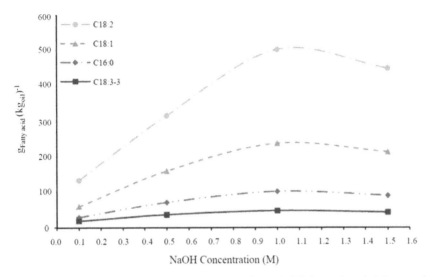

FIGURE 6.1 Optimization of catalyst concentration, NaOH in methanol (ultrasound system at 80% amplitude and 0.9 cycles per ton).

power dissipation. Amplitude values higher than 80% were not applied in this research based on previous studies that postulated a large number of cavitation bubbles generated in the solution for higher amplitude values, with no increase of the conversion of fatty acids. Many of these bubbles will coalesce forming larger, more long-lived bubbles which will certainly act as a barrier to the transfer of acoustic energy through the liquid [7, 24].

6.3.3 EFFECTS OF CYCLE

The instrument was used in ultrasonic wave cycle mode: this implies that the cycle setting was possible to be adjusted with a pulse setting to enable rhythmic processing of media (on for few seconds followed by few seconds off). For example, "0.3" means that the mixture is sonicated for 0.3 s and then sonication stops for 0.7 s. Hence in cycle mode, the ratio of sound-emission time to cyclic pause time can be adjusted continuously from 0 to 1 s cycles per ton with the objective to obtain the maximum extraction yield. The effect of ultrasonic wave cycle on the transesterification process was studied at 0.3, 0.5, 0.7 and 0.9 s cycles per ton. In

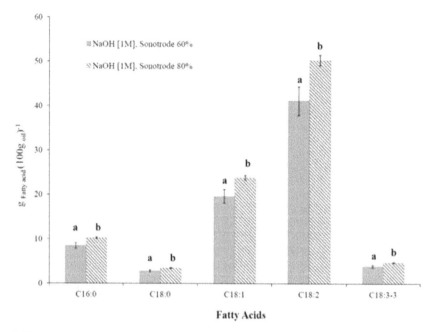

FIGURE 6.2 Effects of ultrasonic amplitude in the ultrasonic-assisted derivatization method. (Mean ± SD, n = 3). Different letter over the same fatty acid indicates statistically significant different value (p < 0.05).

Figure 6.3, the efficiency of transesterification of fatty acids was increased with significant differences with the increase in ultrasonic wave cycle. The maximum values of FAME (830 ± 30 g total fatty acid (kg oil)$^{-1}$) were achieved for 0.9 s cycles per ton. This ultrasonic wave cycle value was therefore used in all subsequent experiments. The maximum cycle was never used in this study to prevent damage of the sonic probe as recommend by other investigators [6].

6.3.4 EFFECTS OF REACTION TIME

Sonication increases the speed of chemical reaction in the transesterification process [17]. Figure 6.4 shows the effects of reaction time on the derivatization of fatty acids carried out at different times at 80% amplitude and 0.9 s cycles per ton. The maximum values of FAME were achieved

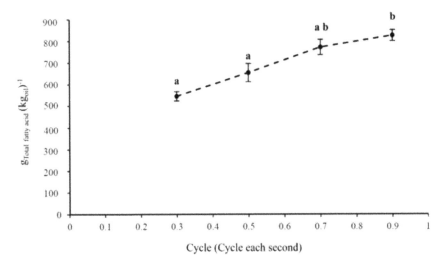

FIGURE 6.3   Effects of ultrasonic wave cycle on the ultrasonic-assisted derivatization method using NaOH in methanol (1 M) and 80% of ultrasonic amplitude (Mean ± SD, n = 3). Different letter over the cycle value indicates statistical significant different values (p < 0.05).

around 1.2–1.3 minutes (p < 0.05). However, the reaction time selected was 1.2 minutes (840 ± 20 g total fatty acid (kg oil)–1) because the aim was to reduce the reaction time, and avoid preheating of samples and possible degradations. In addition after this maximum, a drastic decrease in the amount of FAME was observed. Similar results were obtained by Singh et al. [37] due to cracking followed by oxidation of the FAME to lower-chained organic fractions. Based on the results in this study, the ultrasonic cavitation provides the necessary activation energy to start the reaction considerably reducing the derivatization time compared with other studies reported which use external heating and/or mechanical stirring [18].

6.3.5 ANALYSIS OF ALMOND OILS

The developed method in the present work was carried out by using different extracted almond oils from Marcona and American Butte cultivars and determining their major fatty acid profile by GC-MS. For this purpose, each sample was analyzed in triplicate by using the optimal conditions

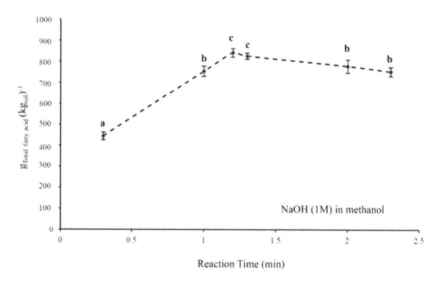

FIGURE 6.4 Effects of reaction time in the transesterification process by ultrasonic irradiation at 80% amplitude, NaOH in methanol (1 M) and 0.9 s cycles per ton (Mean ± SD, n = 3). Different letter over the reaction time value indicates statistical significant different value (p < 0.05).

found for ultrasonic-assisted derivatization method (1M NaOH in MeOH, 80% of ultrasonic amplitude, 0.9 for ultrasonic wave cycle value and 1.2 min of reaction time). The main analytical parameters typically used to ensure reliable results were determined. Accuracy of the proposed method was assessed and the linearity was also evaluated using seven calibration standards injected in triplicate. A linear behavior was observed over the explored concentration range with acceptable determination coefficient values (r^2 ranged from 0.9988 to 0.9997). LOD and LOQ were calculated using a signal-to-noise ratio of 3:1 and 10:1, respectively. The results for both parameters varied from 0.04 to 0.3 mg Kg^{-1} for LOD and from 0.14 to 1.0 mg Kg^{-1} for LOQ.

The same major fatty acids were detected in all almond oil samples after analysis by GC–MS as expected. As a result, a total of five major FAME from palmitic, stearic, palmitoleic, oleic and linoleic acids were identified and quantified, which are indicated in the Figure 6.5 and were related to their corresponding fatty acid (C16:0, C16:1, C18:0, C18:1 and C18:2). Figure 6.5 shows the full-scan chromatogram of American Butte and Marcona almond oil samples obtained at optimized ultrasonic-assisted derivatization method.

As it can be seen in Figure 6.5 when comparing the major fatty acid profile obtained for American Butte and Marcona cultivars, a different behavior was observed related to the American almond samples. Figure 6.6 shows the amount of major FAME detected for both samples. American Butte samples presented significant differences compared with Marcona almond cultivar. Lower amounts of palmitic, palmitoleic, stearic and oleic acids and a higher content of linoleic acid leading to statistical differences between both almond cultivars were obtained ($p < 0.05$). These results are consistent with previous studies indicating that linoleic acid content in American Butte samples is higher than in Spanish almond samples which make American Butte cultivar more prone to lipid oxidation [2, 3].

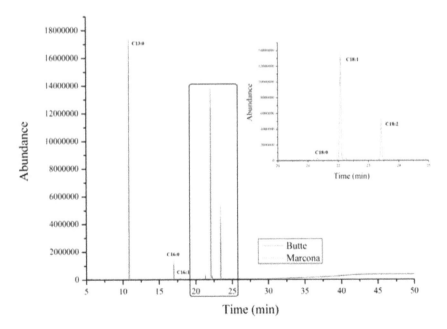

FIGURE 6.5 GC–MS chromatogram for major fatty acids obtained from American butte and Marcona almond oils (by using optimum ultrasonic-assisted derivatization conditions: 80% amplitude, NaOH in methanol (1 M), 0.9 s cycles per ton and 1.2 min).

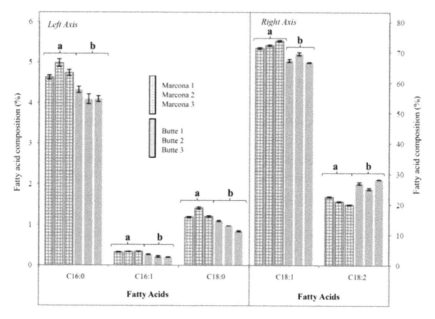

FIGURE 6.6 Comparison of major fatty acid profile (%) obtained for Marcona and American butte almond oils analyzed by GC-MS. (Mean ± SD, n = 3). Different letter over the same fatty acid indicates statistically significant different values ($p < 0.05$).

6.4   CONCLUSIONS

A fast, accurate and applicable method was developed for determining of the major fatty acid profile in oils based on ultrasonic-assisted derivatization method via sonotrode followed by GC-MS analysis. One of the main objectives was to reduce the reaction time and use of NaOH as catalyst reagent. Furthermore, with this procedure, it may be possible to eliminate thermal heating and the use of relatively hazardous reagents used in other procedures already. The developed method was satisfactorily applied to analysis of almond oils from two different cultivars (Spanish Marcona and American Butte) achieving suitable and good analytical parameters of validation. American Butte showed a higher linoleic acid content, in comparison with Spanish almond cultivar. These results can be used as a preliminary approach to avoid adulteration practices in food products with guarantee of origin as in nougats. In conclusion, ultrasonic-assisted derivatization can be a valuable tool for the transesterification of fatty acids of different oils, aiming to obtain the fatty acid profile at laboratory scale.

6.5   SUMMARY

In order to investigate new methods to carry out a fast determination of the major fatty acid profile in edible oils an ultrasonic-assisted derivatization followed by gas chromatography-mass spectrometry (GC-MS) has been developed. Effects of different operating parameters (catalyst concentration, amplitude, cycle and reaction time) were investigated. The best conditions for the derivatization process were achieved using sodium hydroxide in methanol (1M) as derivatization reagent, 80% ultrasonic amplitude and 1.2 min of reaction time. Major fatty acid profiles were determined by using the optimized method (GC-MS) from different edible oils. The results demonstrated that the ultrasonic-assisted derivatization method can be used to determine the major fatty acid profile in a fast and simple way due to reduce considerably the analysis time and to carry out a classification between Spanish and American almond cultivars.

ACKNOWLEDGMENTS

Authors thank Prof. Maria del Carmen Garrigós (University of Alicante) for useful discussion as well as Colefruse S. A> (San Juan, Alicante) for providing the almond samples. Moreover, Marina Ramos thanks University of Alicante (Spain) for UAFPU2011-48539721S predoctoral research grant.

KEYWORDS

- Almond
- Almond oil
- American Butte almond oil

- **Amplitude of ultrasonic processor**
- **Classification**
- **Derivatization**
- **Fatty acid**
- **Fatty acid methyl esters**
- **GC-MS**
- **Major fatty acid profile**
- **Marcona almond oil**
- **Optimized method**
- **Reaction time**
- **Sodium hydroxide**
- **Sonication**
- **Sonotrode**
- **Soybean oil**
- **Transesterification**
- **Ultrasonic irradiation**
- **Ultrasonic wave cycle**
- **Ultrasound**
- **Vegetable oil**

REFERENCES

1. Almond Board of California, (www.almondboard.com). Last access June 2015.
2. Beltrán, A., Ramos, M., Grané, N., Martín, M. L., & Garrigós, M. C. (2011). Monitoring the oxidation of almond oils by HS-SPME-GC-MS and ATR-FTIR: Application of volatile compounds determination to cultivar authenticity. *Food Chemistry*, 126, 603–609.
3. Beltrán Sanahuja, A., Prats Moya, M., Maestre Pérez, S., Grané Teruel, N., & Martín Carratalá, M. (2009). Classification of Four Almond Cultivars Using Oil Degradation Parameters Based on FTIR and GC Data. *Journal of the American Oil Chemists' Society*, 86, 51–58.
4. Brondz, I. (2002). Development of fatty acid analysis by high-performance liquid chromatography, gas chromatography, and related techniques. *Analytica Chimica Acta*, 465, 1–37.

5. Caballo-López, A., & Luque De Castro, M. D. (2003). Continuous ultrasound-assisted leaching of phenoxyacid herbicides in soil and sediment with in-situ sample treatment. *Chromatographia, 58,* 257–262.
6. Canals, A., Remedio, H. del, & Hernández, M. del R. (2002). Ultrasound-assisted method for determination of chemical oxygen demand. *Analytical and Bioanalytical Chemistry, 374,* 1132–1140.
7. Canals, A., & Del Remedio Hernández, M. (2002). Ultrasound-assisted method. *Analytical and Bioanalytical Chemistry, 374.*
8. Delgado-Povedano, M. M., & Luque de Castro, M. D. (2013). Ultrasound-assisted extraction and in situ derivatization. *Journal of Chromatography, 1296,* 226–234.
9. Demirbas, A. (2008). Comparison of transesterification methods for production of biodiesel from vegetable oils and fats. *Energy Conversion and Management, 49,* 125–130.
10. Dittmar, T., Dimmig, T., Ondruschka, B., Heyn, B., Haupt, J., & Lauterbach, M. (2003). Production of fatty acid methyl esters from rapeseed oil and spent fat in batch operation—Part 1 of a three-part series on the manufacture of biodiesel. *Chemie Ingenieur Technik, 75,* 595–601.
11. Djenni, Z., Pingret, D., Mason, T., & Chemat, F. (2013). Sono–Soxhlet: In Situ Ultrasound-Assisted Extraction of Food Products. *Food Analytical Methods, 6,* 1229–1233.
12. Gallego-Gallegos, M., Liva, M., Olivas, R. M., & Cámara, C. (2006). Focused ultrasound and molecularly imprinted polymers: A new approach to organotin analysis in environmental samples. *Journal of Chromatography A, 1114,* 82–88.
13. Grane Teruel, N., Prats Moya, M., Berenguer Navarro, V., & Martin Carratala, M. (2001). A possible way to predict the genetic relatedness of selected almond cultivars. *Journal of the American Oil Chemists' Society, 78,* 617–619.
14. Hanh, H. D., Dong, N. T., Starvarache, C., Okitsu, K., Maeda, Y., & Nishimura, R. (2008). Methanolysis of triolein by low frequency ultrasonic irradiation. *Energy Conversion and Management, 49,* 276–280.
15. Hingu, S. M., Gogate, P. R., & Rathod, V. K. (2010). Synthesis of biodiesel from waste cooking oil using sonochemical reactors. *Ultrasonics Sonochemistry, 17,* 827–832.
16. Hristozov, D., Domini, C. E., Kmetov, V., Stefanova, V., Georgieva, D., & Canals, A. (2004). Direct ultrasound-assisted extraction of heavy metals from sewage sludge samples for ICP-OES analysis. *Analytica Chimica Acta, 516,* 187–196.
17. Kumar, D., Kumar, G., & Singh, C. P. (2010). Fast, easy ethanolysis of coconut oil for biodiesel production assisted by ultrasonication. *Ultrasonics Sonochemistry, 17,* 555–559.
18. Kumar, D., Kumar, G., & Singh, C. P. (2010). Ultrasonic-assisted transesterification of Jatropha curcus oil using solid catalyst, Na/SiO_2. *Ultrasonics Sonochemistry, 17,* 839–844.
19. Lee, D.-S., Noh, B. S., Bae, S. Y., & Kim, K. (1998). Characterization of fatty acids composition in vegetable oils by gas chromatography and chemometrics *Analytica Chimica Acta, 358,* 163–175.
20. Li, H. Z., Zhang, Z. J., Hou, T. Y., Li, X. J., & Chen, T. (2015). Optimization of ultrasound-assisted hexane extraction of perilla oil using response surface methodology. *Industrial Crops and Products, 76,* 18–24.

21. Liao, J., Qu, B., Liu, D., & Zheng, N. (2015). New method to enhance the extraction yield of rutin from Sophora japonica using a novel ultrasonic extraction system by determining optimum ultrasonic frequency. *Ultrasonics Sonochemistry, 27,* 110–116.

22. Lifka, J., & Ondruschka, B. (2004). Influence of Mass Transfer on the Production of Biodiesel. *Chemical Engineering & Technology, 27,* 1156–1159.

23. Lin, J. R., & Yen, T. F. (1993). An upgrading process through cavitation and surfactant. *Energy & Fuels, 7,* 111–118.

24. Mahamuni, N. N., & Adewuyi, Y. G. (2009). Optimization of the synthesis of biodiesel via ultrasound-enhanced base-catalyzed transesterification of soybean oil using a multifrequency ultrasonic reactor. *Energy and Fuels, 23,* 2757–2766.

25. Marriott, P. J., Shellie, R., & Cornwell, C. (2001). Gas chromatographic technologies for the analysis of essential oils. *Journal of Chromatography A, 936,* 1–22.

26. Mason, T. J. (1999). Sonochemistry: current uses and future prospects in the chemical and processing industries. *Philosophical Transactions of the Royal Society of London. Series A: Mathematical, Physical and Engineering Sciences, 357,* 355–369.

27. Niñoles, L., Clemente, G., Ventanas, S., & Benedito, J. (2007). Quality assessment of Iberian pigs through backfat ultrasound characterization and fatty acid composition. *Meat Science, 76,* 102–111.

28. Omar, T. A., & Salimon, J. (2013). Validation and application of a gas chromatographic method for determining fatty acids and trans fats in some bakery products. *Journal of Taibah University for Science, 7,* 56–63.

29. Paiva, E. J. M., da Silva, M. L. C. P., Barboza, J. C. S., de Oliveira, P. C., de Castro, H. F., & Giordani, D. S. (2013). Non-edible babassu oil as a new source for energy production–a feasibility transesterification survey assisted by ultrasound. *Ultrasonics Sonochemistry, 20,* 833–838.

30. Pan, Z., Qu, W., Ma, H., Atungulu, G. G., & McHugh, T. H. (2011). Continuous and pulsed ultrasound-assisted extractions of antioxidants from pomegranate peel. *Ultrasonics Sonochemistry, 18,* 1249–1257.

31. Rodrigues, S., Fernandes, F. A. N., de Brito, E. S., Sousa, A. D., & Narain, N. (2015). Ultrasound extraction of phenolics and anthocyanins from jabuticaba peel. *Industrial Crops and Products, 69,* 400–407.

32. Rosa, M. T. M. G., Silva, E. K., Santos, D. T., Petenate, A. J., & Meireles, M. A. A. (2015). Obtaining annatto seed oil miniemulsions by ultrasonication using aqueous extract from Brazilian ginseng roots as a biosurfactant. *Journal of Food Engineering, 168,* 68–78.

33. Ruiz-Rodriguez, A., Reglero, G., & Ibañez, E. (2010). Recent trends in the advanced analysis of bioactive fatty acids. *Journal of Pharmaceutical and Biomedical Analysis, 51,* 305–326.

34. Sánchez, Á., Cancela, A., Maceiras, R., & Alfonsin, V. (2014). Lipids extraction from microalgae for biodiesel production. In: *Proceedings of 2014 International Renewable and Sustainable Energy Conference, IRSEC.*

35. Sanz-Landaluze, J., Bartolome, L., Zuloaga, O., González, L., Dietz, C., & Cámara, C. (2006). Accelerated extraction for determination of polycyclic aromatic hydrocarbons in marine biota. *Analytical and Bioanalytical Chemistry, 384,* 1331–1340.

36. Schuchardt, U., Sercheli, R., & Vargas, R. M. (1998). Transesterification of vegetable oils: a review. *Journal of the Brazilian Chemical Society, 9,* 199–210.

37. Singh, A. K., Fernando, S. D., & Hernandez, R. (2007). Base-catalyzed fast transesterification of soybean oil using ultrasonication. *Energy and Fuels,* 21, 1161–1164.
38. Stavarache, C., Vinatoru, M., & Maeda, Y. (2006). Ultrasonic versus silent methylation of vegetable oils. *Ultrasonics Sonochemistry,* 13, 401–407.
39. Stavarache, C., Vinatoru, M., & Maeda, Y. (2007). Aspects of ultrasonically assisted transesterification of various vegetable oils with methanol. *Ultrasonics Sonochemistry,* 14, 380–386.
40. Stavarache, C., Vinatoru, M., Nishimura, R., & Maeda, Y. (2005). Fatty acids methyl esters from vegetable oil by means of ultrasonic energy. *Ultrasonics Sonochemistry,* 12, 367–372.
41. Sun, Q., Li, A., Li, M., Hou, B., & Shi, W. (2015). Advantageous production of biodiesel from activated sludge fed with glucose-based wastewater. *Huanjing Kexue Xuebao/Acta Scientiae Circumstantiae,* 35, 819–825.
42. Talpur, F. N., Bhanger, M. I., Rahman, A. U., & Memon, G. Z. (2008). Application of factorial design in optimization of anion exchange resin based methylation of vegetable oil and fats. *Innovative Food Science and Emerging Technologies,* 9, 608–613.
43. Wang, W., Ma, X., Xu, Y., Cao, Y., Jiang, Z., Ding, T., Ye, X., & Liu, D. (2015). Ultrasound-assisted heating extraction of pectin from grapefruit peel: Optimization and comparison with the conventional method. *Food Chemistry,* 178, 106–114.

CHAPTER 7

PRINCIPLES OF ULTRASONIC TECHNOLOGY FOR TREATMENT OF MILK AND MILK PRODUCTS

ASAAD REHMAN SAEED AL-HILPHY,[1] DEEPAK KUMAR VERMA,[2] ALAA KAREEM NIAMAH,[3] SUDHANSHI BILLORIA,[4] and PREM PRAKASH SRIVASTAV[5]

[1]*Assistant Professor, Department of Food Science, College of Agriculture, University of Basrah, Basra City, Iraq; Mobile: +00-96 47702696458; E-mail: aalhilphy@yahoo.co.uk*

[2]*PhD Research Scholar, Department of Agricultural and Food Engineering, Indian Institute of Technology, Kharagpur – 721302, West Bengal, India; Mobile: +91-74071-70260, +91-93359-93005, Telephone: +91-32222-81673, Fax: +91-3222-282224, E-mail: deepak.verma@agfe.iitkgp.ernet.in, rajadkv@rediffmail.com*

[3]*Assistant Professor, Department of Food Science, College of Agriculture, University of Basrah, Basra City, Iraq; Mobile: +00-96 47709042069; E-mail: alaakareem2002@hotmail.com*

[4]*Research Scholar, Department of Agricultural and Food Engineering, Indian Institute of Technology, Kharagpur – 721302, West Bengal, INDIA, Mobile: +91-8768126479, E-mail: sudharihant@gmail.com*

[5]*Associate Professor, Department of Agricultural and Food Engineering, Indian Institute of Technology, Kharagpur – 721302, West Bengal, India; Mobile: +91-9434043426, Telephone: +91-3222-283134, Fax: +91-3222-282224, E-mail: pps@agfe.iitkgp.ernet.in*

CONTENTS

7.1   INTRODUCTION

Ultrasonication is the science of sound waves above the audible range of human beings. Frequency of the sound determines its tone and pitch. Low frequency sound waves have bass tone whereas high frequency produces treble tones. Frequencies greater than 18 kHz are usually considered as ultrasonic and cannot be heard by human ear. Usually, Ultrasonication (US) treatment has been in the range of 10 to 1000 W/cm² and frequencies of 20 to 100 kHz. It has a pronounced effect on physical and chemical properties of liquids [14]. In 1944, the first patent was granted in Switzerland for using US.

   In food processing, the role of ultrasonication is very important because of the consumer interest in minimizing quality losses to the processed foods. The US device operates through intense sound pressure waves in liquids (such as milk). Pressure waves pass through the liquid, leading to the formation of tiny bubbles that eventually collapse. This process is called cavitation. Cavitation process generates a shock wave that contains sufficient energy to destroy the covalent bonds, the chemical compounds and then inhibit the processes within cells [16]. Therefore, US is currently recognized as a progressive technology in the food processing and has a wide range of applications due to safe, nontoxic and environmentally friendly characteristics of sound waves. Normally ultrasonication of high frequency is used in milk homogenization process.

This chapter discusses principles and applications of innovative and eco-friendly ultrasonic technology for treatment of milk and milk products.

7.2 ULTRASONICATION: A BRIEF HISTORY

Acoustics is the science of sound which deals with the study of all mechanical waves in solids, gases and liquids including areas like sound, US, infrasound and vibration. The major developments in the history of ultrasonic are summarized in Table 7.1. On the other hand, American national standards institute classifies ultrasound as a sound at frequencies greater than 2000 Hz.

7.3 TECHNOLOGY OF ULTRASONIC TREATMENT

7.3.1 GOALS OF ULTRASONIC TREATMENT

The applications of ultrasonic treatments in milk and milk products as an innovative and eco-friendly technology are noteworthy. This technology is a rapidly growing field in the area of research, generating interest among scientists to create novel and interesting methodologies compared to other classical techniques. The use of ultrasonic treatment with various goals is increasing day-by-day also in milk industry for various reasons including analysis and modification of milk products. Some of the important goals of ultrasonication in milk industry are outlined below:

1. To increase the stability and consistency of emulsion obtained from separation of fat and water phases during the homogenization of milk.
2. To offer a net advantage in terms of productivity, selectivity and yield, with enhanced quality, processing time, reduced physical and chemical hazards.
3. To reduce the fat globule size in milk during the milk homogenization.
4. To use as a non-destructive analytical technique for process control and quality assurance with particular reference to structure,

TABLE 7.1   Major Developments in the History of Ultrasonication

Year	Major inventions/discoveries/developments
1794	Lazzaro Spallanzani discovered Echolocation in bats, when he demonstrated that bats hunted and navigated by in audible sound and not vision.
1876	Frances Galton invented the ultrasonic whistle.
1877	Rayleigh's "Theory of Sound" laid foundation for modern acoustics.
1893	Frances Galton manufactured a whistle, producing US.
1903	Lebedev and coworkers developed complete ultrasonic system to study absorption of waves.
1912	Richardson files first patent for an underwater echo ranging sonar
1915	Langevin originated modern science of ultrasonics through work on the "Hydrophone" for submarine detection.
1917	US technology was applied to detect submarines.
1920	Use of US to inactivate microbes was reported later on 1920s [25].
1922	Hartmann developed the air-jet ultrasonic generator.
1925	Pierce developed the ultrasonic interferometer.
1926	Boyle and Lehmann discovered the effect of bubbles and cavitation in liquids by US.
1927	Wood and Loomis described effects of intense US.
1928	Sokolov proposed use of US for flaw detection.
1930	Debye and Sears and Lucas and Biquard discover diffraction of light by US.
1930	Harvey reported on the physical, chemical, and biological effects of US in macromolecules, microorganisms and cells.
1931	Mulhauser obtained a patent for using two ultrasonic transducers to detect flaws in solids.
1937	Sokolov invented an ultrasonic image tube.
1938	Pierce and Griffin detect the ultrasonic cries of bats.
1939	Pohlman investigated the therapeutic uses of ultrasonic.
1940	Firestone, in the United States and Sproule, in Britain, discovered ultrasonic pulse-echo metal-flaw detection.
1942	Dussik brothers made first attempt at medical imaging with US.
1944	Lynn and Putnam successfully used US waves to destroy brain tissue of animals.
1944	The first patent for ultrasonic emulsification was in Switzerland.
1945	Start of the development of power ultrasonic processes.
1948	Start of extensive study of ultrasonic medical imaging in the United States and Japan.
1954	Jaffe discovered the new piezoelectric ceramics lead titanate-zirconate.

Source: Harvey and Loomis [25]; http://www.ob-US.net/ultrasonics_history.html

composition and physical state of milk and milk products which are considered as physicochemical properties.

7.3.2 PRINCIPAL MECHANISM OF ULTRASONIC TREATMENT

US is a mechanical energy generated by electrical energy supplied to a piezoelectric material referred to as transducer, which then converts it to mechanical vibration at specific frequency transmitting through the fluid. Mechanical vibration at increasing frequencies is called as sound energy. The energy dissipated is calculated from the following equation [35]:

$$E_d = kf^2A^2 \qquad\qquad (1)$$

where, E_d = energy dissipated, k = constant (If it have some value please mention here), f = frequency, and A = amplitude.

The normal human sound range is from 16–20,000 Hz. Beyond this upper limit, the mechanical vibration is called ultrasonic as depicted in Figure 7.1 [17, 18].

Ultrasonic waves hit the surface of material and generate a force, when the force is perpendicular to the surface. It results in a compression wave that moves through the food; also if the force is parallel to the surface it produces a shear wave. However, the waves become attenuated as they

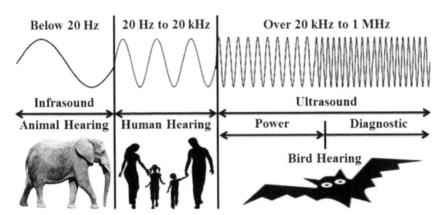

FIGURE 7.1   Illustration of ultrasonic frequency.

move through the food. Ultrasonication produces a very rapid localized change in pressure and temperature, which causes shear disruption, cavitation, thinning of cell membranes, localized heating and free radical production with a lethal effect on microorganisms [20] and reducing the size of fat globules [2]. The medium (liquid milk) responds to the diffusion of ultrasonic waves and sustains them by vibrating elastically. Hecht [26] and Knorr et al. [33] stated that there are two types of elastic vibrations in the medium: Condensation and rarefaction. During condensation, the medium particles are compressed, and an increase in pressure and density of the medium occurs. In addition, rarefaction causes decrease in the density and pressure of the medium particles [1, 22, 26].

Effect of ultrasonic on fluid is to impose an acoustic pressure and hydrostatic pressure acting on the medium. McClements [40] and Muthukumaran et al. [43] stated that the acoustic pressure follows a sinusoidal wave dependent on time, distance, frequency and maximum pressure amplitude of the waves. Acoustic pressure can be calculated from the following equation:

$$P_a = P_{a.(max)} \sin(2\pi f t) \tag{2}$$

The value of f is calculated by the following equation by McClements [40]:

$$f = \frac{1}{t} \tag{3}$$

where, is the maximum pressure amplitude of the wave and refers to the power input of the transducer, f is the sinusoid frequency, and t is the time period of the sinusoidal wave.

Leighton [35] stated that acoustic streaming pressure wave induces motion and mixing within the fluid at low intensity, but at high intensity the local pressure in the expansion phase of the cycle falls under the vapor pressure of the liquid, this causes tiny bubbles to grow. On the other hand, it generates negative transient pressure in the fluid and increasing bubble growth and producing new cavities by the tensioning effect on the fluid [37]. Compression cycle causes shrinking of bubbles (increasing surface area) and their content are absorbed back into the liquid. Not all of the vapors are absorbed back into the liquid and the bubble continues to grow over the number of cycles. Moholkar et al. [41] explained that critical size of the bubble wall matches that of the applied frequency of the sound waves causing the bubble to implode during a single compression cycle, the process is called cavitation. The bubble size can be calculated by the following equation [35]:

$$f.R = 3 \qquad (4)$$

where, f = frequency (in MHz), and R = bubble radius (in microns).

Suslick [58] and Laborde et al. [34] stated that imploding bubbles can be dramatic with the temperature of 2000–5000 K and pressures of 300–1200 bars [21, 51]. This implosion in turn produces very high shear energy waves and turbulence in the cavitation zone. In addition, Suslick [58] found that heat, pressure and turbulence (which is used to accelerate the mass transfer in chemical reaction) create new reaction pathways in chemical reactions and dislodge the particle or even create different products from those obtained under conventional conditions.

7.3.3 TYPES OF ULTRASONIC TREATMENT

Ultrasonic treatment is an emerging technology for research in the milk and milk products. This technology generally deals with the frequency of sound waves [8, 15, 48]. According to the frequency, US can be classified into three following frequency ranges. All of these ranges are characterized by important criteria viz. sound power (W), sound intensity (W·m^{-2}) or sound energy density (Ws·m^{-3}) [20, 33, 40].

- The power ultrasound ranged between 16–100 kHz (Table 7.2);
- High frequency ultrasound ranged between 100–1 MHz (Table 7.3);
- Diagnostic ultrasound ranged between 1–10 MHz (1 MHz = 1 million cycles per second).

7.4 USE OF ULTRASONIC TREATMENT IN MILK HOMOGENIZATION

Raw milk is an emulsion, which is composed of fat globules, various solids and water. Cow milk and coconut milk at 25°C and atmospheric pressure

TABLE 7.2 Effects of Low-Intensity Ultrasonic Treatment in Milk and Milk Products [38]

Application	Effects
On living cells	1. Stimulation of activity
	2. Sono chemical destruction
On enzymes	1. Stimulation of activity
	2. Controlled denaturing
'Jet' impact effect on surface	1. Improved impregnation
	2. Improved extraction
Miscellaneous	1. Crystallization and freezing
	2. Emulsification

TABLE 7.3 Effects of High-Intensity Ultrasonic Treatment in Milk and Milk Products

Application	Mechanisms	Benefits
Crystallization	Nucleation and modification of crystal formation	Formation of smaller crystal formation during freezing
Emulsification/ Homogenization	High shear micro-streaming	Cost effective emulsion formation
Filtration/ Screening	Disturbance of the boundary layer	Increased flux rates, reduced fouling
Separation	Agglomeration of components of pressure nodal points	Adjunct for use in nonchemical separation procedures
Viscosity Alteration	Reversible and non-reversible structural modification via vibrational and high shear micro-streaming. Sono-chemical modification involving crosslinking and restructuring	Nonchemical modification for improved processing traits, reduced additives, differentiated functionality
Defoaming	Airborne pressure waves causing bubble collapse	Increased production throughput, reduction or elimination of anti-foam chemicals and reduced wastage in bottling lines
Enzyme and microbial inactivation	Increased heat transfer and high shear. Direct cavitation damage to microbial cell membranes	Enzyme inactivation adjunct at lower temperatures for improved quality attributes
Fermentation	Improved substrate transfer and stimulation of living tissue, enzyme processes	Increasing production of metabolites, acceleration of fermentation processes
Heat Transfer	Improved heat transfer through acoustic streaming and cavitation	Acceleration of heating, cooling and drying of products at low temperature

coagulates after six hours and divert to cream, skim milk and water. Over the course of time, fat globules start separating and rise as a cream layer and skim milk as a lower layer. Homogenization prevents separation of these components [4]. Also, homogenization process rearranges the milk density and reduces the diameter of fat globules to produce uniform size and increase the surface tension of the fatty membrane [11]. Nowadays, the use of ultrasonic treatment has become popular for homogenization of milk because of various benefits [3, 12, 23].

High amplitude of ultrasonication and longer exposure time give a greater effect on the degree of homogenization. The capillary wave mechanism contributes to fat globules disruption. Decreasing fat globule size is enabled only if the fat globules diameter is larger than the oscillation wavelength. Cavitation is the most accepted mechanism for ultrasonic emulsification. This mechanism is dependent on the imploding bubbles, which produce powerful shock waves in the milk surrounding the ultrasonic probe and jets of high velocity. This micro jet effectively results in disruption of fat globules [36]. US treatment with high power has an important effect on the milk homogenization compared with traditional homogenization.

Homogenization is used to prevent creaming. Physical structure of milk is affected by homogenization and has several advantages like: reducing the fat globule size, resulting in no cream-line formation, whiter and attractive color, reduced oxidation of fat, rich flavor, better taste and better stability of cultured milk products [61].

7.4.1   PHENOMENON OF FAT BREAKING IN MILK DURING HOMOGENIZATION WITH ULTRASONIC TREATMENT

Milk and milk products have been consumed since the domestication of mammals [52] in which fat is a source of energy. Fat is known as the most important and reactive component in milk, which has the ability to easily react with proteins and phosphors. The fats in fresh milk are wrapped by proteins and phosphors as depicted in Figure 7.2.

The milk constituents have variation in their density and this density has to be rearranged by a particular standard process, known as homogenization. US assisted homogenization process achieves the homogeneity

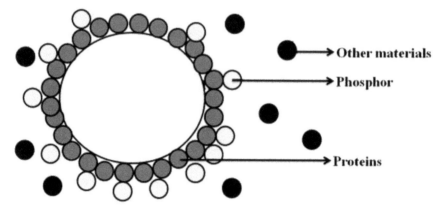

FIGURE 7.2 The structure of fat wrapped by proteins and phosphors in fresh milk: Adopted from Ketaren [31].

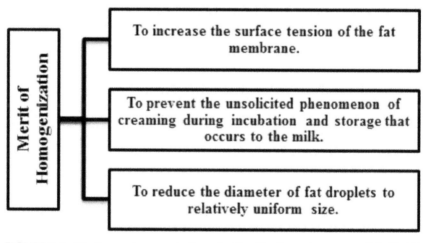

FIGURE 7.3 Merits over homogenization of milk with ultrasonic treatment [11, 64, 65].

of the milk by the modification in the particle density of the components [16], which is commonly applied in the dairy industry. This technology is considered as an important pre-treatment process due to many of its merits (Figure 7.3) in the processing of most dairy products such as yogurt, ice cream, etc. [3, 12, 23, 53].

7.5   EFFECTS OF ULTRASONIC TREATMENT

7.5.1   HOMOGENIZER FACTOR (OR MILK HOMOGENIZATION)

Homogenization factor refers to the inverse of homogenization efficiency [19, 61]. Homogenization factor of milk should be in the range of 1 to 10%. The homogenized milk is put in volumetric bottle with a capacity of 150 ml and kept in a refrigerator for 48 hr. The fat content in the upper layer of the bottle should not exceed 1/10th and the bottom of bottle should be 9/10. The fat content is determined with the Gerber method [32]. The homogenization factor can be calculated from the following equation:

$$H_I = \left[\frac{(a-b)}{a}\right] \times 100 \qquad (5)$$

where, a and b refer to the fat content of a sample of the upper and bottom layers.

Figure 7.4 illustrates that homogenization factor was significantly ($p < 0.05$) reduced with increasing time in the ultrasonic treatments and

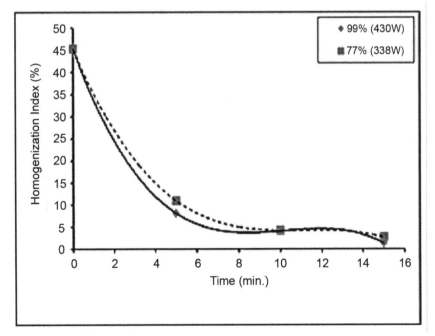

FIGURE 7.4   Homogenization factor versus time for different ultrasonic treatments: Adopted from Al-Hilphy et al. [2].

power of 430 and 338 W. Reducing homogenization factor means high homogenization efficiency. A minimum homogenization factor was 2 and 3% at 430 W and 338 W, respectively for 15 min of time duration. The following empirical equations can be used for calculating homogenization factor (%) at 338 W and 430 W:

$$H_1(430 \text{ W}) = -0.0427t^3 + 1.3053t^2 - 12.904t + 45.333 \qquad (6)$$

$$H_1(338 \text{ W}) = -0.0299t^3 + 1.0014t^2 - 11.134t + 45.333 \qquad (7)$$

where, t = time duration (in minute).

The correlation coefficient (R^2) was 0.999, for the two equations. The highest homogenization efficiency and the smallest fat globule diameter are 3.22 and 0.725 µm, respectively at 450 W power level and 10 min of time duration, respectively. The diameter of fat globules at a power level of 180 W for 10 min is similar to those of traditional homogenization [19].

7.5.2 PHYSICAL PROPERTIES OF MILK

7.5.2.1 Size

The ultrasonication (US) is mainly used for size reduction of fat globules by homogenization. Although size reduction is possible only if the size of fat globules is larger than the oscillation wavelength, and for oil in water emulsions like milk, it is 10 µm [2]. The mechanism which is said to cause disruption of fat globules is called *capillary wave's mechanism*. The size of the particles decreases with an increase in the power levels of US and also it has a significant effect on decrease in size with the increase in time. Shanmugam et al. [54] reported higher impact of 41 W power level than at the 20 W power level with the maximum reduction in size during the initial 15 min and the prolonged treatment showed gradual decrease in size reduction. The reduction in size of the particles is attributed to the shear forces generated during acoustic cavitation. High intensity US is an effective technique to control the size and shape of the whey protein particles. The accurate and proper selection of the variables can determine the mean size, polydispersity as well as the shape of the protein particles [23].

7.5.2.2   pH

The pH of milk is not much affected by the US process. Yanjun et al. [66] studied the effect of US on pH of milk after power US (PUS) pre-treatment with the probe but did not find any significant differences (p > 0.05) between the control (untreated) and US treated samples for varied time durations. Their results were similar to those of Jambrak et al. [29], who also could not observe any significant differences (p > 0.05) between the pH of control and sonicated samples at 20 kHz. However, results of some other researchers Walstra et al. [64] and Bermudez-Aguirre et al. [5] contradicted with the Yanjun et al. [66] and Jambrak et al. [29] and showed the lower values of pH of milk after the treatment. Walstra et al. [64] attributed the decrease in pH of milk to the enzymatic hydrolysis of the phosphoric esters.

7.5.2.3   Temperature

During the process of US, high intensity sound waves are passed through the solutions at frequency more 29 kHz, which leads to acoustic cavitation. During acoustic cavitation, micro bubbles that are present in the solution grow in size until a maximum critical size is reached when they violently implode generating localized temperature hot spots exceeding 5000 K and pressures of several thousand bars. At the time of cavitation, sufficient shear forces are created to break polymer chains apart [68]. Although there is an increase in temperature, which is very high, this rise in temperature is temporary and is localized to the site of cavitation and explosion.

7.5.2.4   Viscosity

Viscosity is a very important operation in dairy industry, which governs the efficiency, throughput and viability of a product and/or process. Viscosity often poses problems during ultrafiltration and spray drying of milk. US can be used to generate strong shear forces to reduce viscosity of dairy products potentially improving the efficiency of both concentration and spray drying processes. Zisu et al. [69] treated concentrated skim milk with high intensity US with low frequency though acoustic cavitation process. Batch sonication for 1 min at 40–80 W and continuous treatment delivering an applied energy density of 4–7 J mL^{-1}, reduced the viscosity of medium-heat skim milk concentrates containing 50–60% solids. Viscosity reduction of approximately 10% was achieved, which improved to >17% in highly viscous thickened material. Shear thinning behavior was also seen after sonication at shear rates below 150 s^{-1}. The US treatment could only delay the rate of thickening despite being able to lower the viscosity of concentrated skim milk up to 50% solids. But when US was used during the concentration process, the reduction in viscosity of skim milk concentrates could be achieved rapidly [69].

The US was also performed for the whey protein concentrate [69]. To achieve viscosity reductions at flow rates in excess of 300 ml/min, a 4 kW ultrasonic unit was used. Sonication treatment was found to reduce the solution viscosity at all power levels. The drop in viscosity was comparable or greater than that achieved using the 1 kW unit at the slower flow rate of 300 ml /min. The flow rate affects the time the protein system is exposed to US and also influences the sonication efficiency. Even at high amplitude of 84%, low viscosity reductions were observed at a fast flow rate of 6000 mL/min because the residence time was short and sonication efficiency was poor. The exposure of each ml of solution to energy affects the energy density parameter which has a strong correlation with the reduction in viscosity. Larger viscosity reductions can be achieved by the recirculation of the retina through the ultrasonic field at the residence time of the product increases with circulation.

7.5.2.5   Emulsion Stability Index

Emulsions are stable suspensions of immiscible liquids, where stability can be obtained by dispersion of a very fine liquid (usually in lesser quantities) called *dispersed phase*(in another liquid called *continuous phase*). An emulsion can be regarded as stable emulsion, if the droplets of the dispersed phase do not coalesce, rise or settle down over a period of time. Factors affecting the stability of an emulsion are: interfacial surface forces, size of dispersed phase (fat globules in the case of milk), viscous properties of the continuous phase and density differences between two phases. As almost all of these factors are influenced during the US of milk, the process also has a significant effect on the emulsion stability index. Sfakianakis and Tzia [52] reported that high intensity US > 20 kHz and amplitude >100 W resulted in the stability of milk as an emulsion. Cavitation is the most accepted mechanism for US based emulsification. This mechanism is based on implosion of bubbles that are produced during US which produces powered shock waves in the milk surrounding the ultrasonic probe and jets of high velocity.

7.5.2.6   Absorbance/Turbidity

US treatment was found to affect the turbidity of the milk samples, but was found to be consistent with the overall changes in particle size data of milk components soluble particles of ultra centrifuged supernatant, fat and casein in the study conducted by Shanmugam et al. [54]. The changes in turbidity of the sonicated samples were attributed to the changes in whey protein and its aggregates. Soluble particles showed greater reduction in magnitude as compared to other particles under study. The milk samples were visibly clearer in sonicated samples than the unsonicated ones with an increase in sonication time from 15–60 min and similar trends were found at power level of 41 W.

7.5.3   MILK FAT

Iswarin and Permadi [28] found that ultrasonic homogenization with high amplitude has an important effect on coconut milk homogenization. Also, it reduced the fat globule size. Diameter of fat globules can be calculated from the following equation [24]:

$$f(d) = \frac{1}{\sqrt{2\pi}\ln(\sigma)} exp\left(-\left[\frac{\ln(d)-\ln(\bar{d})}{\sqrt{2}\ln(\sigma)}\right]^2\right) \tag{8}$$

where, d is the diameter of fat globules,   is the mean value, and   is the variance.

Increasing the power to 100 W led to an increase of homogenization degree and decrease in fat globule size in cow milk [6, 7]. Table 7.4 shows the fat percentage, immediately after the treatment and 48 h after the treatment. The fat globules in homogenized milk by using ultrasonic are distributed uniformly in milk. The percentages of fat in the upper and bottom layers of un-treated milk sample were 7.4 and 6.5, respectively, but after 48 h, the percentages of fat in the upper and bottom layers of un-treated samples were 7.5 and 6.5, respectively. The results showed that the percentage of fat in the upper and bottom layers of homogenized milk samples by using ultrasonic treatments were better than un-treated samples, because of reduction in fat globule size by ultrasonic treatments. The degree of homogenization of milk was significantly ($p < 0.05$) increased with increasing power and time. Luque de Castro and Priego-Capote [36]

TABLE 7.4 Fat Percentage of Milk Immediately After the Treatment and 48 h After the Treatment [2]

Treatments	Place	No treatment	Direct after treatment	After 48 h
Fresh milk	Upper	7.4	–	7.5
	Lower	6.4	–	6.5
338 W at 5 min.	Upper	–	7.5	7.5
	Lower	–	7.3	6.5
338 W at 10 min.	Upper	–	7.4	7.5
	Lower	–	7.3	7.0
338 W at 15 min.	Upper	–	7.5	7.7
	Lower	–	7.4	7.5
430 W at 5 min.	Upper	–	7.2	7.5
	Lower	–	7.0	6.9
430 W at 10 min.	Upper	–	7.3	7.6
	Lower	–	7.1	7.3
430 W at 15 min.	Upper	–	7.1	7.3
	Lower	–	7.0	7.2

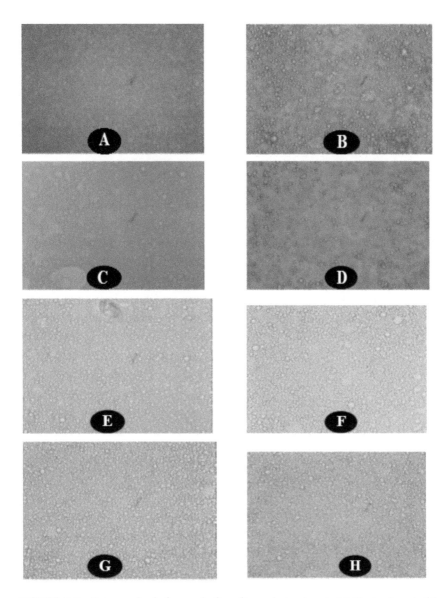

FIGURE 7.5   Fat granules before and after ultrasonic treatment: (a) No-treatment, (b) treatment at 338 W for 5 min, (c) treatment at 338 W for 10 min, (d) treatment at 338 W for 15 min, (e) treatment at 430 W for 5 min, (f) treatment at 430 W for 10 min, (g) treatment at 430 W for 15 min, (h) treatment by only homogenizer: Adopted from Al-Hilphy et al. [2].

stated that high amplitude ultrasonic treatment has a great effect on milk homogenization compared with traditional homogenization. Figure 7.5 shows that the increasing power and time produced small fat globules (Figure 7.5b–h) compared with non-homogenized milk samples (Figure 7.5a).

7.5.4. MILK'S MICROBES

7.5.4.1 Beneficial Microbes

Generally, US method is associated with damage and inhibition of microbial cells, but also shows some signs of the beneficial effects of the sonication on the microbial cells [13]. The treatment of milk with low-frequency US, significantly increases total bacterial count, coliform bacteria, staphylococci and enterococci [27].

The treatment of yoghurt with high-power US at low frequencies (about 20–24 kHz) leads to the reduction in fermentation time [39]. Few components of milk such as lactose and milk fat can reduce the effects of US on bacterial cells [10]. US treatment (20 kHz at 10 and 20 min) gave best yield of vit.-B_{12} (cobalamin) by *Propionibacterium freudenreichii* subsp. *Shermanii* after growth in milk whey [57]. Around 10% increase in the rate of fermentation (30 kHz and 2–8 W) and improvement in quality of the final product of the US treated samples of whey was observed [55].

7.5.4.2 Spoilage Microbes

US treatment can reduce the temperature required for pasteurization of milk and dairy products and improve microbial, chemical and sensory characteristics [45]. The US treatment had a pronounced effect on gram-negative bacteria such as *Pseudomonas fluorescens* and *E. coli* while the effect does not appear on gram-positive bacteria such as *Staphylococcus epidermidis* and *S. aureus* [47]. The US treatment also led to the reduction in numbers of pathogenic bacteria in milk. The viable counts of *Pseudomonas fluorescens* and *E. coli* were reduced 100% after treatment at 6 and 10 min, respectively [9]. The D value of microbial load varied significantly when used heat and US treatments as shown in Table 7.5.

TABLE 7.5   Inactivation of Microorganisms by Using Heat, US and Pressure [49]

Organism	Temperature (°C)	D value (in minute)			Manosonication/ manothermosonication
		Heat	US	Thermoso- nication	
Aspergillus flavus	55	17.40	–	5.06	–
	60	2.60		1.20	
Cronabacter sakazakii	56	0.86	–	–	0.28
Enterecoccus faecium	62	11.2	30	1.8	–
Escherichia coli K12	61	0.79	1.01	0.44	0.40 (300 kPa)
Lactobacillus acidophilus	60	70.5	–	43.3	–
Listeria innocua	63	30	–	10	–
Listeria monocytogenes	Ambient	–	4.3	–	1.5 (200 kPa) 1.0 (400 kPa)
Penicillium digitatum	50	25.42	–	9.59	–
Saccharomyces cerevisiae	60	3.53	3.1	0.73	–
Staphylococcus aureus	50.5	19.7	–	7.3	–
Yersenia entercolitica	30	–	1.52	–	0.2 (600 kPa)

7.5.5   ANTIOXIDANT ACTIVITY

The US treatment has no effect on antioxidant activity of skim milk and no effect of sulfhydryl groups (R-SH) on increased antioxidant activity of sonicated skim milk [60].

7.5.6   DIFFERENT MILK PRODUCTS

Ultrasonic wave method was used in milk and milk products leading to the change in chemical reactions and their pathway. US has proven to be an extremely useful tool in the promotion of rates of reaction in a range of interactive systems. It led to an increase of the conversion processes with great success, return to improve, changing the course of the interaction and/or begin to interact in biological and chemical electrochemical systems [62]. Sound waves were used in the food for the first time in 1960 for cleaning and characterization of certain nutrients in some foods [15].

7.5.6.1 Yogurt

Recently, US was introduced in the fermented dairy industry to improve the chemical characteristics and rheological properties of the final product, compared with thermal treatment. Yoghurt, which is a product of milk, was treated with US and showed an effect on the physical and textural characteristics (20 kHz, 50–500 W, 1–10 min) such as: increased viscosity, lower syneresis and improved water holding capacity of the final product [65].

Some studies have shown that US improves acidification characteristics of lactobacilli, thus reducing the production time while accelerating the hydrolysis of lactose, which stimulates the effect of sweetening in the milk without increasing the calorie content [56]. US was also used in soy yoghurt production, and the low frequencies of US increased the survival percentage of *Bifidobacteria* and decreased fermentation time [46].

7.5.6.2 Cheese

US was used (20 kHz frequency) in the cheese industry resulting in increased yield and improved activity of proteolytic enzymes and reduced process time [63]. US techniques have been used for the classification of defects in the cheese, on the basis of the differences in the spectrum of cheese with and without defect. The US was also used by Orlandini and Annibaldi [44] for the same subject in Parmesan cheese. The changes during the ripening of Tortadel Casar cheese treated with the ultrasonic waves (500 kHz and 1 MHz) were studied by measuring acidity, heat clot and follow-up phases of cheese ripening [30]. In this study, effects of US treatment on starter bacteria in hard cheese resulted in an increase in lactose dehydrogenase enzyme activity during the period of ripening cheese and autolysis of lactococci cells more than with lactobacilli cells [59].

7.5.6.3   Ice Cream

US techniques used in ice-cream products resulted in reducing the size of borates snow and increases the rate of heat transfer [67], thereby reducing the time of freezing process. It also yields a better quality ice cream products when used in ice cream manufacturing [42].

7.6   PROSPECTIVE FUTURE AND RESEARCH OPPORTUNITIES

Ultrasound is a novel technology that has a large importance in food processing such as milk homogenization and elimination of bacteria. Continuous large-scale apparatus can be manufactured for milk processing and thermo-sonication of milk. There are a lot of opportunities for exploitation of ultrasonication in milk homogenization because its energy consumption is lower compared with traditional methods. On the other hand, the connection between ultrasonic and oscillating magnetic treatment can be used in the future to reduce energy consumption in milk processing. Low-frequency US can be used for significant increase of probiotic bacteria in bio-production.

7.7   SUMMARY

Ultrasound is among one of the most promising new technologies that can be used in foods such as milk and dairy products. It is cheap compared to the other techniques. It can be used in many ways depending on the type of frequency, type of food and time period. High frequency US can eliminate microorganisms in milk products and maintain the chemical characteristics and rheological properties of the final product. Low-frequency US increases the number of beneficial microbes in starter cultures and probiotic bacteria.

KEYWORDS

- Acoustic pressure
- Capillary wave's mechanism
- Cavitation
- Continuous phase
- Dairy products
- Dispersed phase
- Emulsions
- Enzymatic hydrolysis
- Fat globules
- Food engineering
- Food processing
- Frequency
- Functional properties
- High frequency
- High-intensity ultrasonic treatment
- Homogenization factor
- Homogenization of milk
- Hydrostatic pressure
- Low-intensity ultrasonic treatment
- Microbial inhibition
- Nizo method
- Nizo value
- Non-thermal technology
- Physical properties of milk
- Power ultrasound
- Sonication
- Sonication efficiency
- Sonication treatment
- Sound waves
- Ultra centrifuged

- **Ultrasonic frequency**
- **Ultrasonication**
- **Ultrasound**
- **Ultrasound technology**
- **Ultrasound treatment**

REFERENCES

1. AHSMD, (2002). *American Heritage Stedman's Medical Dictionary*. Boston, MA: Houghton Mifflin.
2. Al-Hilphy, A. R. S., Niamah, A. K., & Al-Timimi, A. B. (2012). Effect of ultrasonic treatment on buffalo milk homogenization and numbers of bacteria. *World Journal of Dairy & Food Sciences*, 7(2), 185–189.
3. Ashok, K., M., Bhaskaracharya, R., Kentish, S., Lee, J., Palmer, M., & Zisu, B. (2010). The ultrasonic processing of dairy products, an overview. *Dairy Science & Technology*, 90(2–3), 147–168.
4. Bennion, M. (1980). *The Science of Food*. 1st Ed. John Wiley & Sons, New York.
5. Bermudez-Aguirre, D., Mawson, R., & Barbosa-Canovas, G. V. (2008). Microstructure of fat globules in whole milk after thermosonication treatment. *Journal of Food Science*, 73(7), 325–332.
6. Bosiljkov, T. B., Tripalo, M., Brni, M. D., Ježek, Karlovi, S., & Jagušt, I. (2011). Influence of high in density ultrasound with different probe diameter on the degree of homogenization (variance) and physical properties of cow milk. *African Journal of Biotechnology*, 10, 34–41.
7. Butz, P., & Tauscher, B. (2002). Emerging technologies: chemical aspects. *Food Research International*. 35, 279–284.
8. Cameron, M., McMaster, L. D., & Britz, T. J. (2009). Impact of ultrasound on dairy spoilage microbes and milk components. *Dairy Science & Technology*, 89, 83–98.
9. Chandrapala, J., Oliver, C., Kentish, S., & Ashok Kumar, M. (2012). Ultrasonic in food processing food quality assurance and food safety. *Trends in Food Science & Technology*, 26, 88–98.
10. Charley, H. (1982). Food Science. 2nd Ed., Macmillan, USA.
11. Chemat, F., Huma, Z., & Khan, M. K. (2011). Applications of ultrasound in food technology: Processing, preservation and extraction. *Ultrasonics Sonochemistry*, 18(4), 813–835.
12. Chisti, Y. (2003). Sonobioreactors: using ultrasound for enhanced microbial productivity. *Trends in Biotechnology*, 21(2), 89–93.
13. Chow, R. C. Y., Blindt, R. A., Chivers, R. C., & Povey, M. J. (2003). The sonocrystallization of ice sucrose solutions: Primary and secondary nucleation. *Ultrasonics*, 41(8), 595–604.

14. Cucheval, A., & Chow, R. C. Y. (2008). A study on the emulsify cation of oil by power ultrasound. *Ultrasonics Sonochemistry*, 15(5), 916–920.

15. Demirdöven, A., & Baysal, T. (2009). The use of ultrasound and combined technologies in food preservation. *Food Reviews International*, 25(1), 1–11.

16. Dhankhar, P. (2014). Homogenization Fundamentals a review. *IOSR Journal of Engineering*. 4(5), 1–8.

17. Elmehdi, H. M., Page, J. H., & Scanlon, M. G. (2003). Using ultrasound to investigate the cellular structure of bread crumb. *Journal of Cereal Science*, 38, 33–42.

18. EPADU, (2015). (http://epadu.csp.org.uk/).

19. Ertugay, M. F., Şengul, M., & Şengul, M. (2004). Effect of ultrasound treatment on milk homogenization and particle size distribution of fat. Turk. *Journal of Veterinary & Animal Sciences*, 28, 303–208.

20. Fellows, P. J. (2000). Food processing technology principles and practice. Wood Head Publishing Limited, England. pp. 563.

21. Feng, H., Barbosa-Cánovas, G. V., & Weiss, J. (2011). Ultrasound Technologies for Food and Bioprocessing, Springer New York, Dordrecht Heidelberg London.

22. Gallego-Juárez, J. A., Elvira-Segura, L., & Rodríguez-Corral, G. (2003). A power ultrasonic technology for deliquoring. *Ultrasonics*, 41, 255–259.

23. Gordon, L., & Pilosof, A. M. R. (2010). Application of high-intensity ultrasounds to control the size of whey proteins particles. *Food Biophysics*, 5(3), 203–210.

24. Gregory, J. (2006). Particles in water, Taylor & Francis Group, London.

25. Harvey, E. N., & Loomis, A. L. (1929). The destruction of luminous bacteria by high frequency sound waves. *Journal of Bacteriology*, 17, 373–376.

26. Hecht, E. (1996). Physics: Calculus. Pacific Grove, CA, Brooks/Cole, pp. 445–450, 489–521.

27. Huhtanen, C. N. (1968). Effect of low-frequency ultrasound and elevated temperatures on isolation of bacteria from raw milk. *Applied Microbiology*, 16 (3), 470–475.

28. Iswarin, S. J., & Permadi, B. (2012). Coconut milk's Fat breaking by means of ultrasound. *International Journal of Basic & Applied Sciences*, 12(1), 1–8.

29. Jambrak, A. R., Mason, T. J., Lelas, V., Herceg, Z., & Herceg, I. L. (2008). Effect of ultrasound treatment on solubility and foaming properties of whey protein suspensions. *Journal of Food Engineering*, 86, 281–287.

30. Jiménez, A., Crespo, A., Piedehierro, J., Montaña, R. M., Patricia, G. M., Paniagua, J. M., José, R. M., & Antolín, A. (2010). Preliminary study to assess ultrasonic characteristics of Tortadel Casar – Type cheese. Proceedings of the 20th International Congress on Acoustics, ICA. pp. 1–8.

31. Ketaren, S. (1986). Pengantarteknologiminyak dan Lemak Pangan. Jakarta: UI Press. Hal. pp. 3–17.

32. Kleyn, D. H., Lynch, J. M., Barbano, D. M., Bloom, M. J., & Mitchell, M. W. (2001). Determination of fat in raw and processed milks by the Gerber method: collaborative study. *Journal of AOAC International*, 84(5), 1499–1508.

33. Knorr, D., Zenker, M., Heinz, V., & Lee, D. (2004). Applications and potential of ultrasonics in food processing. *Trends in Food Science & Technology*, 15, 261–266.

34. Laborde, J. L., Bouyer, C., Caltagirone, J. P., & Gerard, A. (1998). Acoustic bubble cavitation at low frequencies. *Ultrasonics*, 36, 589–594.

35. Leighton, T. G. (1994). The Acoustic Bubble, Academic Press, San Diego.

36. Luque de Castro M. D., & Priego-Capote, F. (2007). Ultrasound assistance to liquid–liquid extraction: A debatable analytical tool. *Analytical Chimica Acta*, 583, 2–9.

37. Mason, T. J. 1998. Power ultrasound in food processing—The way forward. In: *Ultrasound in Food Processing* by Povey, M. J. W., & Mason, T. J. (Eds.). Blackie Academic and Professional, London, pp. 105–126.

38. Mason, T. J., Paniwnyk, L., & Lorimer, J. P. (1996). The use of ultrasound in food technology. *Ultrasonics Sonochemistry*, 3, S253–S260.

39. Masuzawa, N., & Odhaira, E. (2002). Attempts to shorten the time of lactic fermentation by ultrasonic irradiation. *Japanese Journal of Applied Physics*, 41, 3277–3278.

40. McClements, J. D. (1995). Advances in the application of ultrasound in food analysis and processing. *Trends in Food Science & Technology*, 6, 293–299.

41. Moholkar, V.S., Rekveld, S., & Warmoeskerken, M. M. C. G. (2000). Modeling of the acoustic pressure fields and the distribution of the cavitation phenomena in a dual frequency sonic processor. *Ultrasonics*, 38, 666–670.

42. Mortazavi, A., & Tabatabaie, F. (2008). Study of ice cream freezing process after treatment with ultrasound. *World Applied Sciences Journal*, 4(2), 188–190.

43. Muthukumaran, S., Kentish, S. E., Stevens, G. W., Ashok Kumar, M., & Mawson, R. (2007). The application of ultrasound to dairy ultrafiltration: The influence of operating conditions. *Journal of Food Engineering*, 81, 364–373.

44. Orlandini, I., & Annibaldi, S. (1983). New techniques in evaluation of the structure of Parmesan cheese. *Ultrasonication and X-Rays Science Latiero-Caseria*, 34, 20–30.

45. Patist, A., & Darren, B. (2008). Ultrasonic innovations in the food industry: From the laboratory to commercial production. *Innovative Food Science and Emerging Technologies*, 9, 147–154.

46. Phuc, N. T. M. (2011). High Intensity ultrasound aided milk Fermentation by Bifidobacteria. PhD Thesis, National University of Singapore, pp. 191.

47. Pitt, W. G., McBride, M.O., Lunceford, J. K., Roper, R. J., & Sagers, R. D. (1994). Ultrasonic enhancement of antibiotic action on gram-negative bacteria. *Antimicrobial Agents and Chemotherapy*, 38(11), 2577–2582.

48. Piyasena, P., Mohareb, E., & McKellar, R. C. (2003). Inactivation of microbes using ultrasound: a review. *International Journal of Food Microbiology*, 87, 207–216.

49. Şahin Ercan, S., & Soysal, C. (2013). Use of ultrasound in food preservation. *Natural Science*. 5, 5–13.

50. Salmin, O., Salmin, N. P. N., & Solyankin, D. A. (1997). The patent of the Russian Federation 2104636 Milk 6 A01J11/16. The way of the production of high-fat milk products and the devices of its realization, Appl. 26.04.96, publ. 21.01.

51. Sehgal, C., Steer, R. P., Sutherland, R. G., & Verrall, R. E. (1979). Sonoluminescence of argon saturated alkali metal salt solutions as a probe of acoustic cavitation. *The Journal of Chemical Physics*, 70, 2242–2248.

52. Sfakianakis, P., & Tzia, C. (2014). Conventional and innovative processing of milk for yogurt manufacture; development of texture and flavor: A Review. *Foods*, 3 (1), 176–193.

53. Sfakianakis, P., Topakas, E., & Tzia, C. (2015). Comparative study on high-intensity ultrasound and pressure milk homogenization: effect on the kinetics of yogurt fermentation process. *Food and Bioprocess Technology*, 8(3), 548–557.

54. Shanmugam, A., Chandrapala, J., & Ashokkumar, M. (2012). The effect of ultrasound on the physical and functional properties of skim milk. *Innovative Food Science and Emerging Technologies*, 16, 251–258.

55. Shershenkov, B., & Suchkova, E. (2015). Upgrading the technology of functional dairy products by means of fermentation process ultrasonic intensification. *Agronomy Research*, 13(4), 1074–1085.

56. Shimada, T., Ohdaira, E., & Masuzawa, N. 2004. Effect of ultrasonic frequency on lactic acid fermentation promotion by ultrasonic irradiation. *Japanese Journal of Applied Physics*, 43, 2831–2832.

57. Suchkova, E., Shershenkov, B., & Baranenko, D. 2014. Effect of ultrasonic treatment on metabolic activity of *Propionibacterium shermanii*, cultivated in nutrient medium based on milk whey. Agronomy Research, 12(3), 813–820.

58. Suslick, K. S. (1988). Ultrasound—Its Chemical, Physical, and Biological Effects. Suslick, K. S. (Ed.). VCH Publishers, New York, pp. 336.

59. Tabatabaei, F., & Mortazavi, S. A. (2010). Effects of Ultrasound Treatment on Viability and Autolysis of Starter Bacteria in Hard Cheese. *American-Eurasian Journal of Agricultural & Environmental Sciences*, 8(3), 301–304.

60. Taylor, M. J., & Richardson, T. (1980). Antioxidant activity of skim milk: Effect of sonication. *Journal of Dairy Science*, 63(11), 1938–1942.

61. Teknatext, A. B. (1995). Dairy processing handbook tetra pack processing systems AB. Lund, Sweden, pp. 263–278.

62. Thompson, L. H., & Doraiswamy, L. K. (1999). Sonochemistry science and engineering. *Industrial & Engineering Chemistry Research*, 38(4), 1215–1249.

63. Villamiel, M.; Hamersveld, E. H., & de Jong, P. (1999). Effect of Ultrasound Processing on the Quality of Dairy Products. *Milchwissenschaft*, 54, 69–73.

64. Walstra, P., Wouters, J. T. M., & Geurts, T. J. (2006). Dairy Science and Technology. 2nd Ed. Boca Raton, FL, CRC, Taylor and Francis.

65. Wu, H., Hulbert, G. J., & Mount, J. R. (2000). Effects of ultrasound on milk homogenization and fermentation with yogurt starter. *Innovative Food Science and Emerging Technologies*, 1(3), 211–218.

66. Yanjun, S., Jianhang, C., Shuwen, Z., Hongjuan, L., Jing, L., Lu, L., Uluko, H., Yanling, S., Wenming, C., Wupeng, G., & Jiaping, L. (2014). Effect of power ultrasound pre-treatment on the physical and functional properties of reconstituted milk protein concentrate. *Journal of Food Engineering*, 124, 11–18.

67. Zheng, L., & Sun, D. W. (2005). Ultrasonic Acceleration of Food Freezing. In: *Emerging Technologies for Food Processing* by Sun, D. W. (Ed.). London, UK, Academic Press, Elsevier, pp. 603–626.

68. Zisu, B., Bhaskaracharya, R., Kentish, S., & Ashokkumar, M. (2010). Ultrasonic processing of dairy systems in large scale reactors. *Ultrasonics Sonochemistry*, 17, 1075–1081.

69. Zisu, B., Schleyer, M., & Chandrapala, J. (2013). Application of ultrasound to reduce viscosity and control the rate of age thickening of concentrated skim milk. *International Dairy Journal*, 31, 41–43.

PART III

FOODS FOR SPECIFIC NEEDS

CHAPTER 8

NATURAL COLOR FOR FOODS: A TECHNICAL INSIGHT

NAVNEET SINGH DEORA,[1] AASTHA DESWAL,[2]
and SANJITH MADHAVAN[3]

[1]Research Scientist, Prasan Solutions (India) Pvt. Limited (Nalanda R&D Center), Cochin – 682021, Kerala. Mobile: +91-7042307007, E-mail: navneetsinghdeora@gmail.com

[2]Research Scientist, Prasan Solutions (India) Pvt. Limited (Nalanda R&D Center), Cochin – 682021, Kerala. Mobile: +91-8137892690, E-mail: deswalad@gmail.com

[3]Technical Director, Prasan Solutions (India) Pvt. Limited (Nalanda R&D Center), Cochin – 682021, Kerala. Mobile: +91-8137892690, E-mail: sanjith@prasansolutions.com

CONTENTS

8.1   INTRODUCTION

The era of natural food additives has started, some consumers deliberately choose minimally processed foods over processed ones, and when they have to choose processed food they will generally select one with fewer additives and/or containing natural additives. Although the natural additives do not always represent a benefit compared to chemical ones, in most cases they are believed to be healthier, can carry out various functions in the food, and confer added value (bioactivity, nutraceutical). Natural additives are compounds, groups of compounds, or essential oils from plants that are already used empirically by the population for taste purposes. Fungi, seaweeds, and algae are also interesting sources of natural additives. These natural compounds have been around for some time, but in recent years they have gained more interest from the food industry for direct application or in synergy with other natural or chemical additives [11]. Among the many effects, the most studied natural additive activities are their antimicrobial and antioxidant powers [1, 3, 8, 21, 26].

8.2   HISTORY OF FOOD COLOR

The addition of colorants to foods is thought to have occurred in Egyptian cities, where candy makers around 1500 BC added natural extracts and wine to improve the products appearance [25]. Up to the middle of the 19th century ingredients, such as the spice saffron, from the area local to the production units were added for decorative effect to certain foodstuffs. Following the industrial revolution both the food industry and processed food were developed rapidly. The addition of color, via mineral and metal based-compounds, was used to disguise low quality and adulterated foods. Some more lurid examples are: Red lead (Pb_3O_4) and vermillion (HgS) were routinely used to color cheese and confectionery; Copper arsenate was used to recolor used tealeaves for resale. Such coloring also caused two deaths when used to color a dessert in 1860. Toxic chemicals were used to tint certain candies and pickles. Historical records show that injuries, even deaths, resulted from tainted colorants. In 1856 the first synthetic color (Maurine), was developed by Sir William Henry Perkin and by the turn of the century, unmonitored color additives had spread through USA and Europe in all sorts of popular foods, including ketchup, mustard, jellies, and wine. Sellers at that time offered more than 80 artificial coloring agents, some intended for dyeing textiles, not foods. Many color additives had never been tested for toxicity or other adverse effects. In the beginning of 1900s, the bulk of chemically synthesized colors were derived from aniline (petroleum product) that is toxic. Originally, these were dubbed 'coal-tar' colors because the starting materials were obtained from bituminous coal. Though colors from plant, animal and mineral sources had been used in earlier times, yet the only coloring agents available remained in use early in this century and manufacturers had strong economic incentives to phase them out. Chemically synthesized colors simply were easier to produce, less expensive, and superior in coloring properties. Only tiny amounts were needed. They could be blended easily and did not impart unwanted flavors to foods. But as their use grew, so did safety concerns. This led to numerous regulations throughout the world. For example, USA reduced the permitted list of synthetic colors to seven from 700 being used. However 'adulteration' continued for many years and this, together with more recent adverse press comments on food colors and health, has continued to contribute to concerns of the consumers about color addition to foodstuffs [12, 22, 27].

8.3 CLASSIFICATION OF FOOD COLORS

During past 100 years, following the discovery of the first synthetic dye by Sir William Perkin in 1856 and the subsequent development of the dyestuffs industry, synthetic colors have been added to food. For centuries prior to this, natural products in the form of spices, berries and herbs were used to enhance the color and flavor of food. During this century, the use of synthetic color has steadily increased at the expense of natural colors, principally due to ready availability and lower relative price. In the last 20 years following the delisting of several synthetic colors, notably that of amaranth in the USA in 1976 and that of all synthetic colors by Norway also in 1976, there has been an increase in the use of colors derived from natural sources. Generally three types of organic food colors are recognized in the literature: synthetic colors, nature-identical colors and natural colors.

8.4   NATURAL COLORS

Significant developments have occurred with natural colors since their wider commercialization around 25 years ago [9, 25]. The growth in use of natural colors comes from increasing consumer pressure for natural products in light of their distrust for the food industry, based on unsubstantiated health scares related to additives in general, but especially related to hyperactivity and its perceived association with many azo dyes for instance tartrazine [5]. Color is spread widely throughout nature in fruit, vegetables, seeds and roots. In our daily diets, we consume large quantities of many pigments, especially anthocyanins, carotenoids (nature is thought to produce in excess of 100 million tons per annum of carotenoids, of which more than 600 structures have been identified) and chlorophylls. Our intake from naturally colored processed food is fairly insignificant when compared to this. Pigments from nature vary widely in their physical and chemical properties. Many are sensitive to oxidation, pH change and light and their inherent solubility varies widely [6].

There are currently 13 permitted naturally derived colors within the Europe and 26 colors exempt from certification in the USA [23]. Table 8.1 shows the color code and naturally derived colors. Other colors exempt from certification within the USA are: Ultramarine blue (limited to animal feed), toasted partially defatted cooked cottonseed flour, ferrous gluconate, dried algae meal (limited to chicken feed), carrot oil and corn endosperm oil (limited to chicken feed). These have limited use either because of an application restriction or poor stability. Natural colors were initially considered much less stable, more difficult to use and more expensive than the synthetic colors. It was always thought the color shades achievable would be less vibrant and appealing. It is estimated that worldwide up to 70% of all plants have not been investigated fully and that only 0.5% have been exhaustively studied [28] for color selection. Therefore, it can be concluded that we have only just begun our search for natural food color sources. Unfortunately, however, most pigments fall into the classes

TABLE 8.1   Natural Colors (and Colors of Natural Origin) Listed by the EU [23]

Number	Color
E100	Curcumin
E101	Riboflavin, riboflavin-5'-phosphate
E120	Cochineal, carminic acid, carmines
E140	Chlorophylls and chlorophyllins
E141	Copper complexes of chlorophylls and chlorophyll ins
E150a	Plain caramel
E153	Vegetable carbon
El60a	Mixed carotenes and J3-caroten
El60b	Annatto, bixin, norbixin
El60c	Paprika extract, capsanthin, capsorubin
El60d	Lycopene
El60e	J3-Apo-8'-carotenal (C30)
E161b	Lutein
E161g	Canthaxanthin
E162	Beetroot red, betanin
E163	Anthocyanins

mentioned above, making minor pigment classes rare. Any new pigment source would require safety assessment, which would be costly and time consuming, prior to any FDA petitioning and EU approval for use as a food colorant. The final drawback is that many undiscovered pigments will be in un-prospected land or the sea and commercialization could be an uneconomic prospect.

8.4.1 NATURAL COLOR FORMULATIONS

Food color manufacturers are able to offer a complete spectrum of natural and naturally derived colors through expertise in formulation, and are able to provide easy to use and stable forms that are suitable for use in a wide range of applications. In addition, they offer colors that are free from other additives such as Sulphur dioxide, as well as those that are acceptable to a wider range of communities and in accordance with specific dietary or ceremonial laws, e.g., kosher [11]. Food color manufacturers therefore continue to develop new technologies to meet customer needs and they are very proactive in offering technical and application support for the replacement of synthetic dyes with natural color alternatives.

Formulations can be produced using complex high pressure milling and processing, which give enhanced light-stable colors [19]. Other formulations offer excellent dispersibility and stability to heat, light and oxidation and can be used in a variety of applications. For example, dispersible emulsions have been developed for carotenes, which overcome the oxidative color fading which has previously limited their application. Micro-emulsions have been developed for clarity along with enhanced stability to heat, light and oxidation. Patented encapsulation technologies have been developed to meet the requirements of modern food processing, i.e., improved stability to light, pH and oxidation, and reduced color migration, extension of natural color shades and increased color intensity and brightness. These are available as water-dispersible forms of oil-soluble pigments [10]. Hydrocolloid complexion and cyclodextrin inclusion have all been used to promote stability and dispersibility of oil- and water-soluble color formulations [11, 15].

While the coloring of foodstuffs with natural products is usually viewed as a healthier option to synthetic dyes, the development of natural color formulations may also require the use of other food additives such as antioxidants, emulsifiers and carriers, i.e., 'additives within additives.' It is arguable therefore that the removal of a synthetic dye (E-number) from a foodstuff ingredients list is not necessarily a healthier option if it is replaced by a natural color along with one or more E-numbers to aid its application. Another issue that consumers have with natural colors is that in some cases, the natural color source is not in itself regarded as a foodstuff. Cochineal, for example, is derived from an insect.

The insolubility of some natural colors in water, moderate solubility in fats and oils and susceptibility to oxidation impede the direct use of the relatively coarse particles, which also limits their coloring ability. Processes have been described for the production of nano particulate active substance dispersions to overcome these limitations [4]. The technological requirements for these formulations are particularly high, for example carotenoid use in the coloring of aqueous media. However, the nano-particulate nature of the products is stated to realize a wide diversity of coloring properties associated with improved bioavailability. A molecular-disperse solution of a carotenoid is prepared with or without an emulsifier and/or edible oil, in a volatile, water-miscible organic solvent at elevated temperature, with the addition of an aqueous solution of a protective colloid. The hydrophilic component is then transferred into the aqueous phase leaving the hydrophobic phase of the carotenoid as a nanodisperse phase. Examples of carotenoid permitted food colorings, which can be used in this type of product are well characterized, widely available and occur in both natural or synthetic forms, e.g., β-carotene, bixin, β apo-80-carotenal, the ethyl ester of β-apo-80-carotenoic acid and lycopene.

The applications of nanotechnology in the food sector are only recently emergent, but they are predicted to grow rapidly in the coming years. According to Chaudry et al. [5], many of the world's largest food companies are reported to have been actively exploring the potential of nanotechnology for use in food or food packaging. Among many other food additive functionalities, applications in this area already span development of improved color. However, the rapid proliferation of nanotechnologies in a wide range of consumer products has also raised a number of safety, environmental, ethical, policy and regulatory issues. The interactions of nano sized materials at the molecular or physiological levels and their potential effects and impacts on consumer's health and the environment are main concerns, which arise from the lack of knowledge. The nanotechnology-derived foods are also new to consumers and it remains unclear how public perception, attitudes, choice and acceptance will impact the future of such applications in the food sector.

8.4.2 SAFETY EVALUATION

All countries need to have access to reliable risk assessment of chemicals in food, but not all have the expertise and funds available to carry out separate risk assessments on large number of chemicals. Under the auspices of the European Commission (EC), this responsibility lies with the *European Food Safety Authority* (EFSA), whose role is to assess and communicate on all risks associated with the food chain including food additives [13, 24]. The EFSA Panel on food additives and nutrient sources added to food deals with questions of safety in the use of food additives, nutrient sources and other substances deliberately added to food, excluding flavorings and enzymes. EFSA are responsible for evaluating the data in order to calculate acceptable daily intake (ADI) values for all additives, which is the amount of an additive that can be taken in daily over a lifetime without damaging health. It is expressed in relation to body weight (BW) in order to allow for different body size, such as for children of different ages.

The Joint FAO/WHO Expert Committee on Food Additives (JECFA) performs a similarly vital role in providing a reliable and independent source of expert advice in the international setting, thus contributing to the setting of standards on a global scale. To date, JECFA has evaluated more than 1500 food additives. The ADI values are then used to calculate the maximum permitted levels of additives in specific foodstuffs. In line with all food additives, food color manufacturers and the food industry have to demonstrate not only a technological case for need for the color (or a particular formulation) but it must also undergo stringent toxicity testing before consideration for inclusion on the permitted list [24]. However, the degree of safety evaluation required of a synthetic coloring materials designed for food use is currently prohibitively expensive and the less stringent testing designated for natural compounds per se has obvious economical attractions. As part of the substantial data package required for the approval of a new additive, industry must provide sufficient information on the potential uses and levels of use for their respective competent authorities. As part of the assessment of the continuing acceptability of individual additives, estimates are made of their toxicology and potential intakes, and EFSA is then asked to advise the EC whether the use of any particular additive needs to be restricted. The interpretation of the test results and formal safety assessment carried out by EFSA ensures that all the tests have been carried out in accordance with the published guidelines. It also makes sure that the results and conclusions of the studies are scientifically valid. It is not unusual for a complex case that requires new trials and additional data to take several years, which has obvious economic implications for the food industry. Moreover, the safety of all additives will be reviewed every 10 years as a matter of routine and the safety of any additive can be reviewed again in the light of new toxicological data that might have become available.

The EC law also requires additive manufacturers to demonstrate that there is a genuine need for the product. In the 2010 EFSA Management Plan Activity 2 (evaluation of products, substances and claims subject to authorization), the planned actions include 25 opinions on applications with respect to the re-evaluation of food colors. Article 12 of EC/1333/2008 states that: *"When a food additive is already included in a Community list and there is a significant change in its production methods or in the starting materials used, or there is a change in particle size, for example through nanotechnology, the food additive prepared by those new methods or materials shall be considered as a different additive and a new entry in the Community lists or a change in the specifications shall be required before it can be placed on the market."*

The most significantly aspect related to the use of nanoscale food additives may be perhaps in the re-evaluation of safety assessment. Whether or not developments in nanotechnology constitute new scientific information may be for EFSA to assess in the first instance. The current EU purity specification for TiO_2, for example, does not prescribe criteria related to particle size, which clearly is a principal issue with nanotechnology. This additive was last evaluated in 1977, but was scheduled for re-assessment in 2010.

In cases where EFSA (or JECFA) consider that the use of an additive is safe for use over the period of time required to generate and evaluate further safety data, they will assign a temporary ADI. There is also a category of ADI 'Not Specified,' which is applied to additives generally of very low toxicity, where the maximum possible dietary intake of the additive arising from its use at levels necessary to achieve the desired effect is not considered to represent a hazard to health. In some cases the ADI's allocated by EFSA and JECFA may differ. This can be because expert groups differ on judging how each toxic effect should be weighted and in deciding which no-effect levels and safety factors to apply, but often it is simply due to evaluations being carried out at different times and hence are based on different data sets. At present, all ADI's used by national and international authorities are based on the highest intake in mg/kg/day, which does not give rise to observable adverse effects.

The fact that an ADI can be developed for a substance does not, however, mean that its use in food will be automatically permitted. It is a matter for EFSA to decide firstly whether there is a demonstrable need for the additive, and secondly whether it is necessary to place restrictions on the use of an additive to ensure that dietary exposure to it remains within acceptable limits as defined by the ADI. In some cases, such restrictions may make it impractical to use the additive at all. Interestingly, and perhaps due to market forces, there is an increasing tendency to categorize additives as natural or artificial and to make assumptions about their safety accordingly. However, there is no inherent reason why chemicals present in nature or derived thereof should be safer than any others. EFSA and JECFA advise that all additives, whatever their origin, need to be examined for both need and safety-in-use.

Food colors from natural sources tend to exhibit variable composition because of the inherent variability of their source materials and their different methods of extraction [16]. Therefore, there is a requirement to continually improve specifications and to have available suitable analytical methods for natural food color additives because their consumption is both widespread and increasing. Specific dietary advice and other strategies, to ensure that consumers can maintain a safe and adequate diet in terms of additive intake, may only be established using relevant scientific knowledge.

Consequently, there is a clear need for analytical methods to support the purity of specifications, provide intake data on additives, and enforce EU Regulations. Many of the ADI's for natural and nature-identical colorings are designated as 'acceptable' due to their historical use as food ingredients. However, once they have been isolated from their source materials, many natural coloring materials are particularly susceptible to oxidation, photo-induced degradation and isomerization, and may be exposed to any number of agents, which may affect their stability. The processing of natural products may give rise to various artifacts as well as degradation products, which may be carried through to the final color formulation and thence into a foodstuff where they may be considered undesirable. Moreover, once added to a foodstuff, a coloring material may be further 'processed,' e.g., by cooking or by mixing with other foods which may affect its stability depending upon the stabilizing effects of the food matrix.

Suitable analytical methods are required in order to carry out surveillance for additives in food, especially those for which no suitable methods of analysis currently exist, and to ensure the ADIs are not exceeded, especially for young children. In order to build on the systems already being used, related research work on the fate of color additives in food must also be carried out. This is usually achieved through the development and application of analytical methods to the measurement of additives and their degradation products in foods.

8.4.3   LEGISLATION

Natural colors are widely permitted throughout the world. However, there is no universally accepted definition of this term and many countries exclude from their list of permitted colors those substances that have both a flavoring and a coloring effect. Thus spices are generally not regarded as colors. Sweden, for example, states that *"turmeric, paprika, saffron and sandalwood shall not be considered to be colors but primarily spices, providing that none of their flavoring components have been removed"* [22]. Italian legislation states that *"natural substances having a secondary coloring effect, such as paprika, turmeric, saffron and sandalwood, are not classed as colors but must be declared as ingredients in the normal way"* [20]. Similar comments are included in the food legislation of The Netherlands, Switzerland and Norway.

The European Union (EU) permits a wide range of colors, some of which are of natural origin and these are listed in Table 8.1. It is, however, important to note that lycopene is not widely available commercially and four of the colors are only available commercially as nature-identical products, namely: E101 riboflavin, riboflavin-S'-phosphate, E160e l3-apo-8'-carotenal and E161g canthaxanthin [18]. I3-Carotene is available in both natural extract and nature identical forms, the latter of which is more widely used. Council Directive 94/36IEC specifies conditions of use for permitted colors and is fully implemented effective from 20th June 1996. The directive stipulates the purposes for which individually approved colors may be used and in many cases it also limits the levels at which they may be applied.

Generally the directive is less restrictive in its treatment of natural colors many of which can be used at quantum satis, a flexibility that is not granted to any of the azo-dyes. The USA has a different set of natural colors. These do not require certification (therefore do not have FD & C numbers) and are permanently listed. One of the advantages of using natural colors is that they are generally more widely permitted in foodstuffs than synthetic colors. It should be remembered that color usage may be controlled in three distinct ways as follows:

- National legislation lists: those colors that may be used in foods.
- Color use within a country is limited by the type of food that may be colored.
- The maximum quantity of color that can be added to a food may also be specified.

Thus, for example, beetroot extract is permitted in Sweden. However, its use is limited to specified food products only, such as sugar confectionery, flour confectionery and edible ices. There is a maximum dose level limit of 20 mg/kg as betanin in the first two categories of food and 50 mg/kg in edible ices. The 20 mg/kg is equivalent to 0.4% of beetroot extract containing 0.5% betanin, which is the standard strength for such an extract. Its use, however, is not currently permitted in dry mix desserts or milk shakes. Although now that Sweden is a member of the ED its food legislation will need to change and beetroot extract will become a quantum satiscolor. It is therefore essential to consult the legislation relating to the particular food product before using any color. It is obviously not sufficient just to check the simplified list of permitted colors since this relates solely to one aspect of color regulation. Figure 8.1 summarizes the chronological development of EU legislation on food colors.

8.4.4   LABELING OF COLORS

Colors used in food products, like other additives, must be declared in the ingredients list by category name, color, plus either the name of the color or the E-number. Colors present in mixtures with other additives (e.g.,

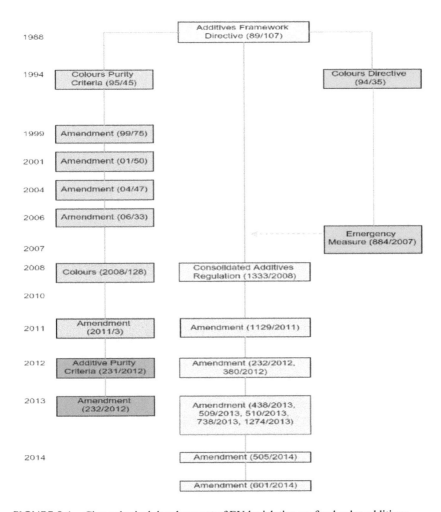

FIGURE 8.1 Chronological development of EU legislation on food color additives.

flavor) are not excluded from this and must be declared, because they will have a coloring function in the final product. Within the UK, it tends to be a mixture of the declaration possibilities, depending on the printing area available as well as the food manufacturers and retailers own policies. The consumer still has a negative opinion of the E numbers following unsubstantiated popular press articles linking colors to unwanted behavior in some people (mainly children) in the early 1980s, and because of this many foods name the color added.

8.4.5 US LEGISLATION

In the USA, the use of food colors is governed by the Code of Federal Regulations (CFR). This is divided into 50 titles and title 21 is assigned to the Food & Drug Administration. Parts 70–82 list color additives. These are divided into two categories: de-certified color additives and FD&C colors. These are synthetically produced organic molecules that have had their purity checked by the FDA. There are seven of these, which are water soluble dyes and six insoluble Lake Colorants exempt from certification [14]. These are derived from animal, vegetable, mineral origin or are synthetic duplicates of naturally existing colors. As such they contain complex mixtures of numerous components. From a regulatory point of view, there is no such thing as a *natural color*. Although, it is generally accepted that colorants exempt from certification are usually naturally derived.

8.4.6 LABELING OF COLORS IN THE USA

In terms of labeling. the terms natural color and food color are not permitted as they may indicate that the color occurs naturally, when it does not [14]. There are numerous options available: artificial color, artificial color added, and color added. These terms do not indicate any real benefit when using naturally derived colors. A preferred option is colored with x or x (color) naming the color source, e.g., Annatto. If the name of the specific color is not included, the label declaration must then say artificially colored or artificial color added.

8.4.7 NATURE IDENTICAL COLORS

Nature identical colors have been developed to match their counterparts in nature. The most common pigments that are synthesized are carotenoids consisting of conjugated hydrocarbons, and as such they are prone to oxidative attack and a subsequent loss of color. Color formulations have been developed with antioxidant systems (e.g., tocopheryl and ascorbyl palmitate), to reduce this effect.

Nature identical, β-carotene, has a large portion of the colorant market (around 17% of the global and 40% of the European market, the annual output is thought to exceed 500 tons) and was first marketed in 1954. Its principal use is in yellow fats (margarines, low fat spreads etc.), soft drinks, and confectionery and bakery products. Two main manufacturers supply NI β-carotene, Hoffman la Roche (Basel, Switzerland) and BASF (Ludwigshafen, Germany), within their vitamin portfolio. For most food and drink applications, the challenge to these suppliers is to provide oil and water dispersible forms (especially for the yellow fat and soft drink markets). This is achieved using methods such as emulsification and pigment suspension as follows [4]:

- Emulsion formulations consist of very fine β-carotene containing oil droplets dispersed throughout an aqueous phase. These can be spray dried to give water dispersible powders.
- Oil dispersible suspensions are made using micronization to form nano sized β-carotene crystal suspended in vegetable oil. Dispersing this suspension through a protective colloid polymer matrix, which coats and stabilizes the pigment, makes water dispersible beadlet forms. This suspension is then spray dried to form high strength powders.

By using these methods, water and oil dispersible nature identical colors with pigment contents from 1 to 10% have been developed. The shade achieved is dependent on the formulation and processing used and varies from golden yellow to a red/orange shade.

8.4.8   USE OF BIOTECHNOLOGY

Biotechnology can allow the efficient mass production of colorants [17]. Plant cell and tissue culture, microbial fermentation and gene manipulation have all been investigated with respect to pigment production. However, extensive safety testing of such products is required before they are given clearance as safe food additives. There is also the obstacle of research and development investment and manufacturing facilities. Plant tissues are often considered to be an effective alternative method for the production of natural pigments [2, 17]. Carotenoids, anthocyanin and betalains have already been produced in plant cell cultures. Continuous production using currently available techniques appears to be impossible because most pigments are not excreted by the cells but stored within them. To date no food grade pigments have been shown to be producible on large-scale plant cell culture processes. Thus the development seems to be worth pursuing only in the case of plants that cannot be successfully cultivated or propagated. The pigments isolated from cell cultures would also display the same instability as those isolated from naturally grown plants. Single cell algae and fungi are better options for new biotechnologically derived colorants.

8.4.9. PROBLEMS, OPPORTUNITIES, AND FUTURE PERSPECTIVES

8.4.9.1   Future Outlook

The aim of color manufacturers—whether the color is synthetic, nature identical or naturally derived—is to constantly support and train the food industry in the correct selection and application of color. The addition of color is often thought of as last on the list in the development process. Time pressures and ingredient rationalization often mean that the most suitable color is not used, which can cause problems in the future, both in manufacturing, lack of consumer appeal and potential new product failure. Color suppliers will continue to mirror the flavor industry by offering formulations and pre-blends along with a comprehensive technical advice and sample service.

8.4.9.2   Future Developments in EU Legislation

Undoubtedly at some time in the future, the colors directive will be subject to amendment, but this seems a distant prospect at present. The current preoccupation with GM foods is resulting in two potential changes affecting additives. A proposal is already under discussion, which will enforce the need for food additives developed from a new source material to be subjected to review by the Scientific Committee on Food. New sources of starting material will include GM crops and as time goes on is likely to include some colors. The extent of such an evaluation is unclear at present, as discussions are still at an early stage, but it seems eminently sensible that new source materials or new production processes (this may affect the fermentative source of b-carotene) should require a safety evaluation. Another proposal, which is expected, is that compulsory labeling of GM ingredients will be extended to food additives, including colors, if they are derived from GM sources. There is an undertaking in the Colors Directive that the European Commission should report to the European Parliament within 5 years of the adoption of the directive on changes in the colors market and levels of use and consumption.

The deadline will not be met but the impression is that the UK is more advanced in this exercise than the majority of member states. It is not yet clear whether consumption patterns will confirm that intakes of colors remain within acceptable limits. The exercise has the potential to provoke amendments to the Color Directive if any high intakes are found.

8.5   CONCLUSIONS

Significant developments in natural colors have occurred during the past 10 years and this is likely to continue in the area of stabilizing the currently permitted range of pigments by the development of the formulation and processing technology as well as the continued searches for untapped sources of permitted pigments. Developments are only likely to cease when colors such as a heat and acid stable vegetarian natural red color has been developed or alternatively a stable non-pH dependent natural blue shade. Large-scale production of food colorants in bioreactors (based mostly on cell cultures of bacteria and other microorganisms including microalgae or fungi) may open up new perspectives in food industry. Fermentation laboratories work in sterile conditions with predictable yield and without climatic influences in an eco-friendly way. However, more research is needed to further optimize the pigment composition and yield by finding the best parameters for pigment production. There is also further work to complete a full range of GM free colors to meet current consumer/retailer concerns, especially in the nature identical colorarea. The growing functional food ingredients market is likely to see natural pigments used for their health giving rather than their coloring properties. This is a very exiting area, which should be realized in the future.

The increasing demand in natural food color additives cannot currently be fully satisfied. The lifetime of commercial products has considerably decreased. For instance, it is considered that the formulations of cosmetic products are totally renewed every 2 years, whereas 50% of the processed food recipes are revised every year. For these reasons, new natural coloring agents and/or new natural sources of colorants, or new formulations, are intensively studied. Research is continuing on examining the structure and stability of natural colors from a variety of fruits, vegetables, and flowers in order to find new and possibly more stable sources of color. Food color manufacturers are active in this area in order to provide food manufacturers with a wider selection of colors with different stabilities.

8.6 SUMMARY

In the last decade, it has been observed that food industry is shifting to the use of more natural colors due to customer demand for cleaner labels and healthier products. One of the primary drives for the increased use of natural colors in food is consumer demand for more transparent communication in product labeling and recipes. Since July 2010 in the EU, all food products containing such colors have to carry a warning label indicating that the product may have an adverse effect on activity and attention in children. For this reason, the natural colors business in Europe as well as around the world is experiencing a considerable growth, resulting in an increase of products claiming "no artificial" or "natural Color." This trend is likely to continue as natural is a sustainable benefit for consumers.

In spite of this trend, the technical replacement of artificial with natural colors is not straightforward. Therefore the food industry is required to overcome problems related to process and shelf life of products formulated with natural coloring solutions. These problems arise from the fact that natural colors, and in particular coloring foods, are not as pure as artificial ones but contain other components such as proteins, sugars, etc. Thus higher dosages are required, having an impact not only on the formulation cost, but also on the chemical and/or physical properties of the food matrix.

This book chapter explores some novel, natural sources of food colors as well as the stability of colors in different food system. Recent developments of new and improved colors will be reviewed in terms of current consumer concerns such as the incorporation of more natural products into food.

KEYWORDS

- Additives
- ADI
- Emulsion
- EU
- Food color
- Formulation
- JECFA
- Market trends
- Natural color
- Nature identical
- Oil dispersion
- Oxidation
- Plant pigments
- Regulations
- Stability
- Synthetic
- UK
- US

REFERENCES

1. Ankri, S., & Mirelman, D. (1999). Antimicrobial properties of allicin from garlic. *Microbes and Infection*, 1(2), 125–129.
2. Arad, S. M., & Yaron, A. (1992). Natural pigments from red microalgae for use in foods and cosmetics. *Trends in Food Science & Technology*, 3, 92–97.
3. Bagamboula, C., Uyttendaele, M., & Debevere, J. (2003). Antimicrobial effect of spices and herbs on Shigella sonnei and Shigella flexneri. *Journal of Food Protection*, 66(4), 668–673.
4. BASF, (1997). A vitamin and colourant. *Food Ingredients and Analysis International*, 19, 28–30.
5. Bridle, P., & Timberlake, C. (1997). Anthocyanins as natural food colors—selected aspects. *Food chemistry*, 58(1), 103–109.
6. Calvo, C., & Salvador, A. (2000). Use of natural colorants in food gels. Influence of composition of gels on their color and study of their stability during storage. *Food hydrocolloids*, 14(5), 439–443.
7. Chaudhry, Q. (2008). Applications and implications of nanotechnologies for the food sector. *Food Additives and Contaminants,* 25(3), 241–258.
8. Collins, F. W. (1989). Oat phenolics: avenanthramides, novel substituted N-cinnamo-ylanthranilate alkaloids from oat grouts and hulls. *Journal of Agricultural and Food Chemistry*, 37(1), 60–66.
9. Delgado-Vargas, F., Jiménez, A., Paredes-López, O. (2000). Natural pigments: carotenoids, anthocyanins, and betalains—characteristics, biosynthesis, processing, and stability. *Critical Reviews in Food Science and Nutrition*, 40(3), 173–289.
10. Downham, A., & Collins, P. (2000). Coloring our foods in the last and next millennium. *International Journal of Food Science & Technology,* 35(1), 5–22.
11. Galaffu, N., Bortlik, K., & Michel, M. (2015). An industry perspective on natural food color stability. In: *Color Additives for Foods and Beverages*, M.J. Scotter, (Ed.), Woodhead Publishing: Oxford. p. 91–130.
12. Gandul-Rojas, B., Roca, M. A., & Gallardo-Guerrero, L. Detection of the color adulteration of green table olives with copper chlorophyllin complexes (E-141ii colorant). *LWT-Food Science and Technology*, 46(1), 311–318.
13. Hallagan, J., Allen, D., & Borzelleca, J. (1995). The safety and regulatory status of food, drug and cosmetics color additives exempt from certification. *Food and Chemical Toxicology*, 33(6), 515–528.
14. Harp, B. P., & Barrows, J. N. (2015). *US Regulation of Color Additives in Foods*. In: *Color Additives for Foods and Beverages*, M. J. Scotter, (Ed.), Woodhead Publishing: Oxford. p. 75–88.
15. Henry, B. (1996). Natural food colors. In: *Natural Food Colorants*. Springer. p. 40–79.
16. Hutchings, J. B. (2011). *Food Color and Appearance*. Springer Science & Business Media.
17. Mapari, S. A. (2005). Exploring fungal biodiversity for the production of water-soluble pigments as potential natural food colorants. *Current Opinion in Biotechnology*, 16(2), 231–238.

18. Østerlie, M., & Lerfall, J. (2005). Lycopene from tomato products added minced meat: Effect on storage quality and color. *Food Research International,* 38(8), 925–929.

19. Overseal, A. (1999). *Nutri-Seal^TM Functional Food Ingredients.* Swadlincote: Overseal Foods Ltd., p. 2–15.

20. Parmar, M., & Phutela, U. G. (2015). Biocolors: The New Generation Additives. *Int. J. Curr. Microbiol. App. Sci.,* 4(7), 688–694.

21. Pillai, P., & Ramaswamy, K. (2012). Effect of naturally occurring antimicrobials and chemical preservatives on the growth of *Aspergillus parasiticus. Journal of Food Science and Technology*, 49(2), 228–233.

22. Scotter, M. J. (2011). Emerging and persistent issues with artificial food colors: natural color additives as alternatives to synthetic colors in food and drink. *Quality Assurance and Safety of Crops & Foods*, 3(1), 28–39.

23. Scotter, M. J. (2011). Methods for the determination of European Union-permitted added natural colors in foods: a review. *Food Additives and Contaminants*, 28(5), 527–596.

24. Scotter, M. J. (2015). Overview of EU regulations and safety assessment for food colors. In: *Color Additives for Foods and Beverages*, M.J. Scotter, (Ed.), Woodhead Publishing: Oxford. pp. 61–74.

25. Solymosi, K. (2015). Food color additives of natural origin. In: *Color Additives for Foods and Beverages*, M. J. Scotter, (Ed.), Woodhead Publishing: Oxford. pp. 3–34.

26. Sreeramulu, D., & Raghunath, M. (2011). Antioxidant and Phenolic Content of Nuts, Oil Seeds, Milk and Milk Products Commonly Consumed in India. *Food & Nutrition Sciences*, 2(5).

27. Tripathi, M., Khanna, S. K., & Das, M. (2007). Surveillance on use of synthetic colors in eatables vis-a-vis prevention of food adulteration act of India. *Food Control*, 18(3), 211–219.

28. Wissgott, U., & Bortlik, K. (1996). Prospects for new natural food colorants. *Trends in Food Science & Technology*, 7(9), 298–302.

CHAPTER 9

POTENTIAL USE OF PSEUDOCEREALS: BUCKWHEAT, QUINOA AND AMARANTH

SHRUTI PANDEY[1] and B. V. SATHYENDRA RAO[2]

[1]Research Scientist, Department of Grain Science & Technology, CSIR – Central Food Technological Research Institute, Mysore, 570020, India. Mobile: +91-9449859778, E-mail: shruti@cftri.res.in

[2]Senior Principal Scientist, Department of Grain Science & Technology, CSIR – Central Food Technological Research Institute, Mysore, 570020, India. Mobile: +91-9986846780, E-mail: sathyendra@cftri.res.in

CONTENTS

9.1   INTRODUCTION

Cereal and cereal products are one of the most important staple foods. About two billion tons of cereals are produced in the world annually. The major cereals in the world are: wheat, rye, barley, oats, maize, rice, millet and sorghum. All cereals are members of the grass family. Although many plants from the family *chenopodiaceae* are used for human nutrition (e.g., spinach, beet), yet only three plants, Buckwheat, Quinoa and Amaranthus have gained importance as grains, so called pseudo-cereals, worldwide. Botanically they are assigned to the dicotyledonous, but they all produce starch – rich seeds that can be used like cereals. Pseudocereals are dicotyledonous species which are not closely related to each other or to the true cereals monocotyledonous (e.g., wheat, rice, barley, etc.). The name pseudocereal derives from their production of small grain-like seeds that resemble in function and composition those of the true cereals.

Pseudocereals are non-grasses that are used in much the same way as cereals (true cereals are grasses). Their seed can be ground into flour and otherwise used as cereals. All three pseudocereals (Buckwheat, Quinoa and Amaranthus) have advantageous nutritional properties and are very well able to increase the range of starch rich plants for human nutrition [48].

This chapter discusses potential use of buckwheat, quinoa and amaranthus as pseudocereals in our daily diet, processing as well as nutritional profile.

9.2   BUCKWHEAT

Buckwheat (*Fagopyrum esculentum*) belonging to the family Buckwheat belongs in the branch of *Angiospermatophyta* (angiosperms) in the class of *Dicotylenodopsyda* (dicotyledons). Its exact taxonomic place is Polygonales order, *Polygonaceae* family, subfamily *Polygonoideae* and *Fagopyrum* genus (Figure 9.1). Polygonaceae is a moisture loving, cool-climate, annual cereal crop [7]. It is a native of Central Asia, cultivated in China and other Eastern countries as a bread-corn. There are two well-known varieties of buckwheat mainly *Fagopyrum esculentum Moench* and *Fagopyrum tataricum* (L.) Gaertn. *Fagopyrum esculentum Moench* is widely grown in India as compared to the other variety.

It is silvery grey to brown or black in color [90]. It takes about 90–100 days to harvest the crop and can grow well in fertile land [44].

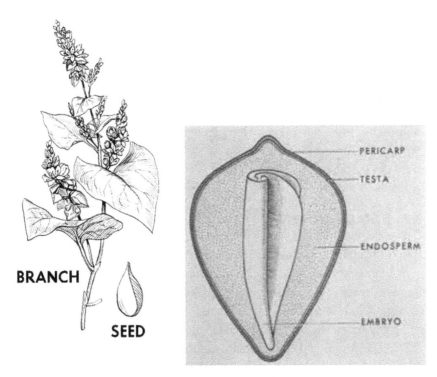

FIGURE 9.1   Buckwheat plant and structure of the seed.

The seed coat is green or tan, which darkens the buckwheat flour. The buckwheat fruit is triangular in shape and measures about 4–9 mm long [62]. The outer layer of the achene is a dark brown or black fibrous hull (pericarp). *Dehulled achenes* are called groats. Buckwheat groats bear structural resemblance to traditional cereal grains. The kernel is made up of a testa, an aleurone layer, an embryo, and a central endosperm [106]. The bran of the buckwheat is thin or thick depending on variety and varies in its adhesion to pericarp and endosperm. For, this reason, different varieties are more or less difficult to mill. The 1000-grain weight may vary from 10–20 g, depending mainly on the hull thickness. Its renewed popularity stems from its many bioactive components, which have been shown to provide various health benefits much sought after in natural foods.

9.2.1 NUTRITIONAL COMPOSITION OF BUCKWHEAT SEED

It was found that the whole buckwheat and buckwheat groats were high in retrograded starch, protein, dietary fiber, soluble carbohydrate, fagopyritols and significant levels of zinc, copper and manganese. It is also a good source of thiamine and rutin. Essential amino acids comprised 36.75% of total protein, lysine, histidine, valine; and leucine was higher and methionine, tryptophan and phenyl alanine were lower in groat than in whole buckwheat proteins [41].

9.2.2 STARCH IN SEEDS

Starch is the main component in buckwheat seed. Starch granules are 4–15 µm in diameter [91]. Physicochemical properties of buckwheat starch were investigated [40]. The results revealed that the average starch granules were 4.3–11.4 µm in size (average 7.8 µm). The starch had water-binding capacity value of 103.7%, blue value of 0.35 and amylase content of 25%. The initial and final gelatinization temperatures were 61–65°C, respectively. Amylograph data showed that the starch had an initial pasting temperature of 64°C. Regular steaming of buckwheat to produce infant and dietetic foods decreases amounts of reducing sugars because of their participation in browning reactions [36]. Very strong steaming however may actually increase total-reducing sugars, presumably because of starch hydrolysis. Similarly, starch hydrolysis may increase maltose formation. It was also found that hydrothermal treatment of buckwheat lowered starch content due to partial dextrinization. The treatment affected some of the physicochemical properties of the starch: Hygroscopicity increased; and viscosity, amylase content, and molecular weight decreased. Partial pyro dextrinization and gelation enhanced susceptibility to enzymatic hydrolysis, appearance of the starch changed and a thick glossy mass formed on drying [45].

9.2.3 NON-STARCH POLYSACCHARIDE

The presence of water-soluble polysaccharide in buckwheat endosperm has been reported [27]. A study revealed the isolation of component (polysaccharide A_1) and its structural elucidation [5]. The molecular weight was 240,000–260,000. Polysaccharide A_1 consisted of xylose, mannose, galactose and glucuronic acid. The results suggested that the main chain of this polysaccharide consisted of glucuronic acid, mannose and galactose and the former two occupied branching positions with xylose and galactose residues as non-reducing ends.

9.2.4   PROTEINS IN BUCKWHEAT

The protein content in common buckwheat varies from 7 to 21%, depending on the cultivar and environmental factors during growth. Most currently grown cultivars yield seeds with 11–15% protein on a whole seed basis. The principal protein of buckwheat is globulin [91]. Based on the feeding experiments, it was postulated that the proteins in buckwheat are the best-known source of high biological value proteins in the plant kingdom, having 92.3% of the value of non-fat milk solids and 81.4% of whole egg solids. The proteins of buckwheat have excellent supplementary value to the cereal grains [89]. The estimated value of protein efficiency ratio of buckwheat proteins is 1.8 compared with 1.2 for normal maize, 2.3 for opaque-2 maize, 1.7 for polished rice, 1.9 for oats, 1.5 for wheat and 2.5 for wheat germ.

9.2.5   AMINO ACIDS IN BUCKWHEAT

Amino acid composition of whole buckwheat revealed a higher amount of albumins and globulins and lower storage proteins than the whole grain proteins. In addition, the concentration of essential amino acids in the buckwheat aleurone granules was higher [85]. The proteins were rich in lysine (6.1%) and contained less glutamic acid and proline and more arginine and aspartic acid than cereal proteins. The 56% of glutamic and aspartic acids were in the form of amides [92].

9.2.6   LIPIDS IN BUCKWHEAT

Seeds of common buckwheat contain 1.5–3.7% of total lipids. The highest concentration is in the embryo at 7–14% and the lowest is in the hull at

0.4–0.9%. Among these, 81–85% are neutral lipids, 8–11% are phospho-lipids and 3–55% are glycolipids. Buckwheat oil contains 16–20% satu-rated fatty acids, 30–45% oleic acid and 31–41% linoleic acid [13]. It was reported that palmitic oleic, linoleic and linolenic acids account for about 95% of the buckwheat fatty acids. The 11 fatty acids have been detected in the whole seeds of buckwheat [12]. It has been reported that hydrother-mal treatment of buckwheat grain decreased free and bound lipids and increased firmly bound lipids [11].

9.2.7   POLYPHENOLS

It is well known that grains of buckwheat can be stored for a long time due to the antioxidants present in them [38]. The concentration of natural anti-oxidants may show strong variation depending on several factors including variety, location and environmental conditions. The three major classes of phenolics are flavonoids, phenolic acids and condensed Tannins. Three of the numerous classes of flavonoids are found in buckwheat: flavonols, anthocyanins and C-glucosyl-flavones. Rutin (quercetin-3-rutinoside), a well-known flavonol diglucoside used as a drug for treatment of vascular disorders, occurs in the leaves, stems, flowers and fruit of buckwheat. The rutin content in buckwheat varieties may range from 12.6–35.9 mg/100 g dry weight basis [42].

According to few researchers, location influences the rutin concentra-tion in the seed while the growing season has a significant influence on the total flavonoid in the hulls. Other reported flavonols are quercitin (quer-cetin 3-rhamnoside) and hyperin (quercetin 3-galactoside). At least three red pigments have been found in the hypocotyls of buckwheat seedlings. One of these is cyanidin, the other two are presumed to be glycosides of cyanidin. The C-glycosylflavones present in buckwheat seedling cotyle-dons are vitexin, isovitexin, orientin and isoorientin. The phenolic acids of buckwheat seed are the hydro benzoic acids, synigic, *p-hydroxy-benzoic, vanillic and p-coumaric acids*. Soluble oligomeric condensed tannins are present in common buckwheat seeds, which, along with the phenolic acids, provide astrigency and affect color and nutritive value of buckwheat products.

9.2.8 PROCESSING AND USES OF BUCKWHEAT

Hulling the buckwheat is an important task as the structure of buckwheat grain is three edged, with rounded pyramidal form. The hull is thick or thin, depending on the variety, and covers the bran tightly or loosely and for this reason the different varieties can be hulled more or less easily [37]. Under the hull is a silver color membrane, the germ is on the lower rounded part of the grain. It can be hulled with the natural humidity content of the grain (dry hulling) or after hydrothermal pre-treatment of the buckwheat.

During the dry procedure the sand, weeds and immature grains and foreign materials are removed, then the buckwheat is classified according to the size, and is hulled at its natural humidity content. The buckwheat fractions which are classified according to size are separately hulled with the adjustment of the aperture of the hulling machine. Various special hulling machines can be used for buckwheat hulling, machines with beater blades, stone segmented, impact style, or a machine with rotating hulling stone, where the lower hulling stone is usually covered with cork or gum. The hulled and classified main product is placed again on the riddle where flour fraction is separated from the whole grains. The hulls are used as packing material or as fuel. The bran is used for swine feeding. However this process results in losses. A large proportion of the soft and fragile inner part of the grain breaks during hulling [47].

The other method hydrothermal treatment before hulling is carried out in two steps: the humidity content of the cleaned grain material is increased by adding of water or by steaming to 22% of water content of the dry weight. The product is then heat treated typically to 150–164°C for 10–20 min [24]. The heat treatment can be made by roasting, steaming or a combination of the two methods. During cooling and drying, as a result of the high tension generated in the hull, the hull splits and can be removed easily. After cooling and cold conditioning, the grains are separated by sieving into four fractions of various sizes, and then hulled. The dehulled groats are sized and the hulls and flour separated. The whole groat may be cut, sieved and the hulls still adhering removed. The dehulling results in browning of the seeds. The dehulled groats are then milled coarsely using corrugated or smooth rollers. The flour produces can be used for bakery and confectionery products. It is also used for human consumption, as a vegetable crop and honey crop.

9.3 QUINOA

Quinoa is a broad leaf plant used like the cereals. Quinoa is a native of the Andes and dates back more than 5000 years. It was called *"the mother grain"* by the Incas and considered sacred. Quinoa is a nutrient power-house packed with wholesome protein, essential fatty acids, minerals and vitamins and therefore is recognized as a super food. Quinoa belongs to the class Dicotyledoneae, family Chenopodiaceae, genus Chenopodium, and species quinoa. The species *Chenopodium quinoa* includes both domesticated and weedy forms [108, 107]. The genus *Chenopodium* includes about 250 species [17].

Quinoa is an annual herbaceous plant found in the Andean region of South America, grown at altitudes ranging from 0 to 4,000 m above sea level. However, it is known that the best production in the region is achieved between altitudes 2,500 and 3,800 m with 250–500 mm of annual precipitation and 5–14°C of mean annual temperature. According to the agro-ecological characteristics of their area of origin as well as morphological and physiological features [94, 95], recognized five major quinoa groups are: (1) valley quinoa, (2) plateau quinoa, (3) salt flat quinoa, (4) sea-level quinoa and (5) subtropical quinoa. Four genotypic groups of quinoa have been identified on the basis of their genotypic by environment interaction: G1, inter-Andean valley varieties; G2, Peruvian Plateau varieties; G3, Bolivian Plateau varieties and G4, sea-level central Chilean varieties [14]. Genetic variability of quinoa is also expressed by the diversity of color of the stems, inflorescences and grains; shape and size of the inflorescences; proteins content; saponin content and the presence of oxalate crystals in their leaves. The analysis of genetic markers (RAPD) of four populations of Bolivian cultivated quinoa also demonstrated both intra- and inter-population variations [71, 72]. The main producing countries are Bolivia, Peru, and Ecuador, it is also cultivated in USA, China, Europe, Canada, and India. Production of Quinoa has increased in the last 20 years, especially in Bolivia.

Quinoa plants do best in sandy, well-drained soils with a low nutrient content, moderate salinity, and a soil pH of 6 to 8.5. The seedbed must be well prepared and drained to avoid waterlogging. In the Andes, the seeds are normally broadcast over the land and raked into the soil. Sometimes it is sown in containers of soil and transplanted later.

This gluten free pseudocereal is rich in lysine, the essential amino acid, making it a more complete protein. Quinoa contains about 1.8% phospholipids. It is also rich in iron and magnesium and provides fiber, vitamin E, copper and phosphorus, as well as some B vitamins, potassium, and zinc. Several health-imparting properties as antioxidant, anti-obesity, hypolipemic and anti-diabetic have been validated. Quinoa has been effectively incorporated in a wide range of food products. Possibility of producing highly nutritious grains even under ecologically extreme conditions provides immense promises for tackling looming food insecurity.

9.3.1 STRUCTURE AND COMPOSITION

Quinoa is a gynomonoecious plant with an erect stem, and bears alternate leaves that are variously colored due to the presence of betacyanins. The leaves exhibit polymorphism; the upper leaves being lanceolate while the lower leaves are rhomboidal [32]. A well-developed, highly ramified taproot system is present [29], penetrating as deep as 1.5 m below the surface protecting the plant against drought conditions. The plants reach a height of 0.20 to 3 m depending on genotype, environmental conditions and soil fertility.

The inflorescence is a panicle, 15–70 cm in length and rising from the top of the plant and in the axils of lower leaves. It has a principal axis from which secondary axis arises and is of two types: *amaranthiform* and *glomerulate*. An important feature of quinoa is the presence of hermaphrodite as well as unisexual female flowers [32, 83]. The fruit is an achene, comprising several layers, viz. perigonium, pericarp and episperm [75], from outwards to inside, and may be conical, cylindrical or ellipsoidal. Fruits, dry, one-seeded and derived from an upper ovary are interpreted as achenes. The thin pericarp which covers the seed makes the fruit a utricle (Figure 9.2).

FIGURE 9.2 Medial longitudinal section of quinoa seed (Top: left & right figures) showing the pericarp (PE), seed coat (SC), hypocotyl-radical axis (H), cotelydons (C), endosperm (EN) (in the micropylar region only), radicle (R), funicle (F), shoot appendix (SA) and perisperm (P), [67].

Fruits are discoidal to lenticular, range from 1.6 to 2.3 (less frequently up to 2.7) mm in diameter and have a central starchy perisperm (more or less vitreous or floury) and peripheral embryo. The alveolate pericarp adheres to the seed. Seed shape is discoidal to lenticular with truncate edge; its diameter ranges from 1.5 to 2.2 (less frequently 2.6) mm. Seed tegument is thin, alveolate to smooth [61]. The different colors of the perigonium (green, red, and yellow), pericarp (white-yellow or more or less orange, pink, red or black) and episperm (white-translucent, light brown or black) are responsible of the various colors that can be present in the quinoa inflorescence. Perisperm is rich in starch while endosperm and embryo tissues are rich in lipid bodies, protein bodies with globoid crystals of phytin and proplastids [4, 67]. In quinoa seeds, like amaranthus, the embryo or germ is campylotropous and surrounds perisperm like a ring and together with the seed coat represent the bran fraction. This also contains most of the ash, fiber, and saponins [56, 102].

Saponins in quinoa are the principle antinutritional factors concentrated in the external layers of the seed coat. Saponins are basically glycosidic triterpenoids with glucose constituting about 80% of the weight. They are bitter in taste, form foam in water solution and have toxic effects (haemolytic power). These substances protect the grain against birds, rodents and insects during the maturation stage and storage period [46]. The quantity of saponins is highly variable among different quinoa varieties and in accordance with the saponin concentration, quinoa varieties are classified as 'sweet quinoa' containing <0.11% of saponins and 'bitter quinoa' containing >0.11% saponins. Saponin content is affected by soil-water deficit, high water deficit lowering the saponin content [86]. Saponin content also differs in different growing stages, low saponin is found in the branching stage and high in the blooming stage. In any case, the saponins have to be removed before human consumption.

9.3.2 MILLING OF QUINOA

Yields are maximized when 170–200 kg (370–440 lb) of N/hectare is available. The addition of phosphorus does not improve yield. In eastern North America, it is susceptible to a leaf miner that may reduce crop success and which also affects the common weed and close relative Chenopodium album, but C. album is much more resistant. Quinoa grain is usually harvested by hand and rarely by machine, because the extreme variability of the maturity period of most Quinoa cultivars complicates mechanization. Harvest needs to be precisely timed to avoid high seed losses from shattering, and different panicles on the same plant mature at different times. The seed yield (often around 3 t/ha up to 5 t/ha) is comparable to wheat yields in the Andean areas. In the United States, varieties have been selected for uniformity of maturity and are mechanically harvested using conventional small grain combines. The plants are allowed to stand until they are dry and the grain has reached moisture content below 10%. Handling involves threshing the seed heads and winnowing the seed to remove the husk. Before storage, the seeds need to be dried in order to avoid germination. Dry seeds can be stored raw until washed or mechanically processed to remove the pericarp to eliminate the bitter layer containing saponins.

Saponins are traditionally removed from quinoa by washing the grains in running water. Saponin removal can be performed as a wet (humid) or dry process or preferentially in combination [34, 58]. The wet method is traditionally used by the peasants, which entails successive washing of the grains using friction by hand or a stone to eliminate the episperm. In one of the traditional dry milling methods, a perforated stone of about 50 cm in diameter is used, into which quinoa, preheated in a thick sand layer is placed. Quinoa grain and sand are then rubbed with the feed [96]. A method of soaking the grains for 30 min and agitation at 70°C was developed for effective reduction of saponins. One of the dry milling methods includes brushing of the preheated seeds. Mechanical dry dehulling involves "pearling" the grain to remove the pericarp as bran. High saponin cultivars require more abrasion than low saponin types. On commercial scale, saponins are removed by abrasive dehulling [74], but in this method, some saponin remains attached to the perisperm [10]. Studying the influence of pearling process on phenolic and saponin content [3] have shown that at 20% pearling degree saponing content (bitterness) was perceptible and at 30% pearling, the grains could be considered 'sweet.'

Thus dry milling methods remove about 30 to 40% of the seed as bran fraction. Moisture tempering the grain from 8 to 16% did not improve milling yield or mill fraction composition. The mill fraction typically contained less than 0.3 mg/g of saponin and residual bran was left on the grain. Hot water washing of grains has been attempted for extraction of saponins. Lye washing and mechanical abrasion has also been attempted. Dry milling method reduces the vitamin and mineral content to some extent, the loss being significant in case of potassium, iron and manganese [78], free and bound phenolic acids [3]. Use of slightly alkaline water rather than neutral water has been recommended to debitter quinoa [114]. Extrusion and roasting are also processes that could potentially reduce bitterness. The extremely high loss of grain via mechanical milling and need to use excess water for washing are areas of concern which needs more research work.

9.3.3 NUTRITIONAL ASPECTS OF QUINOA

Quinoa is known for its high protein content and quality, i.e., a balanced amino acid spectrum with high contents of lysine and methionine. Quinoa, along with Amaranth and buckwheat is recommended for celiac disease patient diets since these are gluten free. In addition, these are also recommended as base ingredients for baby food recipes as an alternative to rice due to low allergenicity. Proximate compositions and varietal variations of quinoa are available from Bio Food Comp 2.1 [25] and USDA nutrient database [100]. Assessing the nutritional composition data available, there is still a need for high-quality analytical data on nutritional composition of quinoa [103].

Total protein content of Quinoa (16.3%, d.b.) is higher compared to rice and cereal grains [57, 99]. Essential amino acid balance is excellent because of a wider amino acid range than in cereals and legumes [76], with higher lysine (5.1–6.4%) and methionine (0.4–1%) contents [65]. Quinoa proteins have higher histidine content than barley, soy, or wheat proteins. Albumins and globulins represent the main storage proteins in quinoa. Quinoa is considered to be a gluten-free grain because it contains very little or no prolamin [101]. Individual seed storage proteins in quinoa have been specifically characterized [19] by isolating and characterizing the 11S seed storage protein, chenopodin, an 11S hexamer having a structure similar to glycinin [8]. Chenopodin has a high content of glutamine (glutamic acid, asparagines), aspartic acid, arginine, serine, leucine, and glycine. According to the FAO reference protein [24], chenopodin meets the requirements for leucine, isoleucine, and phenylalanine, β-tyrosine. The other major protein (35% of total protein) is a 2S-type protein with a molecular mass of 8–9 kDa. The amino acid composition of this protein showed that it is high in cysteine, arginine, and histidine. Quinoa globulin is made of monomers each of which consists of a basic and an acidic polypeptide, with molecular mass of 20–25 and 30–40 kDa, respectively, linked by disulfide bonds [19, 1]. It is reported that the protein efficiency ratio (PER) and the protein quality of cooked quinoa was like that of casein [52]. Studies have indicated that PDCAAS values of a few varieties of quinoa was 1, 0.9 for buckwheat, 0.8 for amaranth and the not more than 0.4 for rice [57].

Quinoa has oil content (7% dry basis) higher than corn (4.9% dry basis) and lowers than soy (20.9% dry basis) [43, 99]. Because of the fatty acid profile similar to that of corn and soybean oils, quinoa has been considered as an alternative oil seed. Quinoa lipids contain high amounts of neutral lipids [69]. Triglycerides are the major fraction present, accounting for over 50% of the neutral lipids. Diglycerides are present in whole seeds contribute to 20% of the neutral lipid fraction. The most abundant of the total polar lipids are Lysophosphatidyl ethanolamine (45%) and phosphatidyl choline (12%). Fatty acid composition of quinoa lipids is also characterized as follows: total saturated 19–12.3%, mainly palmitic acid; total monounsaturated 25–28.7%, mainly oleic acid, and total polyunsaturated 58.3%—chiefly linoleic acid (about 90%) [55, 63, 73, 79, 99, 110]. This, however, is found to be similar to that reported for other cereal grains. Quinoa shows a high content of linolenic acid (3.8 to 8.3%), which is related to a reduction of biological markers associated with many degenerative diseases such as cancer, osteoporosis, and cardiovascular disease, inflammatory and autoimmune diseases. Oil fraction of quinoa has a high degree of unsaturation: unsaturated/saturated ratio (4.9–6.2). Despite higher fat content, higher degree of unsaturation, quinoa lipids are generally stable because of vitamin E (α-tocopherol), 0.59–2.6 mg/100 g in the seeds which acts as a natural defense against lipid oxidation [117]. About 33.9–58.4 mg/100 g of squalene is found in the lipid fraction of quinoa [79, 35]. The levels of phytosterols from quinoa were β-sitosterol 63.7

mg/100 g, campesterol 15.6 mg/100 g, and stigma sterols 3.2 mg/100 g, levels higher than in pumpkin seeds, barley, and maize [61].

Carbohydrate content of Quinoa varies from 61 to 73% in different varieties. Starch is the major component of quinoa carbohydrates, and is present between 32% and 69.2% [99]. Total dietary fiber of quinoa is close to that of cereals (7–9.7% db), and the soluble fiber content is reported between 1.3% and 6.1% (db). There are about 3% of simple sugars [73]. The amylose content of quinoa starch varies between 3% and 20% [33, 49, 66, 70, 93, 105]. The amylose fraction of quinoa starch is low. The number-average degree of polymerization of quinoa amylose (900) is lower than that of barley (1,700) [93]. Amylose has an average of 11.6 chains per molecule. Amylopectin content in quinoa starch is 77.5% [97]. The amylopectin fraction is high and has a unique chain length distribution as a waxy amylopectin, with 6700 glucan units for the amylopectin fraction of quinoa starch [101]. Quinoa glucans were classified as amylopectin-type short-chain branched glucan [99].

Granule size of quinoa starch is reported to be 1–2 μm in diameter, smaller than that of rice or barley. X-ray diffraction studies indicate that relative crystallinity of quinoa starch can be described as higher than that of normal barley starch, lower than that of amaranth starch, and similar to that of waxy barley starch [70, 93]. Gelatinization onset and peak temperatures of quinoa starches ranged from 44.6 to 53.7°C and from 50.5 to 61.7°C, respectively, and the gelatinization enthalpies from 12.8 to 15 J/g of dry starch [49]. Rapid Visco Analysis (RVA) shows the normal pasting feature of cereal and root starches [70].

Quinoa carbohydrates have beneficial hypoglycemic effects and induce lowering of free fatty acids. Studies made in individuals with celiac disease showed that glycemic index of quinoa was slightly lower than that of gluten-free pasta and bread. Quinoa induced lower free fatty acid levels than gluten-free pasta and significantly lower triglyceride concentrations compared to gluten-free bread [15]. In vitro digestibility (amylase) of raw quinoa starch was reported at 22%, while that of autoclaved, cooked, and drum-dried samples was 32%, 45%, and 73%, respectively [77]. Since granule size is small, quinoa starch can be used to produce a creamy, smooth texture that exhibits properties similar to fats [50]. Quinoa starch also has excellent stability under freezing and retrogradation processes [2].

Quinoa is similar to that of cereals containing significant content of thiamin (0.29–0.36% mg/100 g), riboflavin (0.30–0.32% mg/100 g), niacin (1.24 -1.52 mg/100 g), vitamin B6 (0.487 mg/100 g) and total folate (0.18 mg/100 g). The riboflavin content in 100 g contributes 80% of the daily needs of children and 40% of those of adults [60]. The niacin content does not cover the daily needs, but is beneficial in the diet. Thiamin values in quinoa are lower than those in oat or barley, but those of niacin, riboflavin, vitamin B6, and total folate are higher [99, 101].

Quinoa is also rich in micronutrients such as minerals and vitamins. The main Minerals are potassium, phosphorus, and magnesium. According to the National Academy of Sciences [60], the magnesium, manganese, copper, and iron present in 100 g of QS cover the daily needs of infants and adults, while the phosphorus and zinc content in 100 g is sufficient for children, but covers 40–60% of the daily needs of adults. The potassium content can contribute between 18% and 22% of infant and adult requirements, while the calcium content can contribute 10% of the requirements. However, the mineral content of QS is higher than that of cereals like oat (except phosphorus) or barley, especially that of potassium, magnesium, and calcium but are affected by the pearling process.

Six flavonol glycosides which can serve as a good source of free radical scavenging agents have been isolated [113]. Tannin content for quinoa is reported as 0.051% db, a value comparable to that of amaranth. The reported contents (db) were 251.5 mg/g of ferulic acid, 0.8 mg/g of p-coumaric acid, and 6.31 mg/g of caffeic acid showing higher antioxidant activity compared to grains like rice and buckwheat [31].

Saponins possess a broad variety of biological effects: analgesic, anti-inflammatory, antimicrobial, antioxidant, antiviral, and cytotoxic activity, effect on the absorption of minerals and vitamins and on animal growth, hemolytic and immune stimulatory effects, increased permeability of the intestinal mucosa neuroprotective action, and reduction of fat absorption. Five quinoa saponins (glycosides of oleanolic acid and hedera genin) have shown some antifungal activity on Candida albicans [83] and higher antifungal activity against Botrytis cinerea with alkali-treated quinoa saponin also has been established [109]. Saponins also have commercial–industrial importance as they are used in the preparation of soaps, detergents, and shampoos. However, the biological properties of quinoa saponins require further study.

9.3.4   USES

Quinoa is an excellent example of a functional food, with superior nutritional and nutraceutical values that can lower the risk of various diseases. Functional properties are provided by the unique proteins, essential amino acids, fatty acids, vitamins, minerals, phytohormones and antioxidants. Functional properties like solubility, water-holding capacity, gelation, emulsification and foaming allows its diversified uses. Lower granular size, freeze stability, etc., of the quinoa starch provides functional properties that can be exploited with novel uses.

However, quinoa can be consumed as a rice replacement, as a hot breakfast cereal, can be boiled in water to make infant cereal food. Seeds can be ground and used as flour, or sprouted. Seeds can also be popped. Flour can be mixed with/ substituted for flours of other cereals to make different kinds of products. There are several developments with quinoa flour at a smaller scale, like bread, cookies, muffins, pasta, snacks, drinks, flakes, breakfast cereals, baby foods, beer, diet supplements, and extrudates. Solid state fermentation of quinoa with *Rhizopus Ologosporus* Saito has provided a good quality tempeh [19]. Quinoa provides an opportunity to develop gluten-free cereal-based products [28]. In view of the nutritional and nutraceutical advantages offered it is important to increase and promote production and consumption of quinoa, develop new functional products that can be available on the market for the ordinary user, and scale them up to industrial level.

9.4   AMARANTH

The word amaranth in Greek means everlasting or non-wilting [59]. It belongs to the Class—dicotyledoneae, Sub class—carophyllidae, Order—caryophyllales, Family—Amaranthaceae, Genus—Amaranthus, Species—at least 60 species, e.g., *A. caudatus, A. cruentus* (Figure 9.3). It grows on rich soil as well on poor soils. It is quite drought resistant and grows from sea level to 3500 m.

Amaranth is an annual, broad-leaved plant. The height of the plant varies between 0.5–3 m in height. The flowers, which attain 90 cm length, consist of many little clusters of flower, which are either totally or partially

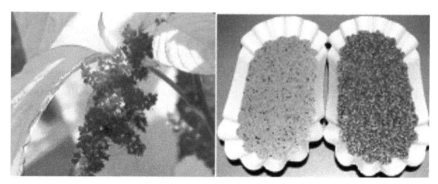

FIGURE 9.3   Amaranth plant and grains.

erect, or hang down. The lentil shaped seeds measure about 1.0 mm in diameter. One thousand grain weighs 0.5–1.0 g. A proportionally large embryo surrounds the starch rich tissue (perisperm) in the form of a ring and lies curved in the inside of the seed coat. The seed color varies from milky white to yellow, golden, red, brown and black [80]. The diameter of the seed is 1 mm, and is light in weight. The embryo encloses the starch rich perisperm like a ring. The embryo is rather large and accounts for 25% of the grain weight. The seed coat is smooth and thin [26].

9.4.1   NUTRITIONAL COMPOSITION OF AMARANTH SEED

Amaranth is characterized by excellent nutritional composition. The content of the nutrients depends on the species, variety, and method of cultivation [30]. The mineral content is twice as high as in cereals. Bran and germ have high content of ash than the perisperm. The 66% of total minerals are found in the bran and germ fractions. They are particularly high in the contents of calcium, magnesium, iron, potassium and zinc. As for vitamins it is a good source of vitamin E, B_2 and C. The insoluble dietary fiber is higher in bran fraction ranging from 19.5–27.9%.

9.4.2   STARCH IN AMARANTH GRAIN

Starch is located in the cells of the perisperm in the form of very small starch granules. The size of the starch granule is around 1 μm with a range of 0.5–1.5 μm [72]. Their shape is angular and polygonal and of unusually uniform size [87]. In the cotyledons, starch granules are assembled in great agglomerates consisting of several thousand single granules [104], with a size of 80 μm. If the protein content of isolated starch is high, the granules can also be cemented together and formed into popcorn ball like structure of a unique spherical shape [112]. X-ray diffraction measurements of amaranth show atypical diffraction pattern of cereal starches [64, 71]. The content of amylose is low in the starch, however the values vary from 0.1–10% even within the same species. The gelatinization and viscosity behavior of the amaranth starch differ considerably from most cereal starches and likely comparable with maize starch. It shows higher GT and viscosities, which even increase during cooling period [51, 64, 81, 98, 111]. Furthermore the low amylase content is responsible for a high water binding capacity, high swelling power, high enzyme susceptibility, and good freeze- thaw and retrogradation stability [78, 84].

9.4.3 CARBOHYDRATES IN AMARANTH SEED

The content of carbohydrate is lower as compared to wheat. Only small amounts of mono- and di- saccharides can be found in amaranth seed in the range of 3–5% [20]. The main components of these low molecular weight carbohydrates are sucrose, raffinose, stachyose and maltose.

9.4.4 PROTEINS IN AMARANTH SEED

The protein content of amaranth lies between 11.7–18.4%. The Biological value is 75. Of the proteins, 65% are located in the germ and seed coat, and 35% in the starch-rich perisperm [80]. Amaranth does not contain gluten and is therefore suitable for diets of persons with celiac disease. The amaranth protein consists of 405 albumins, about 20% globulins, 2–3% prolamins, and 25–30% glutelins [82]. In amaranth 7S and 12S storage globulins are found. The globulins present possess important emulsifying properties [53, 54].

9.4.5 AMINO ACID COMPOSITION

The amino acid composition reveals a high content of lysine, histidine and arginine. Leucine and threonine are the limiting amino acid in amaranth. It shows a high biological value of 75 and therefore it has similar value to milk protein [18].

9.4.6 FATS IN AMARANTH SEED

The fat content is higher as compared to other cereals. The fat is characterized by a high content of unsaturated fatty acids, with a very high content of Linoleic acid followed by oleic acid and palmitic acid [9]. The unsaponifiables of amaranth oil are comprised of squalene. It is a highly unsaturated open chain triterpene, which is used in cosmetic industry.

9.4.7 PHENOLICS IN AMARANTH

Phytic acid, tannins, protease inhibitors and Saponins are present in amaranth. In amaranth two betacyanine compounds can be found, namely amaranthine and iso-amaranthine.

9.4.8 PROCESSING OF AMARANTH

The amaranth seed is very small and because of its botanical peculiarities, the embryo enclosing the starch rich perisperm in the form of a ring, it is not possible to apply the technology and equipment of cereal milling directly without further adaptation. However it is basically possible to obtain flour fractions from amaranth seeds with different chemical compositions and different physical properties through grinding and separation into different fractions. Studies have revealed that, for production of different flour fractions, a strong Scott barley pearler can be used. Within five passes through the pearler, the seed coat and germ were completely separated; a spherical intact starch rich perisperm was left. The bran fraction constituted about 25–26% of the seed weight. Nitrogen, crude fat, dietary fiber and ash were

2.3–2.6 times higher than in the whole seed and vitamins were 2.4–3 times higher [16]. The other processing methods are boiling of the seeds, lime treatment, popping and extrusion cooking can be used for preparing variety of products like pasta, noodles, bakery products, etc.

9.5   CONCLUSIONS

On account of the excellent nutritional composition of the three psuedocereals in this chapter, it gives an excellent alternative to increase the spectrum of foods in our diet. However from the Agriculture side, intense breeding, production and timely harvesting are required. On the other hand intensive research work is required for the processing and value addition of the above crops, according to the taste and consumer habits of the society. Intensive marketing related to these three crops can be done to popularize them and impart knowledge about the nutritional benefits of these crops.

9.6   SUMMARY

Pseudocereals are broadleaf plants (non-grasses) used much the same way as cereals. The starch-rich, gluten-free psuedocereals belonging to the class dicotyledonae, are not morphologically related either to each other closely or to the monocotyledonous true cereals. Pseudocereals include three crops: amaranth and quinoa, of South American origin and buckwheat, a native of China. In countries of their origin, these grains are consumed owing to their nutritive value. Being rich in proteins, essential aminoacids, lipids and minerals they provide an improved dietary composition beneficial to human health.

Amaranth (*Amaranthus* spp.) seeds are small in diameter (0.9 to 1.7 mm), lenticular in shape and its color varies from white, gold, brown, and pink to black. The protein content lies between 11.7–18.4% (65% of which is located in the germ and seed coat, and 35% in the starch- rich perisperm). The amino acid composition reveals a high content of lysine, histidine and arginine. Leucine and threonine are the limiting amino acids in amaranth. Starch is located in the cells of the perisperm in the form of very small starch granules. Phytic acid, tannins, protease inhibitors and saponins are present in amaranth.

Quinoa (*Chenopodium quinoa* Willd.) seeds are 1–2.6 mm in diameter and are round, flattened, and oval-shaped, with colors ranging from pale yellow to pink or black. Quinoa is rich in protein (about 14%), very high for a cereal/pseudocereal and is considered to be a source of complete protein. The highest concentration of proteins and fats are present in the embryo. Quinoa contains 1.8% phospholipids. The starch is located in the perisperm, although small amounts are present in seed coat and embryo.

In amaranth and quinoa seeds, the embryo or germ is campylotropous and surrounds the starch-rich perisperm like a ring and together with the seed coat represent the bran fraction, which is relatively rich in fat and protein. The grains are adapted to different environmental conditions, being cultivated on poor soils and high altitudes.

Buckwheat (*Fagopyrum esculentum*) is a triangular nut, sometimes prominently winged. The seed is 3-edged, 6–9 mm long and rounded form. The bran of buckwheat is thin or thick depending on variety and varies in its adhesion to pericarp and endosperm. The main two varieties cultivated are common buckwheat (*F. esculentum*) and tartary buckwheat (*F. tartaricum*). It contains well-balanced amino acids and proteins of high biological value, significant contents of microelements, resistant starch, total and soluble dietary fibers, vitamins and polyunsaturated fatty acids. The buckwheat starch is nutritionally important to diabetics as it flattens the glycemic curve. The significant contents of rutin, catechins and other polyphenols and their potential antioxidant activity are of great significance. These functional components have health benefits like reducing high blood pressure, lowering cholesterol, controlling blood sugar and preventing cancer risk.

Due to their different morphology and functional properties, known cereal processing methods cannot be applied on processing of pseudocereals without adaptation, but their seed can be ground into flour and used as cereals. Buckwheat can be hulled either in its native form or after hydrothermal pre-treatment. Milling of buckwheat to produce flour can be done either after dehulling the grains or by grinding the whole grains and then sieving them. The flour produced can be used by bakery or confectionery industries. The hydrothermally treated groats of buckwheat after milling are consumed as food after boiling. The first step in quinoa processing is removal of saponins. It can be done by dry hulling using an abrasive dehuller or by washing. However, a combination of both is preferred. The flour of quinoa is used for making bread, cake, cookies etc. Amaranth grain is usually boiled before consumption. The wet cooking process and lime processing improves the protein quality of the grain. Amaranth is also popped before grinding. A variety of products such as pasta, noodles, bakery items, museli etc. can be prepared from amaranth.

On account of nutritional composition, pseudocereals provide an excellent alternative to increase the spectrum of foods in our diet. Intensive research is required in the area of processing and value addition to the crops to meet the taste and consumer habits. The chapter featuring the present status, would provide a detailed account of the physico-chemical and functional properties, processing methods, nutritional and nutraceutical components of the pseudocereals.

KEYWORDS

- Achenes
- Amylase
- Amylopectin
- Betacyanins
- Campylotropous
- Catechins
- Cytotoxic
- Dehuller
- Dicots
- Functional properties
- Glucans
- Glutelins
- Groats
- Hydrothermal treatment

- **Perisperm**
- **Physico chemical properties**
- **Phytic acid**
- **Processing**
- **Prolamins**
- **Psuedocereals**
- **Raffinose**
- **Rutin**
- **Saponins**
- **Stachyose**

REFERENCES

1. Abugoch, L., Romero, N., Tapia, C., Silva, J., & Rivera, M. (2008). Study of some physicochemical and functional properties of quinoa (*Chenopodium quinoa* Wild.) protein isolates. *J. Agric. Food Chem,* 56, 4745–4750.
2. Ahamed, T., Singhal, R., Kulkarni, P., & Pal, M. (1998). A lesser-known grain, Chenopodium quinoa: Review of the chemical composition of its edible parts. *Food Nutr. Bull.,* 19, 61–70.
3. Ana Maria G. C., Giovanna L., & Vito V. Emanuele M., & Maria F. C. (2014). Effect of antioxidant potential of tropical fruit juices on antioxidant enzyme profiles and lipid peroxidation in rats. *Food Chemistry,* 157, 174–178.
4. Ando, H., Chen, Y., Tang, H., Shimizu, M., Watanabe, K., & Miysunaga, T. (2002). Food Components in Fractions of Quinoa Seed. *Food Sci. Technol. Res.,* 8, 80–84.
5. Asano, K., Morita, M., & Fujimaki, M. (1970). Studies on the non-starchy polysaccharides of the endosperm of buckwheat II. The main polysaccharide of water soluble fractions. *Agr. Biol. Chem.,* 34, 15–22.
6. Baker, L. A., & Rayas-Duarte, P. (1998). Freeze-thaw stability of amaranth starch at different storage temperatures and the effects of salts and sugars. *Cereal Chem.,* 75, 301–307.
7. Baker, L. A., & Rayas-Duarte, P. (1998). Retro gradation of amaranth starch at different storage temperatures and effect of salts and sugars. *Cereal Chem.,* 75, 308–314.
8. Barrett, M. (2006). Identification, sequencing, expression and evolutionary relationships of the 11s seed storage protein gene in *Chenopodium quinoa* Willd. Thesis of Master of Science Department of Plant and Animal Sciences, Brigham Young University, pp. 146–168.
9. Becker, R. (1994). Amaranth oil: composition, processing and nutritional qualities. In: *Amaranth-Biology, Chemistry, and Technology,* Paredes-Lopez, O. (Ed.), CRC Press, London, pp. 133–142.

10. Becker, R., & Hanners, G. D. (1991). Composition and nutritional evaluation of quinoa whole grain flour and mill fractions. *Lebensmittel-Wissenschaft Technologie, 23*, 441–444.

11. Belova, Z. A., Nechaev, A. P., Severinenko, S. M., & Yusupova, I. U. (1972). Effect of hydrothermal treatment of buckwheat on its fatty acid content and composition. *Izv. Vyss. Uchebn. Zav. Pishchevaya Tekhnol.*, 38, 301–307.

12. Belova, Z. A., Nechaev, A. P., Severinenko, S. M., & Baikov, V. G. (1971). Forms of lipid and composition of fatty acids in buckwheat grain lipids. *Izv. Vyss. Uchebn. Zav. Pishchevaya Tekhnol.*, 37,143–156.

13. Belova, Z. A., Nechaev, A. P., & Severinenko, S. M. (1969). Lipids in buckwheat. *Maslozhirovaya Promyshlennost*, 35,123–132.

14. Bertero, H. D., De la Vega, A. J., Correa, G., Jacobsen, S. E., & Mujica, A. (2004). Genotype and genotype-by-environment interaction effects for grain yield and grain size of quinoa (*Chenopodium quinoa* Willd.) as revealed by pattern analysis of international multi environmental trials. Field Crops Res., 89, 299–318.

15. Berti, C., Riso, P., Monti, L., & Porrini, M. (2004). In vitro starch digestibility and in vivo glucose Response of gluten-free foods and their gluten counterparts. *Eur. J. Nutr.*, 43, 198–204.

16. Betschart, A. A., Wood, I. D., Shepherd, A. D., & Saunders, R. M. (1981). *Amaranthus cruentus*: milling characteristics, distribution of nutrients within seed components, and the effects of temperature on nutritional quality. *J. Food Science*, 46, 1181–1187.

17. Bhargava, A., Rana, T., Shukla, S., & Ohri, D. (2005). Seed protein electrophoresis of some cultivated and wild species of Chenopodium. *Biol. Plan.*, 49, 505–511.

18. Bressani, R., De Martell, E. C. M., & Godinez, C. M. (1993). Protein quality evaluation of amaranth in adult humans. *Plant Foods Human Nutr.*, 43, 123–143.

19. Brinegar, C., & Goundan, S. (1993). Isolation and characterization of chenopodin, the 11S seed storage protein of quinoa (*Chenopodium quinoa*). J. Agric. Food Chem., 41, 182–185.

20. Cortella, A. R., & Pochettino, M. L. (1990). South American grain chenopods and amaranths: a comparative morphology of starch. *Starch*, 42, 251–255.

21. Del Castillo, C., Winkel, T., Mahy, G., & Bizoux, J. P. (2006). Genetic structure of quinoa (*Chenopodium quinoa* Willd.) from the Bolivian plateau as revealed by RAPD markers. *Genetic Resource Crop Evolution*, 35, 234–239.

22. Del Castillo, C. R. (2008). Diversitégénétique et réponseauxcontraintes du climat: une étude de cas à partir de la biologie de population de quinoa (*Chenopodium quinoa* Willd.) de Bolivie. PhD Thesis, Faculté Universitaire des Sciences Agronomiques, Gembloux, pp. 143–172.

23. Dietrych-Szostak, D., & Oleszek, W. (1999). Effect of processing on the flavonoid content of buckwheat grain. *J. Agric. Food Chem.*, 47, 4384–4387.

24. FAO (1973). Necesidades de energı′ay de proteı′nas. Informe de un Comite′ Especial Mixto FAO/OMS de Expertos, serie Reuniones sobre nutricio′n, No. 52, Roma, FAO. pp. 23–24.

25. FAO/INFOODS (2013). FAO/INFOODS Food Composition Database for Biodiversity version 2.1, Bio Food Comp 2.1, Rome, FAO, pp. 43–49.

26. Franke, W. (1989). Studies on the physic-chemical profile of cereals. *Nutzpflanzenkunde,* 15, 143–152.

27. Fujimaki, M., Igarashi, O., & Asano, K. (1969). Studies on the non-starchy polysaccharides of buckwheat endosperm. Fractionation and constituent sugars of the water soluble, non-starchy polysaccharides of buckwheat. *J. Agr. Chem. Soc. Jpn.,* 43, 625–635.

28. Gallagher, E., Gormleya, T., & Arendt, E. (2004). Review: Recent advances in the formulation of gluten-free cereal-based products. *Trends Food Sci. Technol.,* 15, 143–152.

29. Gandarillas, H. (1979). Botanica. Quinua y Kaniwa. Cultivos Andinos. In: Tapia, M. E. (Ed.), Serie Libros y Materiales Educativos, vol.49. Instituto Interamericano de Ciencias Agricolas, Bogota, Columbia, pp. 20–44.

30. Glowienke, S., & Kuhn, M. (1998). Importance of the Psuedocereal amaranth. II, Importance of growing conditions for the chemical composition of Amaranth. *Getriede Mahl Brot.,* 52, 323–327.

31. Gorinstein, S., Lojek, A., Ciaz, M., Pawelzik, E., Delgado-Licon, E., Medina, O., Moreno, M., Salas, I., & Goshev, I. (2008). Comparison of composition and antioxidant capacity of some cereals and pseudocereals. *Int. J. Food Sci. Technol.,* 43, 629–637.

32. Hunziker, A. T. (1943). Los species alimenticias de Amaranthus Chenopodium cultivadas porlos Indios de America. *Rev. Argen. Agron.,* 30, 297–353.

33. Inouchi, N., Nishi, K., Tanaka, S., Asai, M., Kawase, Y., Hata, Y., Konishi Yue, S., & Fuwa, H. (1999). Characterization of amaranth and quinoa starches. *J. Appl. Glycosci.,* 46, 233–240.

34. Jacobsen, S., Quispe, H., & Mujica, A. (2000). Quinoa: An Alternative crop for saline soils in the Andes. *CIP Progr. Rep.,* 53, 403–408.

35. Jahaniaval, F., Kakuda, Y., & Marcone, M. (2000). Fatty acid and triacylglycerol compositions of seed oils of five amaranthus accessions and their comparison to other oils. *J. Am. Oil Chem. Soc.,* 77, 847–852.

36. Karchik, S. N., Chiba, S., Shimomura, T., & Nishi, K. (1976). Improved method for purification of buckwheat alpha-glucosidase and some kinetic properties. *Agric. Biol. Chem.,* 40, 22–24.

37. Karpati, I., & Banayai, L. (1980). The buckwheat and the tartary buckwheat-Fagopyrum esculentum MONCH, Fagopyrum tataricum (L) GARTN. Culturflora of Hungary. Akademia, Budapest, 23, 321–331.

38. Keli, Y., Dabiao, L., & Genjiu, V. (1992). The quality appraisal of buckwheat germlasm resources in China. In *Proceedings of the 5th Symposium on Buckwheat.* Agriculture Publishing House, Taiyuan, China, pp. 90–97.

39. Khan, F., Zeb, A., Arif, M., Ullah, J., & Wahab, S. (2012).Preparation and evaluation of gluten free ready to serve buckwheat product. *Journal of Agricultural and Biological Science,* 7, 633–637.

40. Kim, S. K., Hahn, T. R., Kwon, T. W., & D-Appolonia, B. L. (1977). Physicochemical properties of buckwheat starch. *Korean J. Food Sci. Technol.,* 9, 138–142.

41. Kirilenko, S. K., & Sarkisova, N. E. (1977). Content of thiamine in the embryo of some buckwheat varieties. *Voprosy Pitaniya,* 4, 91–95.

42. Kitabayashi, H., Ujihara, A., Hirose, T., & Minami, M. (1995). Varietal differences and heritability for rutin content in common buckwheat, Fagopyrum esculentum Moench. *Jpn. J. Breed.,* 45, 75–79.
43. Koziol, M. (1992). Chemical composition and nutritional evaluation of quinoa (*Chenopodium quinoa* Willd), *J. Food Comp. Anal.,* 5, 35–68.
44. Krkoskova, B., & Markova, Z. (2005). Prophylactic components of buckwheat. *Food Research International,* 38, 561–568.
45. Kuzmina, O.V., & Tarzhinskaya, L. R. (1973). The effect of hydrothermal treatment of rice and buckwheat on the properties of starch. *Izv. Vyss. Uchebn. Zav. Pishchevaya Tekhnol.,* 45, 23–25.
46. Laura, M., López A. C., & Axel Emil, Nielsen (2011). Traditional post-harvest processing to make quinoa grains (Chenopodium quinoa var. quinoa) apt for consumption in Northern Lipez (Potosí, Bolivia): ethno archaeological and archaeobotanical analyses, *Archaeol Anthropol Sci.,* 3, 49–70.
47. Leder, F., & Monda, S. (1988). Increasing the nutritive value of flour mill products by natural ways. *Gabonaipar,* 1, 27–31.
48. Li, S. Q., & Zhang, Q. H. (2001). Advances in the development of functional foods from buckwheat. *Critical Reviews in Food Science and Nutrition,* 41, 6–12.
49. Lindeboom, N. (2005). Studies on the characterization, biosynthesis and isolation of starch and protein from quinoa (Chenopodium quinoa Willd.), University of Saskatchewan, pp. 157–186.
50. Lindeboom, N., Chang, P., & Tyler, R. (2004). Analytical, biochemical and physicochemical aspects of starch granule size, with emphasis on small granule starches: A review. *Starch/Staerke,* 56, 89–99.
51. Lopez, M. G., Bello-Perez, L. A., & Paredes-Lopez, O. (1994). Amaranth carbohydrates. In: Amaranth-biology, chemistry, and Technology, Paredes-Lopez, O., Eds., CRC Press, London, pp. 107–132.
52. Mahoney, A., Lopez, J., & Hendricks, D. (1975). Evaluation of the protein quality of quinoa. *J. Agric. Food Chem.,* 23, 190–193.
53. Marcone, M. F. (1999). Evidence confirming the existence of a 7S globulin like storage. *Agric. Food Chem.,* 47, 2234–2238.
54. Marcone, M. F., Beniac, D. R., Harauz, G., & Yada, R. Y. (1994). Quaternary structure and model for the oligomeric seed globulin from *Amaranthus hypochondriacus. J. Agric. Food Chem.,* 42, 2675–2678.
55. Masson & Mella, (1985). Materias grasas de consume habitual y potencial en Chile. (Ed. Universitaria), pp. 23–28.
56. Mastebroek, D., Limburg, H., Gilles, T., & Marvin, H. (2000). Occurrence of sapogenins in leaves and seeds of quinoa (*Chenopodium quinoa* Willd.). *J. Sci. Food Agric.,* 80, 152–156.
57. Mota Carlos, Mariana Santos, Raul Mauro, Norma Samman, Ana Sofia Matos, Duarte Torres, & Isabel Castanheira (2014). Protein content and amino acids profile of pseudocereals, *Food Chemistry,* 23, 234–239.
58. Mujica, A., & Jacobsen, S. (2006). La quinua (*Chenopodium quinoa* Willd.) y sus parientes silvestres. Botanica Economica de los Andes Centrales., pp. 341–345.
59. National academy of Sciences (1984). Amaranth Modern Prospects for An Ancient Crop. NAS, Washington, DC, pp. 234–236.

60. National Academy of Sciences. (2004). Comprehensive DRI table for vitamins, minerals and macronutrients, organized by age and gender. Institute of Medicine. Food and Nutrition Board, Beltsville, MD, pp. 45–49.

61. Ng, S., Anderson, A., Cokera, J., & Ondrusa, M. (2007). Characterization of lipid oxidation products in quinoa (*Chenopodium quinoa*). *Food Chem.,* 101, 185–192.

62. Oomah, B. D., & Mazza, G. (1996). Flavonoids and anti-oxidative activities in buckwheat. *Journal of Agricultural and Food Chemistry,* 44, 1746–1750.

63. Oshodi, A., Ogungbenle, H., & Oladimeji, M. (1999). Chemical composition, nutritionally valuable minerals and functional properties of benni seed, pearl millet and quinoa flours. *Int. J. Food Sci. Nutr.,* 50, 325–331.

64. Paredes-Lopez, O., Bello-Perez, L. A., & Loper, M. G. (1994). Amylopectin: structural, gelatinization and retrogradation studies. *Food Chem.,* 50, 411–417.

65. Prakash, D., & Pal, M. (1998). Chenopodium: Seed protein, fractionation and amino acid composition. *Int. J. Food Sci. Nutr.,* 49, 271–275.

66. Praznik, W., Mundigler, N., Kogler, A., Pelzl, B., & Huber, A. (1999). Molecular background of technological properties of selected starches. *Starch/Staerke,* 51, 197–211.

67. Prego, I., Maldonado, S., & Otegui, M. (1998). Seed structure and localization of reserves in *Chenopodium quinoa. Ann. Bot.,* 82, 481–488.

68. Martin, J. (2003). Protein in A. hypochondriacus seed. *Food Chem.,* 65, 533–542.

69. Przybylski, R., Chauhan, G., & Eskin, N. (1994). Characterization of quinoa (*Chenopodium quinoa*) lipid. *Food Chem.,* 51, 187–192.

70. Qian, J., & Kuhn, M. (1999). Characterization of *Amaranthus cruentus* and *Chenopodium quinoa* starch. *Starch/Staerke,* 51, 116–120.

71. Quain, J. Y., & Kuhm, M. (1999). Characterization of *Amaranthus cruentus* and *Chenopodium quinoa* starch. *Starch,* 51, 116–120.

72. Radosavljevic, M., Jane, J., & Johnson, L.A. (1998). Isolation of amaranth starch by dilute alkaline protease treatment. *Cereal Chem.,* 75, 212–216.

73. Ranhotra, G., Gelroth, J., Glaser, B., Lorenz, K., & Johnson, D. (1993). Composition and protein nutritional quality of quinoa. *Cereal Chem.,* 70, 303–305.

74. Reichert, R. D., Tatarynovich, J. T., & Tyler, R. T. (1986). Abrasive dehulling of quinoa (Chenopodium quinoa): effect on saponin content was determined by an adapted hemolytic assay. *Cereal Chem.,* 63, 471–475.

75. Risi, J., & Galwey, N. W. (1984). The Chenopodium Grains of the Andes: Inca Crops for Modern Agriculture. *Adv. Appl. Biol.,* 10, 145–216.

76. Ruales, J., & Nair, B. M. (1993). Content of fat, vitamins and minerals in quinoa (*Chenopodium quinoa* Willd.) seeds. *Food Chem.,* 48, 131–136.

77. Ruales, J., & Nair, B. M. (1994). Effect of processing on in vitro digestibility of protein and starch in quinoa seeds. *Int. J. Food Sci. Technol.,* 29, 449–456.

78. Ruales, J., & Nair, B. M. (1992). Nutritional quality of the protein in quinoa (*Chenopodium quinoa* Willd) seeds. *Plant Foods Hum. Nutr.,* 42, 1–12.

79. Ryan, E., Galvin, K., O'Connor, T., Maguire, A., & O'Brien, N. (2007). Phytosterol, squalene, tocopherol content and fatty acid profile of selected seeds, grains, and legumes. *Plant Foods Hum. Nutr.,* 62, 85–91.

80. Saunders, R. M., & Becker, R. (1984). Amaranthus: a potential food and feed resource. In: *Advances in Cereal Science and Technology* by Pomeranz, Y., Eds., Vol IV, American Association of Cereal Chemists. St Paul, MN, pp. 357–396.

81. Schoenlechner, R. (1997). Entwicklung und Charakterisierung von Convenience-produk-ten aus amaranth und Quinoa. Dissertation, University of Agricultural Sciences, Vienna, Austria. 32, 234–238.

82. Segura-Nieto, M., Barba dela Rosa, A. P., & Paredes-Lopez, O. (1994). Biochemistry of amaranth proteins. In: *Amaranth-Biology, Chemistry, and Technology*, Paredes-Lopez, O. (Ed.), CRC Press, London, pp. 75–101.

83. Simmonds, N. W. (1965). The grain chenopods of the tropical American highlands. *Annals Botany*, 19, 223–235.

84. Singhal, R. S., & Kulkarni, P. R. (1990). Some properties of *Amaranthus paniculatas* starch pastes. *Starch*, 42, 5–7.

85. Sokolov, O. A., Timchenko, A. V., Semikhov, V. F., Dunaevskii, Ya. E., & Belozerskii, M. A. (1981). Extraction and chemical composition of aleurone from buckwheat. *Fiziol. Rast.*, 28, 1166.

86. Soliz-Guerrero, J. B., Jasso de Rodriguez, D., Rodriguez-Garcia, R., Angulo-Sanchez, J. L., & Mendez-Padilla, G. (2002). Quinoa saponins: concentration and composition analysis. In: *Trends in New Crops and New Uses*. Janick, J. Whipkey, A. (Eds.), ASHS Press, Alexandria, VA, pp. 110–114.

87. Stone, L. A., & Lorenz, K. (1984). The starch of amaranthus-physico chemical properties and functional characteristics. *Starch*, 50, 7–13.

88. Stuardo, M., & San Martin, R. (2008). Antifungal properties of quinoa (*Chenopodium quinoa* Willd.) alkali treated previous term saponins next term against *Botrytis cinerea*. *Ind. Crops Prod.*, 27, 296–302.

89. Sure, B. (1955). Nutritive value of proteins in buckwheat and their role as supplements to proteins in cereal grains. *J. Agric. Food Chem.*, 3, 793–796.

90. Tahir, I., & Farooq, S. (1988). Review article on buckwheat. *Fagopyrum*, 8, 33–53.

91. Taira, H. (1974). Buckwheat. In: *Encyclopedia of Food Technology*, Johnson, A. H., & Peterson, M. J. (Eds.), AVI Publications Corporation, West Port, Connecticut, pp. 139–157.

92. Tamura, S. (1970). Amino Acid Composition of Food in Japan. *Jpn. Agric. Quart.*, 5, 56–58.

93. Tang, H., Watanabe, K., & Mitsunaga, T. (2002). Characterization of storage starches from quinoa, barley and adzuki seeds. *Carbohydr. Polym.*, 49, 13–22.

94. Tapia, M. (1982). The Environment, Crops and Agricultural Systems in the Andes and Southern Peru. IICA. pp. 322–329.

95. Tapia, M. (2000). Cultivos andinos subexplotados y su aporte a la alimentación. FAO-Organizaciones de las Naciones Unidas para la Agricultura y la Alimentación, Santiago. pp. 321–333.

96. Tapia, M. E. (1979). Historia y Distribuciongeographica. Quinua y Kaniwa. Cultivos Andinos. In: *Serie Librosy Materiales Educativos, vol.49*. Tapia, M. E. (Ed.), Instituto Inter americanode Ciencias Agricolas, Bogota, Colombia, pp. 11–15.

97. Tari, T., Annapure, U., Singhal, R., & Kulkarni, P. (2003). Starch-based spherical aggregates: Screening of small granule sized starches for entrapment of a model flavoring compound, vanillin. *Carbohydr. Polym.*, 53, 45–51.

98. Uriyapongson, J., & Rayas-Duarte, P. (1994). Comparison of yield and properties of amaranth starches using wet and dry milling processes. *Cereal Chem.*, 76, 877–883.

99. USDA U.S. Department of Agriculture, Agricultural Research Service (2005). USDA National Nutrient Database for Standard Reference, Release 18. Nutrient Data Laboratory, pp. 23–25.

100. USDA, U.S. Department of Agriculture, Agricultural Research Service (2013), USDA national nutrient database for standard release 26, Nutrient Data Laboratory, Beltsville Human Nutrition Research Centre, Agriculture Research Service, United States Department of Agriculture. pp. 324–334.

101. Valencia-Chamorro S. A. (2003). Quinoa In: *Encyclopedia of Food Science and Nutrition., Vol.8.* Caballero, B. (ed.). Academic Press, Amsterdam, pp. 4895–4902.

102. Varriano-Marston, E., & De-Francisco, A. (1984). Ultra-structure of quinoa fruit (*Chenopodium quinoa* Willd). *Food Microstruct.,* 3, 165–173.

103. Verena Nowak, Juan Du, & Ruth Charrondiere, U. (2015). Assessment of nutritional composition of quinoa (*Chenopodium quinoa* Willd). Food Chemistry, 193, 47–54.

104. Walkowski, A., Fornal, J., Lewandowicz, G., & Sadowska, J. (1997). Structure, physico chemical properties and potential uses of amaranth starch. *Pol. J. Food Nutr. Sci.,* 6, 11–22.

105. Watanabe, K., Peng, L., Tang, H., & Mitsunaga, T. (2007). Molecular structural characteristics of quinoa starch. *Food Sci. Technol. Res.,* 13, 73–76.

106. Wijngaard, H. H., & Arendt, E. K. (2006). Buckwheat. *Cereal Chemistry,* 83, 391–401.

107. Wilson, H. (1988). Quinoa biosystematics. I: Domesticated populations. *Econ. Bot.,* 42, 461–477.

108. Wilson, H. (1981). Genetic variation among South America populations of tetraploid Chenopodium sect. Chenopodium subsect. *Cellulata Syst. Bot.,* 6, 380–398.

109. Wolde, M. G., & Wink, M. (2001). Identification and biological activities of triterpenoid saponins from *Chenopodium quinoa. J. Agric. Food Chem.,* 49, 2327–2332.

110. Wood, S., Lawson, L., Fairbanks, D., Robinson, L., & Andersen, W. (1993). Seed lipid content and fatty acid composition of three quinoa cultivars. *J. Food Comp. Anal.,* 6, 41–44.

111. Yanez, G. A., Messinger, J. K., & Walker, C. E. (1986). *Amaranthus hypochondriacus* starch isolation and partial characterization. *Cereal Chem.,* 75, 212–216.

112. Zhao, J., & Whistler, R. L. (1994). Isolation and characterization of starch from amaranth flour. *Cereal Chem.,* 71, 392–393.

113. Zhu, N., Sheng, S., Li, D., Lavoie, E., Karwe, M., Rosen, R., & Chi-Tang Hi, C. (2001). Antioxidative flavonoid glycosides from quinoa seeds (*Chenopodium quinoa* Willd.). *J. Food Lipids,* 8, 37–44.

114. Zhu, N., Sheng, S., Sang, S., Jhoo, S., Bai, S., Karwe, M., Rosen, R., & Ho, C. (2002). Triterpene saponins from debittered quinoa (*Chenopodium quinoa*) seeds. *J. Agric. Food Chem.,* 50, 865–867.

CHAPTER 10

NUTRACEUTICAL AND FUNCTIONAL FOODS FOR CARDIOVASCULAR HEALTH

RAJ K. KESERVANI,[1] ANIL K. SHARMA,[2] and
RAJESH K. KESHARWANI[3]

[1]*Researcher, School of Pharmaceutical Sciences, Rajiv Gandhi Proudyogiki Vishwavidyalaya, Bhopal (MP), 462036, India, Mobile: +91-7897803904; E-mail: rajksops@gmail.com*

[2]*Researcher, Department of Pharmaceutics, Delhi Institute of Pharmaceutical Sciences and Research, New Delhi, 110017, India, E-mail: sharmarahul2004@gmail.com*

[3]*Assistant Professor, Department of Biotechnology, NIET, NIMS University, Shobha Nagar, Jaipur, Rajasthan, 303121, India, E-mail: rajiiita06@gmail.com*

CONTENTS

10.1 INTRODUCTION

Cardiovascular disease (CVD) involves the heart or blood vessels. Cardiovascular disease includes coronary artery diseases (CAD) such as angina and myocardial infarction (commonly known as a heart attack) [67]. Other CVDs are: stroke, hypertensive heart disease, rheumatic heart disease, cardiomyopathy, atrial fibrillation, congenital heart disease, endocarditis, aortic aneurysms, and peripheral artery disease [29, 67]. The role of dietary factors has been largely investigated by reducing the content of specific food component known to increase risk, that is, sodium or saturated fats [65] or by large longitudinal cohort studies in which baseline dietary intake was related to cardiovascular outcomes [8, 39].

The term Nutraceutical is a hybrid of nutrition and pharmaceutical. Stephen L. DeFelice in 1989 defined it as a food or food product that provides health and medical benefits, including the prevention and treatment of disease [10, 47, 49]. It has been proposed that CVD can be prevented by lifestyle changes, including diet [93]. Early evidence for the role of diet on CVD was from data on trends in food consumption, and ecological studies have shown associations between CVD prevalence and fat intake [65]. Moreover, excessive consumption of foods that are calorie dense, nutritionally poor, highly processed, and rapidly absorbable can lead to systemic inflammation, reduced insulin sensitivity, and a cluster of metabolic abnormalities, including obesity, hypertension, dyslipidemia, and glucose intolerance [82]. More recently, there has been a focus in nutrition research to try and understand the effects of whole foods [27]. An integrated approach combining lifestyle modification with the correct pharmacologic treatment is sought to reduce cardiovascular risk factors, to improve vascular health, and to reduce healthcare expenditure [2].

Vegetable and fruit fibers (with pectin), garlic and oily seeds (walnut, almonds, etc.), and fish oils have lipid-lowering effects in humans, through both inhibition of fat absorption and suppression of hepatic cholesterol synthesis. Homocysteine increases the risk of both cardiovascular and cerebrovascular disorders [64] by enhancing arteriolar constriction and decreasing endothelial vasodilation [11].

This chapter discusses role of nutraceutical and functional foods for cardiovascular health.

TABLE 10.1   Nutraceutical and Functional Foods with Cardiovascular Potential

Mechanism of action	Functional foods	Chemical constituents
Lowering blood cholesterols	• Nuts	• Tocopherol, omega-3 fatty acids
	• Legumes	• Fiber and polyphenols
	• Fruits and vegetables	• Fiber (pectin)
	• Margarine	• Phytosterols
	• Fish oil	• Omega-3 fatty acids
	• Whole grains	• Fiber and phytochemicals
	• Soy proteins	• Genistein and daidzein
	• Dark chocolate	• Flavonoid
Inhibition of LDL-C oxidation	• Fish	• Omega-3 fatty acids
	• Green leafy vegetables, fruits	• Carotenoids
		• Vitamin C
	• Citrus, fruits and vegetables	• Lycopene
		• Polyphenolics and oleic acid
	• Tomato	• Tea Polyphenolics
	• Extra virgin olive oil	
	• Green tea	• Genistein, daidzein, and glycitein
	• Soy proteins	• Flavonoid
	• Dark chocolate	• Polyphenols
	• Pomegranate	
Platelets aggregation	• Grapes and red wines	• Anthocyanins, Catechins, Cyanidins and flavanols, myricetin and quercetin
Decreasing blood pressure	• Fish	• Omega-3 fatty acids
	• Legumes	• Fiber
	• Whole grains	• Fiber and phytochemicals
	• Citrus fruits	• Ascorbic acid
	• Onion and garlic	• Quercetin
	• Green and black teas	• Tea polyphenols
	• Grapes and red wines	• Grape polyphenols
	• Dark chocolate	• Flavonoid

TABLE 10.1   (Continued).

Mechanism of action	Functional foods	Chemical constituents
Antioxidants activity	• Tomatoes	• Lycopene
	• Vegetable oils	• Tocopherol, tocotrienols
	• Soy proteins	• Genistein and daidzein
	• Green leafy vegetables	• Carotenoids
	• Green and black teas	• Tea polyphenols
	• Grapes and red wines	• Anthocyanins, catechins, cyanidins, and flavonols, myricetin and quercetin
Lowering blood triglycerides	• Fish	• Omega 3-fatty acids

10.2   CARDIOVASCULAR POTENTIAL OF NUTRACEUTICAL AND FUNCTIONAL FOODS

Functional foods are *any food or food ingredient that may provide a health benefit beyond the traditional nutrients it contains* [44]. Nutraceutical and functional foods like nuts, legumes, fruits and vegetable dietary fish, soy protein, dark chocolate, citrus fruits, vitamins etc., with their mechanism of action and chemical constituents have cardiovascular potential are listed in Table 10.1 [3, 48–53].

Potential mechanisms for the cardiovascular protective effects of n–3 fatty acids are suggested to be: anti-inflammatory, antithrombotic (reduced platelet aggregability), and antiarrhythmic (reducing the risk of potentially fatal cardiac arrhythmias), lowering of heart rate and blood pressure, hypotriglyceridemic, and improved endothelial function [69].

10.2.1   ROLE OF WHOLE GRAINS IN PREVENTION OF CVD

Most whole grains are abundant sources of dietary fiber and other nutrients, such as minerals and antioxidants, which have shown beneficial effects on human health including improvement of weight loss, insulin sensitivity, and lipid profile, as well as inhibition of systemic inflammation [36, 78, 79]. The Communities study on atherosclerosis risk indicated that consuming three servings of whole grain foods daily was associated with a 28% lower risk of coronary artery disease [94] and there was 7% lower risk of incident of heart failure per each additional whole grain serving [72]. The potential protective role of whole grains was first evaluated in the early 1970s [68]. Based on the results of the prospective Iowa Women's Health Study, it was demonstrated that cereal fiber had different associations with total mortality, depending on whether the fiber came from foods that contained primarily whole grain or refined grain [45]. A more recent meta-analysis based on seven qualifying prospective cohort studies focused on whole grain consumption and cardiovascular outcomes, and it was reported that the inverse association between dietary whole grains and incident CVD was strong and consistent across trials [7, 62, 66, 95].

10.2.2 ROLE OF SOY PROTEIN IN CVD

Soy protein (Figure 10.1) is a protein that is isolated from soybean. It is made from soybean meal that has been dehulled and defatted. Soy protein is generally regarded as being concentrated in protein bodies, which are estimated to contain at least 65–70% of the total soybean protein [63]. Prospective observational studies, in vegetarians [14], Chinese women [100], and in a Japanese population [71], have shown a reduction of total cholesterol and LDL-C as well as of ischemic and cerebrovascular events with a daily soy protein intake of more than 6 g, compared with less than 0.5 g. A large number of clinical studies was summarized in a meta-analysis [4] and confirmed that serum LDL-C concentrations are modified, the effects being related to baseline blood cholesterol levels. The results of this meta-analysis were criticized recently, since more recent studies appeared not to confirm the very powerful cholesterol-reducing effect of soy proteins [89]. Elevated cholesterol is major risk factor for CHD. Each one percent reduction in blood cholesterol is thought to reduce risk of heart disease by approximately 2% [59]. Meta-analyses of the clinical research show that soy protein (~25 g/day) lowers low-density-lipoprotein cholesterol (LDL-C)

FIGURE 10.1   Examples of functional foods in cardiovascular health: (a) Soy protein; (b) Whole grain; (c) Nuts and legumes; (d) Dark chocolate.

by 4 to 6% and also lowers blood triglyceride levels by approximately 5% [37, 46, 83, 99].

10.2.3 ROLE OF NUTS AND LEGUMES IN CVD

Nuts (Figure 10.1) are complex foods containing cholesterol lowering mono- and poly-unsaturated fatty acids, arginine (a precursor to the vaso-dilator nitric oxide), soluble fiber, and several antioxidant polyphenols [88]. Postprandial vascular reactivity is characterized by decreased bio-availability of nitric oxide and increased expression of pro-inflammatory cytokines and cellular adhesion molecules [86]. Fleming et al. studied dietary patterns and concluded that a healthy dietary pattern lowers CHD risk and is complementary to drug therapies by targeting well-established risk factors such as elevated BMI, high blood pressure, and an atherogenic lipid profile [26]. Researchers indicate that this may be one of the reasons a Mediterranean diet supplemented with a handful (30 g) of mixed nuts has been shown to reduce the risk of heart disease by 28% compared to a lower fat diet [25, 90].

Specific aspects of a vegetarian diet, including a lower intake of saturated fat, higher intake of soluble fiber, and increased consumption of whole grains, legumes, nuts, and soy protein, are likely to contribute to its cardiovascular benefits [43]. Several studies have demonstrated an association between whole grain intake and CVD risk, and a recent meta-analysis estimated that a greater intake of whole grains (2.5 servings per day versus 0.2 servings per day) was associated with a 21% lower risk of CVD events [66].

10.2.4 ROLE OF DARK CHOCOLATE IN CVD

One of the more recent studies on cocoa (Figure 1) found cocoa to be an antioxidant-rich super fruit, and demonstrated that cocoa powder offered significantly more antioxidant benefit per gram than the powder form of super fruits: acai, blueberry, cranberry and pomegranate. Cocoa is a flavonoid-rich food that has been recently investigated for its possible role in the prevention of CVD [21, 28]. Administration of dark chocolate in essential hypertensive's reduced ambulatory blood pressure and serum LDL-C levels, whereas white chocolate had no effects [33]. Pereira and coworker indicated that dark chocolate intake improves endothelial function in young healthy people: a randomized and controlled trial; and reported that the daily ingestion of 8 g/day of 70% cocoa chocolate during a month improves the endothelial function of young people, improving the endothelium-dependent vasodilatation [75].

The consumption of dark chocolate acutely decreases wave reflections, that it does not affect aortic stiffness, and that it may exert a beneficial effect on endothelial function in healthy adults. Chocolate consumption may exert a protective effect on the cardiovascular system. Further studies are warranted to assess any long-term effects [98].

10.2.5 ROLE OF FRUITS AND VEGETABLES IN CVD

Fruits like banana, apples, grapes, citrus fruits etc. and vegetables like tomatoes and leafy vegetables, etc., have cardiovascular potentials (Figure 10.2). The potential benefits to vascular health from a tomato-rich diet are often ascribed to high concentrations of lycopene, as tomato products

FIGURE 10.2 Examples of fruits and vegetables for cardiovascular health. (a) Banana; (b) Citrus; (c) Apples; (d) Grapes; (e) Leafy vegetables; (f) Tomatoes.

can account for >80% of the intake of this carotenoid [17]. High lycopene concentrations in blood and adipose tissue correlate with a reduction in CVD incidence [56, 81, 84, 85]; low concentrations are associated with early atherosclerosis [54] and elevated C-reactive protein concentrations [12, 57]. Dohadwala et al. [22] reported that rich diet of grapes can reduce the risk for cardiovascular disease.

Inadequate consumption of fruits and vegetables has been linked with higher incidence of CVD. Fruits and vegetables have been found to decrease susceptibility of LDL particles to oxidation. Several bioactive components are present in fruits and vegetables such as carotenoids, vitamin C, fiber, magnesium, and potassium; and these act synergistically or antagonistically to promote a beneficial effect. Soluble fibers including pectin's from apples and citrus fruits, β-glucan from oats and barley, and fibers from flaxseed and Psyllium are known to lower LDL-C [15, 23]. The mechanisms of their cholesterol-lowering effects are suggested to be the binding of bile acids and inhibition of cholesterol synthesis.

The mechanisms by which fruit and vegetables exert their protective effects are not entirely clear, but these likely include antioxidant and anti-inflammatory effects. Among the possible explanations for this beneficial effect, fruits and vegetables have been found to decrease susceptibility of LDL particles to oxidation [15]. Potassium may also have a protective role on the incidence of CVD as mounting evidence indicates an inverse association between dietary intake of fruits and vegetables and blood pressure [38, 91].

10.2.6 ROLE OF DIETARY FISH IN CVD

People with a high intake of dietary fish and fish oil supplements have a low rate of CVD [40, 58]. Although fish contains various nutrients with potentially favorable effects on health, yet attention has been particularly focused on the omega-3 (n–3) fatty acids. The n–3 fatty acids also include the plant-derived alpha-linolenic acid (ALA, 18:3 n–3), eicosapentaenoic acid (EPA, 20:5 n–3), and docosahexaenoic acid (DHA, 22:6 n–3). Both EPA and DHA are found in oily fish, such as salmon, lake trout, tuna, and herring, and fish-derived products (fish oils). The n–3 fatty acid precursor, α-linolenic acid, is typically found in various plants (e.g., spinach), seeds (nuts and flaxseeds), and oils derived from them. Generally, very little ALA is converted to EPA, and even less to DHA, and therefore direct intake of the latter two is optimal. The American Heart Association's recommendation is to consume at least two 3.5 oz fish meals per week to reduce the risk of CVD, with an emphasis on fatty fish (i.e., salmon, herring, mackerel, sardines) to increase EPA and DHA [61]. Long-chain omega-3 fatty acids may reduce CVD risk through several mechanisms: Lowering effects on lipids, inflammatory markers, and platelets [80]. A recent systematic review limited to cardiovascular events in randomized controlled trials and clinical trials demonstrated several cardiovascular favorable effects when marine omega-3 fatty acids were provided as food or in a supplement for a minimum of six months duration [20].

10.2.7 ROLE OF COFFEE AND TEA IN CVD

The active constituents of coffee apparently responsible for cardio protective effect are diterpenes, such as kahweol and cafestol. Coffee consumption may possibly reduce the risk of myocardial infarction, but data are as yet inconclusive [16, 74]. A dose-response decrease in cardiovascular risk and heart disease mortality was reported for a daily caffeine intake in patients with type-II diabetes [9, 34]. Green tea consumption appears to protect from CVD [96], but results are again inconsistent. It has been reported in a meta-analysis that the incidence of myocardial infarction, among individuals who consumed three cups of tea daily, was not statistically significant and there has been large variability across studies [76].

One population-based study by Tokunaga et al. [97] indicated that men who drink green tea are more likely to have lower total cholesterol than those who do not drink green tea. Tea consumption has been reported to protect against cardiovascular disease (CVD) by reducing blood pressure, blood glucose levels, and body weight [77]. The beneficial effects of tea and coffee (Figure 10.3) consumption could be explained by their high content of vitamins and polyphenols, which are suggested to be negatively associated with chronic diseases [32, 35]. The potential health benefits depend on their antioxidant and anti-inflammatory bioactivity, which may contribute to their protective role against CVD [6].

FIGURE 10.3   Green coffee (Left) and green tea (Right).

10.2.8 ROLE OF ANTIOXIDANT AND VITAMINS SUPPLEMENTATION IN CVD

The term *dietary supplement* can be defined as a product that is intended to supplement the diet with one or more of the following dietary ingredients: a vitamin, a mineral, a herb or other botanical, an amino acid, intended for ingestion in pill, capsule, tablet, or liquid form [53]. It should be noted that nutraceuticals differ from dietary supplements in the following aspects: (i) nutraceuticals must not only supplement the diet but should also aid in the prevention and/or treatment of disease, and (ii) nutraceuticals are used as conventional foods or as sole items of a meal or diet [50, 53]. In view of the detrimental role of free radicals and reactive oxygen species in the pathophysiology of atherosclerosis, supplementation with antioxidants (vitamins A, C, and E, folic acid, β-carotene, selenium, and zinc) was expected to be protective.

A significant cardiovascular benefit of phytochemicals (polyphenols in wine, grapes, and teas), vitamins (ascorbate, tocopherol), and minerals (selenium, magnesium) in foods is thought to be the capability of scavenging free radicals produced during atherogenesis [48, 51, 52]. The powerful antioxidant functions of vitamin C serve to reduce tissue reactive oxygen species concentrations, which in the atherosclerotic condition help prevent endothelial dysfunction, inhibit vascular smooth muscle proliferation, and reduce oxidized LDL cholesterol [1]. In addition as a free radical scavenger, vitamin E is a potent anti-inflammatory agent, especially at high doses [92].

10.2.9 ROLE OF PHYTOCHEMICALS IN CVD

Phytochemicals are biologically active compounds present in plants used for food and medicine (Figure 10.4). A great deal of interest has been generated recently in the isolation, characterization and biological activity of these phytochemicals. Polyphenols have been shown in in vivo studies to exert anti-atherosclerotic effects in the early stages of atherosclerosis development (e.g., decrease LDL oxidation); improve endothelial function and increase nitric oxide release (potent vasodilator); modulate

FIGURE 10.4   Food source of phytochemicals to reduce cardiovascular disease.

inflammation and lipid metabolism (i.e., hypolipidemic effect); improve antioxidant status; protect against atherothrombotic episodes including myocardial ischemia and platelet aggregation [19, 52].

Polyphenols (flavonoids) are the most diverse group of phytochemicals distributed in vegetables, fruits, olive oil, and wine and exhibit wide range of protective roles such as hypolipidemic, antioxidative, antiproliferative, and anti-inflammatory effects to reduce the onset of disease progression [3, 31, 60].

Intake of flavonoids has been associated with decreased cardiovascular mortality and general mortality among elderly Dutch individuals [30]. Several prospective studies have reported inverse associations between flavonoid intake and CVD incidence or mortality [41, 55, 70]. Phytosterols and phytosterols inhibit intestinal absorption of cholesterol [42]. HDL and/or VLDL were generally not affected by stanols/sterols intake. Yet, effects of sterols/stanols on LDLs have been found to be additive to diets and/or cholesterol-lowering drugs [73]. This has been the basis for the development of phytosterols-enriched functional foods.

It has been known for many years that high intake of fruits and vegetables is associated with reduced risk of coronary heart disease [13]. The beneficial effect of fruits and vegetables may be related especially to flavonoids, which are thought to exert their action by inhibiting LDL oxidation and platelet aggregation [18], as well as to inhibit the angiotensin converting enzyme (ACE), a key component in the *renin angiotensin aldosterone system* (RAAS), which regulates blood pressure [5]. Pomegranate juice (PJ) is a rich source of flavonoids and as such it has potent antioxidant activity. The flavonoids in PJ have been linked to a diverse group of polyphenols, including ellagitanins, gallotannins and ellegic acid. Recently, studies suggested that PJ consumption may be beneficial in populations at high risk to develop atherosclerosis and CVD [24, 87].

10.3   CONCLUSIONS

Many functional foods have antioxidant and anti-inflammatory activities, by mechanisms that may require further investigation. Therefore, these functional foods should be incorporated into a healthy diet to provide cardiovascular benefits and hence lower cardiovascular risk. This chapter will also provide the effect of individual bioactive dietary compounds with the effect of some dietary patterns in terms of their cardiovascular protection. Finally, concluded that nuts, legumes, whole grain, green tea and coffee, dark chocolate, fish, fruits and vegetables, phytochemicals, etc., have cardiovascular potential.

10.4   SUMMARY

The relationship between CVD and dietary factors has been a major focus of health research for almost half a century. Epidemiological and clinical studies indicate that the risk of CVD is reduced by the diet rich in fruits, vegetables, unrefined grains, fish and low-fat dairy products, and low in saturated fats and sodium. Other foods such as mono- and polyunsaturated fats, brans, nuts, plant sterols, and soy proteins have all been shown to have a favorable effect on lipid profile and blood pressure. However, nutrition is a very complex research topic and it is not clear whether an individual component of the diet or a combination of nutrients and dietary habits may be responsible for any cardio protective effects. The advances in the knowledge of both the disease processes and healthy dietary components have provided new avenues to develop dietary strategies to prevent and/or to treat CVD. Finally, concluded that nuts, legumes, whole grain, green tea and coffee, dark chocolate, fish, fruits and vegetables, phytochemicals, etc., have cardiovascular potential.

KEYWORDS

- Ascorbate
- Cardiovascular disease
- Carotenoids
- Coffee
- Dark chocolate
- Fish
- Flavonoids
- Free radical scavenger
- Fruit
- Functional foods
- Grapes
- Legume
- Magnesium
- Nutraceutical
- Nuts
- Olive oil
- Omega-3 fatty acids
- Phytochemical
- Selenium

- Soy proteins
- Tea
- Tocopherol
- Vegetables
- Vitamin C
- Whole grain
- β-carotene

REFERENCES

1. Aguirre, R., & May, J. M. (2008). Inflammation in the vascular bed: importance of vitamin C. *Pharmacology and Therapeutics*, 119(1), 96–103.
2. Aldana, S. G., Greenlaw, R., Salberg, A., Merrill, R. M., Hager, R., & Jorgensen, R. B. (2007). The effects of an intensive lifestyle modification program on carotid artery intima-media thickness: a randomized trial. *American Journal of Health Promotion*, 21 (6), 510–516.
3. Alissa, E. M., & Ferns, G. A. (2012). Functional foods and nutraceuticals in the primary prevention of cardiovascular diseases. *J. Nutrit. Metab*, 569486, 1–16.
4. Anderson, J. W., Johnstone, B. M., & Cook-Newell, M. E. (1995). Meta-analysis of the effects of soy protein intake on serum lipids. *New England Journal of Medicine*, 333 (5), 276–282.
5. Aviram, M., & Dornfeld, L. (2001). Pomegranate juice consumption inhibits serum angiotensin converting enzyme activity and reduces systolic blood pressure. *Atherosclerosis*, 158, 195–198.
6. Barnes, S., Prasain, J., D'Alessandro, T., Arabshahi, A., Botting, N., Lila, M. A., Jackson. G., Janle, E. M., & Weaver, C. M. (2011). The metabolism and analysis of isoflavones and other dietary polyphenols in foods and biological systems. *Food Funct*, 2, 235–244.
7. Bazzano, L. A., Serdula, M. K., & Liu, S. (2003). Dietary intake of fruits and vegetables and risk of cardiovascular disease, *Current Atherosclerosis Reports*, 5(6), 492–499.
8. Bernstein, A. M., Sun, Q., Hu, F. B., Stampfer, M. J., Manson, J. E., & Willett, W. C. (2010). Major dietary protein sources and risk of coronary heart disease in women. *Circulation*, 122 (9), 876–883, 2010.
9. Bidel, S., Hu, G., Qiao, Q., Jousilahti, P., Antikainen, R., & Tuomilehto, J. (2006). Coffee consumption and risk of total and cardiovascular mortality among patients with type 2 diabetes. *Diabetologia*, 49(11), 2618–2626.

10. Biesalski, H. K. (2001). Nutraceuticals: the link between nutrition and medicine. In: *Nutraceuticals in Health and Disease Prevention.* Kramer, K., Hoppe, P. P., & Packer, L. (eds.). New York: Marcel Decker, Inc. pp. 1–2.

11. Block, G., Patterson, B., & Subar, A. (1992). Fruit, vegetables, and cancer prevention: a review of the epidemiological evidence. *Nutrition and Cancer*, 18(1), 1–29.

12. Boosalis, M. G., Snowdon, D. A., Tully, C. L., & Gross, M. D. (1996). Acute phase response and plasma carotenoid concentrations in older women: findings from the nun study. *Nutrition,* 12, 475–478.

13. Bors, W., Heller, W., Michel, C., & Saran, M. (1990). Flavonoids as antioxidants: determination of radical-scavenging efficiencies. *Methods. Enzymol,* 186, 343–355.

14. Burslem, J., Schonfeld, G., & Howald, M. A. (1978). Plasma apoprotein and lipoprotein lipid levels in vegetarians. *Metabolism,* 27(6), 711–719.

15. Chopra, M., O'Neill, M. E., Keogh, N., Wortley, G., Southon, S., & Thurnham, D. I. (2000). Influence of increased fruit and vegetable intake on plasma and lipoprotein carotenoids and LDL oxidation in smokers and nonsmokers. *Clinical Chemistry,* 46(11), 1818–1829.

16. Christensen, B., Mosdol, A., Retterstol, L., Landaas, S., & Thelle, D. S. (2001). Abstention from filtered coffee reduces the concentrations of plasma homocysteine and serum cholesterol—a randomized controlled trial. *American Journal of Clinical Nutrition,* 74(3), 302–307.

17. Clinton, S. K. (1998). Lycopene: chemistry, biology, and implications for human health and disease. *Nutr. Rev.,* 56, 35–51.

18. Cook, N., & Samman, S. (1996). Flavonoids-chemistry, metabolism, cardio protective effects and dietary sources. *J. Nutr. Biochem.* 7, 66–76.

19. Davidson, M., Maki, H., & Dicklin K. C. (2009). Effects of consumption of pomegranate juice on carotid intima-media thickness in men and women at moderate risk for coronary heart disease. *American Journal of Cardiology*, 104(7), 936–942.

20. Delgado-Lista, J., Perez-Martinez, P., Lopez-Miranda, J., & Perez-Jimenez, F. (2012). Long chain omega-3 fatty acids and cardiovascular disease: a systematic review. *Br J Nutr.* 107, S201–S213.

21. Ding, E. L., Hutfless, S. M., Ding, X., & Girotra, S. (2006). Chocolate and prevention of cardiovascular disease: a systematic review. *Nutrition and Metabolism*, 3(2), 1–5.

22. Dohadwala, M. M., & Vita, J. A. (2009). Grapes and Cardiovascular Disease. *J Nutr.* 139(9), 1788S–1793S.

23. Erkkilä A. T., & Lichtenstein, A. H., 2006. Fiber and cardiovascular disease risk: how strong is the evidence? Journal of Cardiovascular Nursing, 21(1), 3–8.

24. Esmaillzadeh, A., Tahbaz, F., Gaieni, I., Alavi-Majd, H., & Azadbakht, L. (2004). Concentrated pomegranate juice improves lipid profiles in diabetic patients with hyperlipidemia. *J Med Food.* 7, 305–308.

25. Estruch, R. et al. (2013). Primary Prevention of Cardiovascular Disease with a Mediterranean Diet. *N Engl J Med.* 368, 1279–1290.

26. Fleming, J. A., Holligan, S., & Kris-Etherton, P. M. (2013). Dietary Patterns that Decrease Cardiovascular Disease and Increase Longevity. *J. Clin. Exp. Cardiolog.* S6(6), 1–6.

27. Folsom, A. R., Parker, E. D., & Harnack, L. J. (2007). Degree of concordance with DASH diet guidelines and incidence of hypertension and fatal cardiovascular disease. *American Journal of Hypertension*, 20(3), 225–232.

28. Galleano, M., Oteiza, P. I., & Fraga, C. G. (2009). Cocoa, chocolate, and cardiovascular disease. *Journal of Cardiovascular Pharmacology*, 54(6), 483–490.

29. GBD, (2013). Mortality and Causes of Death, Collaborators (17 December 2014). Global, regional, and national age-sex specific all-cause and cause-specific mortality for 240 causes of death, 1990–2013: a systematic analysis for the Global Burden of Disease (GBD) Study. *Lancet,* 385 (9963), 117–71.

30. Geleijnse, J. M., Launer, L. J., Van Der Kuip, D. A. M., Hofman, A., & Witteman, J. C. M. (2002). Inverse association of tea and flavonoid intakes with incident myocardial infarction: the Rotterdam study. *American Journal of Clinical Nutrition*, 75(5), 880–886.

31. George, T. W., Niwat, C., Waroonphan, S., Gordon, M. H., & Lovegrove, J. A. (2009). Effects of chronic and acute consumption of fruit- and vegetable-puree-based drinks on vasodilation, risk factors for CVD and the response as a result of the eNOS G298T polymorphism, Proceed. *Nutrit. Soci.* 68(2), 148–161.

32. Godos, J., Pluchinotta, F. R., Marventano, S., Buscemi, S., Li, V. G., Galvano, F., & Grosso, G. (2014). Coffee components and cardiovascular risk: beneficial and detrimental effects. *Int. J. Food Sci. Nutr.,* 21, 1–12.

33. Grassi, D., Lippi, C., Necozione, S., Desideri, G., & Ferri, C. (2005). Short-term administration of dark chocolate is followed by a significant increase in insulin sensitivity and a decrease in blood pressure in healthy persons. *American Journal of Clinical Nutrition*, 81(3), 611–614.

34. Greenberg, J. A., Dunbar, C. C., Schnoll, R., Kokolis, R., Kokolis, S., & Kassotis, J. (2007). Caffeinated beverage intake and the risk of heart disease mortality in the elderly: a prospective analysis. *American Journal of Clinical Nutrition*, 85 (2), 392–398.

35. Grosso, G., Bei, R., Mistretta, A., Marventano, S., Calabrese, G., Masuelli, L., Giganti, M. G., Modesti, A., Galvano, F., & Gazzolo, D. (2013). Effects of vitamin C on health: a review of evidence. *Front Biosci (Landmark Ed)*, 18, 1017–1029.

36. Harland, J. I., & Garton, L. E. (2008). Whole-grain intake as a marker of healthy body weight and adiposity. *Public Health Nutr.,* 11, 554–563.

37. Harland, J. I., & Haffner, T. A. (2008). Systematic review, meta-analysis and regression of randomized controlled trials reporting an association between an intake of circa 25 g soya protein per day and blood cholesterol. *Atherosclerosis*, 200(1), 13–27.

38. He, J., Klag, M. J., Whelton, P. K., Chen, J. Y., Qian, M. C., & He, G. Q. (1995). Dietary macronutrients and blood pressure in southwestern China. *Journal of Hypertension*, 13(11), 1267–1274.

39. He, K., Merchant, A., Rimm E. B., et al. (2003). Dietary fat intake and risk of stroke in male US healthcare professionals: 14-year prospective cohort study. *British Medical Journal*, 327(7418), 777–781.

40. He, K., Song, Y., Daviglus, M. L., et al. (2004). Accumulated evidence on fish consumption and coronary heart disease mortality: a meta-analysis of cohort studies. *Circulation*, 109(22), 2705–2711.

41. Hertog, M. G. L., Hollman, P. C. H., Katan, M. B., & Kromhout, D. (1993). Intake of potentially anticarcinogenic flavonoids and their determinants in adults in The Netherlands. *Nutrition and Cancer*, 20(1), 21–29.

42. Hicks, K. B., & Moreau, R. A. (2001). Phytosterols and phytostanols: functional food cholesterol busters. *Food Technology*, 55, 63–67.

43. Hu, F. B. (2003). Plant-based foods and prevention of cardiovascular disease: an overview. *Am. J. Clin. Nutr.* 78(3), 544S–551S.

44. International Food Information Council (IFIC) (1999). *Functional Foods Now.* Washington, DC.

45. Jacobs, D. R., Pereira, M. A., Meyer, K. A., & Kushi, L. H. (2000). Fiber from whole grains, but not refined grains, is inversely associated with all-cause mortality in older women: the Iowa Women's Health Study. *Journal of the American College of Nutrition*, 19 (3), 326S–330S.

46. Jenkins, D. J., Mirrahimi, A., Srichaikul, K., Berryman, C. E., Wang, L., Carleton, A., Abdulnour, S., Sievenpiper, J. L., Kendall, C. W., & Kris-Etherton, P. M. (2010). Soy protein reduces serum cholesterol by both intrinsic and food displacement mechanisms. *J. Nutr.,* 140(12), 2302S–2311S.

47. Kalra, E. K. (2003). Nutraceutical—Definition and Introduction. *AAPS Pharm. Sci.,* 5, 1–2.

48. Keservani, R. K., Kesharwani, R. K., Sharma, A.K., Vyas, N., & Chadoker, A. (2010). Nutritional Supplements: An Overview. *Int. J. Curr. Pharm. Rev. Res.* (IJCPR). 1, 59–75.

49. Keservani, R. K., Kesharwani, R. K., Vyas, N., Jain, S., Raghuvanshi, R., & Sharma, A. K., (2010). Nutraceutical and functional food as future food: A Review. *Der Pharmacia. Lett.* 2, 106–116.

50. Keservani, Raj K., Kesharwani, Rajesh K., & Sharma, Anil K. (2015). Pulmonary and Respiratory Health: Antioxidants and Nutraceuticals. In: *Nutraceuticals and Functional Foods in Human Health and Disease Prevention. U.K.,* D. Bagchi, H. G. Preuss and A. Swaroop eds. CRC Press, Taylor and Francis, 17, pp. 279–298.

51. Keservani, Raj K., Kesharwani, Rajesh K., Sharma, Anil K., & Jarouliya, U. (2015). Dietary Supplements, Nutraceutical and Functional Foods in Immune Response (Immunomodulators), In: *Nutraceuticals and Functional Foods in Human Health and Disease Prevention. U.K.,* D. Bagchi, H. G. Preuss, & A. Swaroop (eds.). CRC Press, Taylor and Francis, 20, pp. 343–360.

52. Keservani, Raj K., & Sharma, Anil K. (2014). Flavonoids: emerging trends and potential health benefits. *J Chin. Pharm. Sci.,* 23 (12), 815–822.

53. Keservani, Raj K., Singh, S., Singh, V., Kesharwani, Rajesh K., & Sharma, Anil K. (2015). Nutraceuticals and Functional Foods in the Prevention of Mental Disorder, In: *Nutraceuticals and Functional Foods in Human Health and Disease Prevention. U.K.,* D. Bagchi, H. G. Preuss, & A. Swaroop (eds.). CRC Press, Taylor and Francis, 15, pp. 255–270.

54. Klipstein-Grobusch, K., Launer, L. J., Geleijnse, J. M., Boeing, H., Hofman, A., & Witteman, J. C. (2000). Serum carotenoids and atherosclerosis—The Rotterdam Study. *Atherosclerosis* 148, 49–56.

55. Knekt, P., Järvinen, R., Reunanen, A., & Maatela, J. (1996). Flavonoid intake and coronary mortality in Finland: a cohort study. *British Medical Journal*, 312(7029), 478–481.

56. Kohlmeier, L., & Kark, J. D. (1997). Lycopene and myocardial infarction risk in the EURAMIC Study. *Am J Epidemiol.*, 146, 618–626.
57. Kritchevsky, S. B., Bush, A. J., Pahor, M., & Gross, M. D. (2000). Serum carotenoids and markers of inflammation in nonsmokers. *Am J Epidemiol.*, 152, 1065–1071.
58. Kromhout, D., Feskens, E. J. M., & Bowles, C. H. (1995). The protective effect of a small amount of fish on coronary heart disease mortality in an elderly population. *International Journal of Epidemiology*, 24(2), 340–345.
59. Law, M. R., Wald, N. J., Wu, T., Hackshaw, A., & Bailey, A. (1994). Systematic underestimation of association between serum cholesterol concentration and ischaemic heart disease in observational studies: data from the BUPA study. *BMJ*, 308(6925), 363–366.
60. Liu, R. H. (2003). Health benefits of fruit and vegetables are from additive and synergistic combinations of phytochemicals. *American J. Clin. Nutrit.*, 78(3), 517S–520S.
61. Lloyd-Jones, D. M., Hong, Y., Labarthe, D., et al. (2010). American Heart Association Strategic Planning Task Force and Statistics Committee: Defining and setting national goals for cardiovascular health promotion and disease reduction: the American Heart Association's strategic Impact Goal through 2020 and beyond. *Circulation*, 121, 586–613.
62. Lupton, J. R., & Turner, N. D. (2003). Dietary fiber and coronary disease: Does the evidence support an association? *Current Atherosclerosis Reports*, 5(6), 500–505.
63. Lusas, E.W., & Riaz, M.N. (1995). Soy protein products: processing and use. *J Nutr.*, 125(3), 573S–580S.
64. Mattson, M. P. (2003). Will caloric restriction and folate protect against AD and PD? *Neurology*, 60(4), 690–695.
65. McGill, H. C. (1979). The relationship of dietary cholesterol to serum cholesterol concentration and to atherosclerosis in man. *American Journal of Clinical Nutrition*, 32(12), 2664–2702.
66. Mellen, P. B., Walsh, T. F., & Herrington, D. M. (2008). Whole grain intake and cardiovascular disease: a meta-analysis. *Nutrition, Metabolism and Cardiovascular Diseases*, 18(4), 283–290.
67. Mendis, S., Puska, P., Norrving, B., (2011). World Health Organization. Global Atlas on Cardiovascular Disease Prevention and Control (PDF). *World Health Organization in collaboration with the World Heart Federation and the World Stroke Organization.* 3–18.
68. Morris, J. N., Marr, J. W., & Clayton, D. G. (1977). Diet and heart: a postscript. *British Medical Journal*, 2(6098), 1307–1314.
69. Mozaffarian, D. (2008) Fish and n–3 fatty acids for the prevention of fatal coronary heart disease and sudden cardiac death. *American Journal of Clinical Nutrition*, 87 (6), 1991S–1996S.
70. Mukamal, K. J., Maclure, M., Muller, J. E., Sherwood, J. B., & Mittleman, M. A. (2002). Tea consumption and mortality after acute myocardial infarction. *Circulation*, 105(21), 2476–2481.
71. Nagata, C., Takatsuka, N., Kurisu, Y., & Shimizu, H. (1998). Decreased serum total cholesterol concentration is associated with high intake of soy products in Japanese men and women. *Journal of Nutrition*, 128(2), 209–213.

72. Nettleton, J. A., Steffen, L. M., Loehr, L. R., Rosamond, W. D., & Folsom, A. R. (2008). Incident heart failure is associated with lower whole-grain intake and greater high-fat dairy and egg intake in the Atherosclerosis Risk in Communities (ARIC) study. *J Am Diet Assoc.*, 108(11), 1881–1887.

73. Normén, L., Holmes, D., & Frohlich, J. (2005). Plant sterols and their role in combined use with statins for lipid lowering. *Current Opinion in Investigational Drugs,* 6 (3), 307–316.

74. Panagiotakos, D. B., Pitsavos, C., Chrysohoou, C., Kokkinos, P., Toutouzas, P., & Stefanadis, C. (2003). The J-shaped effect of coffee consumption on the risk of developing acute coronary syndromes: the CARDIO2000 case-control study. *Journal of Nutrition*, 133(10), 3228–3232.

75. Pereira, T., Vilas B. M., & Conde, J. (2014). Dark chocolate intake improves endothelial function in young healthy people: a randomized and controlled trial. *Cardio Vasc Syst.*, 2(3), 1–6.

76. Peters, U., Poole, C., & Arab, L. (2001). Does tea affect cardiovascular disease? A meta-analysis. *American Journal of Epidemiology*, 154(6), 495–503.

77. Psaltopoulou, T., Ilias, I., & Alevizaki, M. (2010). The role of diet and lifestyle in primary, secondary, and tertiary diabetes prevention: a review of meta-analyses. *Rev Diabet Stud.*, 7, 26–35.

78. Qi, L., & Hu, F. B. (2007). Dietary glycemic load, whole grains, and systemic inflammation in diabetes: the epidemiological evidence. *Curr Opin Lipidol.*, 18, 3–8.

79. Qi, L., van Dam, R. M., Liu, S., Franz, M., Mantzoros, C., & Hu, F. B. (2006). Whole-grain, bran, and cereal fiber intakes and markers of systemic inflammation in diabetic women. *Diabetes Care*, 29, 207–211.

80. Raatz, S. K., Silverstein, J. T., Jahns, L., & Picklo, M. J. (2013). Issues of fish consumption for cardiovascular disease risk reduction. *Nutrients,* 5, 1081–1097.

81. Rao, A. V. (2002). Lycopene, tomatoes and the prevention of coronary heart disease. *Exp Biol Med (Maywood).* 227, 908–913.

82. Reddy, K. S., & Katan, M. B. (2004). Diet, nutrition and the prevention of hypertension and cardiovascular diseases. *Public Health Nutrition*, 7(1), 167–186.

83. Reynolds, K., Chin, A., Lees, K. A., Nguyen, A., Bujnowski, D., & He, J. (2006). A meta-analysis of the effect of soy protein supplementation on serum lipids. *Am. J. Cardiol.* 98(5), 633–640.

84. Rissanen, T. H., Voutilainen, S., Nyyssonen, K., & Salonen, J. T. (2002). Lycopene, atherosclerosis and CHD. *Exp. Biol. Med.*, 227, 900–907.

85. Rissanen, T. H., Voutilainen, S., Nyyssonen, K., Salonen, R., Kaplan G. A., & Salonen, J. T. (2003). Serum lycopene concentration and carotid atherosclerosis: the Kuopio ischemic heart disease risk factor study. *Am. J. Clin. Nutr.*, 77, 133–138.

86. Ros, E., Núñez, I., Pérez-Heras, A., et al. (2004). A walnut diet improves endothelial function in hypercholesterolemia subjects: a randomized crossover trial. *Circulation,* 109(13), 1609–1614.

87. Rosenblat, M., Hayek, T., & Aviram, M. (2006). Anti-oxidative effects of pomegranate juice (PJ) consumption by diabetic patients on serum and on macrophages. *Atherosclerosis,* 187, 363–371.

88. Sabaté J., & Ang, Y. (2009). Nuts and health outcomes: new epidemiologic evidence. *American Journal of Clinical Nutrition*, 89(5), 1643S–1648S.

89. Sacks, F. M., Lichtenstein, A., Van Horn, L., Harris, W., Kris-Etherton, P., & Winston, M. (2006). Soy protein, isoflavones, and cardiovascular health: An American Heart Association Science Advisory for professionals from the Nutrition Committee. *Circulation*, 113(7), 1034–1044.

90. Sala-Vila, A. (2014). Changes in Ultrasound-Assessed Carotid Intima-Media Thickness and Plaque With a Mediterranean Diet: A Substudy of the PREDIMED Trial. *Arterioscler Thromb Vasc Biol.* 34(2), 439–445.

91. Savica, V., Bellinghieri, G., & Kopple, J. D. (2010). The effect of nutrition on blood pressure. *Annual Review of Nutrition*, (30), 365–401.

92. Singh, U., Devaraj, S., & Jialal, I. (2005). Vitamin E, oxidative stress, and inflammation. *Annual Review of Nutrition*, 25, 151–174.

93. Stampfer, M. J., Hu, F. B., Manson, J. E., Rimm, E. B., & Willett, W. C. (2000). Primary prevention of coronary heart disease in women through diet and lifestyle. *New England Journal of Medicine*, 343(1), 16–22.

94. Steffen, L. M., Jacobs, D. R. Jr., Stevens, J., Shahar, E., Carithers, T., Folsom, A. R. (2003). Associations of whole-grain, refined-grain, and fruit and vegetable consumption with risks of all-cause mortality and incident coronary artery disease and ischemic stroke: the atherosclerosis risk in communities (ARIC) study. *Am J Clin Nutr.*, 78(3), 383–390.

95. Streppel, M. T., Arends, L. R., Van't Veer, P., Grobbee, D. E., & Geleijnse, J. M. (2005). Dietary fiber and blood pressure: a meta-analysis of randomized placebo-controlled trials. *Archives of Internal Medicine*, 165(2), 150–156.

96. Sumpio, B. E., Cordova, A. C., Berke-Schlessel, D. W., Qin, F., & Chen, Q. H. (2006). Green tea, the "Asian Paradox," and cardiovascular disease. Journal *of the American College of Surgeons*, 202(5), 813–825.

97. Tokunaga, S., White, I. R., & Frost, C. (2002). Green tea consumption and serum lipids and lipoproteins in a population of healthy workers in Japan. *Ann Epidemiol.*, 12,157–165.

98. Vlachopoulos, C. et al. (2005). Effect of Dark Chocolate on Arterial Function in Healthy Individuals. *American Journal of Hypertension,* 18, 785–791.

99. Zhan, S., & Ho, S. C. (2005). Meta-analysis of the effects of soy protein containing isoflavones on the lipid profile. *Am J Clin Nutr*, 81(2), 397–408.

100. Zhang, X., Shu, X. O., Gao, Y. T., et al. (2003). Soy food consumption is associated with lower risk of coronary heart disease in Chinese women. *Journal of Nutrition*, 133(9), 2874–2878.

PART IV

FOOD PRESERVATION

CHAPTER 11

NATURAL ANTIOXIDANTS DURING FRYING: FOOD INDUSTRY PERSPECTIVE

NAVNEET SINGH DEORA,[1] AASTHA DESWAL,[2]
and SANJITH MADHAVAN[3]

[1]*Research Scientist, Prasan Solutions (India) Pvt. Limited (Nalanda R&D Center), Cochin – 682021, Kerala. Mobile: +91-7042307007, E-mail: navneetsinghdeora@gmail.com*

[2]*Research Scientist, Prasan Solutions (India) Pvt. Limited (Nalanda R&D Center), Cochin – 682021, Kerala. Mobile: +91-8137892690, E-mail: deswalad@gmail.com*

[3]*Technical Director, Prasan Solutions (India) Pvt. Limited (Nalanda R&D Center), Cochin – 682021, Kerala. Mobile: +91-8137892690, E-mail: sanjith@prasansolutions.com*

CONTENTS

11.1   INTRODUCTION

Antioxidants are the compounds possessing the ability to inhibit oxidation when present in food or biological systems. As per the structural features, antioxidants can scavenge free radicals, inactivate pro-oxidant metals, quench singlet oxygen, or inactivate sensitizers (Figure 11.1). The classification of natural antioxidants is shown in Figure 11.2. Apart from their structural features, the effectiveness of antioxidants also depends on the concentration used, the applied temperature, exposure to light, and type of substrates [35, 85].

A number of antioxidants occur naturally in food (endogenous) where they offer protection against oxidative damage. These endogenous anti-

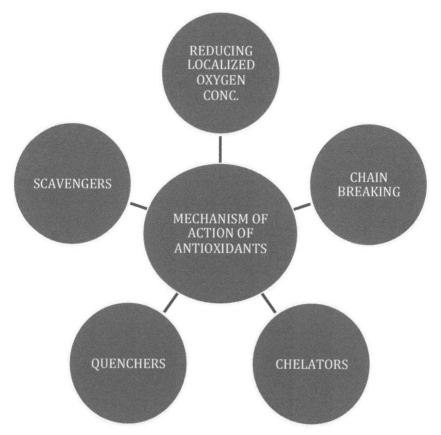

FIGURE 11.1   Antioxidant mechanism.

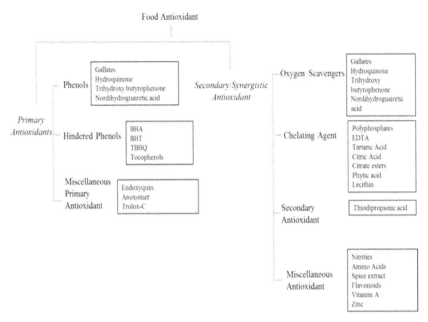

FIGURE 11.2   Different types of antioxidants [43].

oxidants are part of the unsaponifiable components of fats and oils, representing less than 5% of the total lipid composition. However, due to inherent shortcomings in their applicability and performance, a number of synthetic antioxidants such as butylated hydroxytoluene (BHT), butylated hydroxyanisole (BHA), and tert-butyl hydroxyquinone (TBHQ) were prepared. Unfortunately, the use of these common synthetic antioxidants has also been limited due to their failure to meet performance expectations, especially under frying conditions, and their perceived detrimental effect on human health [44, 46]. Consequently, there is a growing involvement in development of new antioxidants with enhanced antioxidant activity and thermal stability, but prepared from natural precursors. The new trend in this direction is leading to adaptation of existing natural antioxidants. A number of these synthetic (or semi-synthetic) antioxidants have been approved by appropriate authorities in various nations around the world [64]. For example, fatty acid esters of vitamin C (ascorbylpalmitate, stearate, oleate and linoleate; E304) have been synthesized to enhance the lipophilicity of vitamin C for stabilization of lipids.

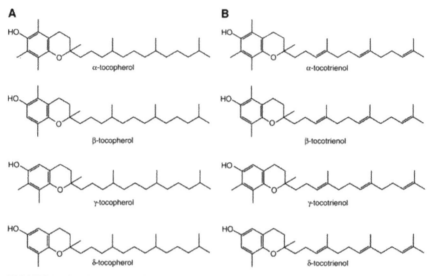

FIGURE 11.3   Structures of tocochromanols.

The chemical assays methods such DPPH, ORAC, FRAP, ABTS, TBA, and β-carotene bleaching are normally used to evaluate the activity of antioxidative compounds. However, these indirect methods are often found out to poorly correlate with the ability of compounds to inhibit oxidation in real food systems due to their inability to account for a number of important variables such as the physical location of the antioxidant and its interaction with other food components.

In real food applications, most of the antioxidant studies have been conducted under ambient or accelerated storage conditions. However, the activity of antioxidative compounds under ambient and accelerate storage tests may not correctly depict their performance during frying, considering the wide differences in operational conditions including: (i) temperatures; (ii) oxygen pressure; (iii) water content; (iv) thermo oxidative degradation of components and their interactions with antioxidative compounds; (v) volatilization, degradation, and thermal inactivation of antioxidants; and (vii) continuous changes in relative locations of antioxidant, oxidant, and substrates due to constant agitation during boiling.

In this chapter, the performance of common natural antioxidants under frying conditions is summarized with the addition of recent trends in the designing of natural antioxidant for frying applications.

11.2   NATURAL ANTIOXIDANTS

11.2.1   Tocochromanols

A number of naturally occurring minor components in edible oils possess antioxidant activity, and are of great importance to both storage and frying stability of oils. The most important and often studied antioxidants are the tocopherols and tocotrienols, collectively referred to as tocochromanols [22]. The structural difference between tocopherols and tocotrienols is in the unsaturation of the phytyl side chain (Figure 11.3). Depending on the position and the number of a methyl substitution on the chromanol ring, four homologues, α, β, γ, and δ, are recognized for each tocopherol and tocotrienol. Of these, the α- and γ-tocopherols are the most abundant in frying oils [46, 51].

More often than not, edible oils rely on tocochromanols for protection against oxidative degradation, mainly because of their excellent radical scavenging activity. It has been reported that lipid peroxy radicals react with tocopherols much faster (10^4 to 10^9 $M^{-1}s^{-1}$) than with lipids (10 to 60 $M^{-1}s^{-1}$) [15]. According to Kamal-Eldin and Appelqvist [31], onetocopherol molecule can protect about 103 to 108 polyunsaturated fatty acid molecules at low peroxide value [31]. A very strong positive correlation was reported between the radical scavenging capacity of different refined oils and the total content of tocochromanols [31, 73]

A number of studies have evaluated the ability of tocochromanols to protect oils under frying conditions (Table 11.1). According to Nogala-Kalucka et al. [58], the addition of 100, 500, or 1000 μg/g of δ-tocopherol to Planta, a commercial blend of hydrogenated canola oil and palm oil resulted in a significant decrease in the peroxide value, p-anisidine value, and hexanal formation during heating at 160°C for 2 h in a Rancimat. Supplementing technical triolein with 100 or 400 μg/g of γ-tocopherol significantly inhibited polymer formation during frying of potato chips at 190°C for 6 h [56]. The effect of α-tocopherol and α-tocotrienol on the performance of antioxidant stripped canola oil at a high temperature was reported by Romero et al. [71]. The pure canola oil triacylglycerols were supplemented with α-tocopherol and α-tocotrienol at various concentrations: 155 and 432 μg/g tocopherol; 138 μg/g tocotrienol; and a mixture of 66 µg/g tocopherol and 72 µg/g of tocotrienol. The samples were heated at 180°C for 18 h in a Rancimat apparatus, and performance assessed by measurement of the amount of total polar components. The authors reported a significant level of protection by both α-tocopherol and α-tocotrienol, although the later was significantly less effective. They also observed that increasing the level of α-tocopherol from 155 to 432 µg/g did not improve its antioxidative activity. In a similar study, Lampi and Kamal-Eldin [37]evaluated the antioxidant activity of α- and γ-tocopherol in purified high oleic sunflower triacylglycerols. The oil with or without α- and γ-tocopherol was heated for 24 h at 185°C in an oven. Thermo oxidative changes in oil were measured by analysis of polymerized materials, and both tocopherols significantly inhibited polymerization of the oil; however, the γ-isomer was more effective. The authors did not observe any synergistic relationship between the two isomers when they were added as a mixture. On the contrary, Warner and Moser [83] observed that samples of purified mid-oleic sunflower oil triacylglycerols containing a mixture of α, γ, and δ tocopherols accumulated significantly lower amounts of the total polar components compared to effectiveness of individual isomers during frying of tortilla chips. Similarly, Barrera-Arellano et al. [11] reported a synergistic interaction between α, β, γ, and δ tocopherols during thermal treatment of purified palm olein and soybean oil triacylglycerols at 180°C for 10 h in a Rancimat as assessed by the amounts of polymers formed. The tocochromanol mixtures isolated from canola, rice bran and palm oils significantly inhibited thermo oxidation of canola oil triacylglycerol during a model frying of formulated food at 185°C as measured by the amount of total polar compounds and the rate of volatile carbonyl compounds and hydroxynonenal formation [5]. Interestingly, no significant difference was observed in the performance of the tocochromanol mixture despite the differences in isomeric composition, suggesting an absence of synergies among tocopherol and tocotrienol isomers. Romero et al. [71] also found no increase in the antioxidant activity when a mixture of α-tocopherol and α-tocotrienol was added to antioxidants stripped canola oil as compared to when only tocotrienol was added.

Normand et al. [59] compared the frying stability of regular and three modified canola oils during a 72-hour actual frying operation, and performance was assessed by the analysis of free fatty acids and total polar

TABLE 11.1 Studies on the Relative Stability of α- and γ-Tocopherols Under Frying Condition

Study	Major Observation	Reference
Frying of potato chips in a blend of palm olein: canola oil (1:1) at 185°C	(a) α-degradation was lower (186–57) than in γ (b) In canola oil, no significant difference with α and γ (c) In palm olein, at 7 h, γ- was more stable than α	[3]
Frying of potatoes, milanesas, and churros in soybean, sunflower and partially hydrogenated oil at 180°C for 7 h/day for 2 days	(a) In soybean, α- was more stable with 67% loss (from 149–49.1) and γ-, 69% loss from 468–145 µg/g (b) In partially hydrogenated mixture of vegetable oil, α- was more stable (17% loss; 101–84) and γ- 66% loss (626–210)	[30]
Frying of French fries and chicken nuggets in RBD palm olein at 180–185°C.	(a) In the heated control, α- was more stable, decreasing from 250 to 50 µg/g at the end of 20 cycle of frying cf to γ-, 100–10 (b) During frying of both French fries, α- wasmore stable with a loss of ~40% after 40 cyclesand total loss of γ- during the same period.	[9]
Frying of tortilla chips in purified mid-oleic sun flower oil fortified with different combination of tocopherols at 180 °C (6 h/d for 1 day)	(a) Added singly at ~850 µg/g, γ was more stable after 6 h (37 vs 35% retention) (b) Added together in equal amounts (400 µg/g, each), no difference in stability after 6 h (40% retention, each) although γ- was more stable at lower frying time ≤3 h) (c) Added together in equal amounts (360 µg/g, each) in the presence of 60 µg/g δ-, no difference in stability at 1 h and 3 h, but more γ- was retained after 6 h (38 vs 32%).	[82]
Frying of French fries in canola oil at 185°C (7 h/d for 7 days) under atmospheric condition, carbon dioxide blanketing, and vacuum	(a) Under standard frying condition, α- was more stable than γ- (37 vs 44 µg/g/d degradation rate) (b) Under carbon dioxide blanketing, α- was more stable than γ- (7 vs 15 µg/g/d degradation rate) (c) Under vacuum α- was more stable than γ- (2 vs 4 µg/g/d degradation rate)	[6]

TABLE 11.1   (Continued)

Study	Major Observation	Reference
Heating of two varieties of extra virgin olive oil at 180°C for 1 h	(a) in all varieties of extra virgin olive oil, γ-tocopherol was more stable (~40–11 µg/g and 11–3 µg/g) compared to α- (165–20 µg/g and 120–9 µg/g)	[50]
Frying of French fries in canola oil at 185 and 215°C (7 h/d for 7 days)	(a) at 185°C, α- (from 214–93 µg/g) more stable than γ- (347–115 µg/g) (b) at 215°C, γ- (347–55 µg/g) more stable than α- (214–10 µg/g) at the end of the 3rd day. (c) All tocopherol were lost at the end of the 6th day at 215°C	[7]
Frying of cassava chips in refined cottonseed oil at 180°C (5 h/d for 5 days)	(a) γ- was more stable withloss of 2.5 µg/g/d compared to α- at 3.7 µg/g/d.	[18]
Heating of genetically modified sunflower oil at 180°C for up to 25 h using a Rancimat	(a) γ- in modified sunflower oil containing only γ was more stable (808–241 µg/g) than α- in sunflower oil containing only α-isomer (826–28 µg/g) (b) γ- was more stable than α in a 50:50 mixture of both sun flower oils after 10 h of heating.	[45]
Frying of potatoes in palm oil at 175 °C for 18 h (6 h/d)	(a) α-tocotrienol was more stable (259–199 µg/g) than γ-tocotrienol (438–75 µg/g)	[73]
Frying of French fries in High oleic/low linolenic canola and partially hydrogenated canola oils at 175 °C	(a) α-tocopherol was more stable than γ-tocopherol	[49]
Heating of refined soybean and partially hydrogenated soybean at 180°C for 10 h in a Rancimat	(a) α- was more stable than γ- for soybean oils with IV 130, 115, and 95. (b) Residual γ- were 27,17, and 12 for soybean 130, 115, and 95, respectively. (c) The corresponding α- contents were ~38, 27, and 18 µg/g. (d) The initial tocopherol ranged from 200–218, α- and 841 –931 µg/g, γ-	[78]

TABLE 11.1 (Continued)

Study	Major Observation	Reference
Heating of palm olein, olive, sun flower, rapeseed and soybean oils at 180°C for 10 h in a Rancimat	(a) In all natural oils γ-tocopherol was more stable (b) In stripped palm and soybean oils supplemented with 250 µg/g each of α- and γ-, the γ-isomer was slightly more stable	[11]
Frying of French fries in regular, hydrogenated, low linoleic, and high linolenic canola oil at 175 °C (12 h/d; 6 d)	(a) In all oils, α-tocopherol was more stable	[59]
Heating of tocopherol fortified triolein, trilinolein and their 1:1 mixture at 180°C for 10 h in a Rancimat	(a) α-tocopherol was less stable than γ-tocopherol	[10]
Frying of wet cotton-balls in soybean, corn and palm olein at 185°C (10 h/d for 3 d)	(a) In soybean oil, α- was more stable (79–7 µg/g) than γ- (474–9 µg/g) (b) In corn oil, α- was more stable (123–93 µg/g) than γ- (585–206 µg/g)	[75].
Heating of rapeseed oil, sunflower oil, and tocopherol fortified high-oleic sunflower oil at 180°C in an oven for up to 24 h	(a) In all oils, γ-tocopherol was more stable than α	[37]
Frying of French fries in regular, hydrogenated, low linoleic, and high linoleniccanola oil at 180°C (8 h/d; 5 d)	(a) In all oils, α-tocopherol was more stable	[42]
Frying of potato chips in rapeseed oil at 162°C for 6 d (12 batches/d)	(a) γ- was more stable (t1/2,7–8 frying operations) than α- (t1/2, 4–5 frying operations).	[24]
Frying of French fries, with and without coating, in a mixture of soybean and canola oils at 180°C (32 frying operation; 17 min/frying operation)	(a) α- was more stable (83% retention) than γ- (50% retention) during tempura-frying (b) α- was more stable (48% retention) than γ- (25% retention) during frying without coating	[51]

TABLE 11.1   (Continued)

Study	Major Observation	Reference
Heating of soybean, canola, rice bran, corn, cottonseed, and safflower oil at 180°C with continuous water spraying and gentle agitation of oil for 10 h	(a) α-tocopherol was more stable in all oils	[87]

components. The rate of tocopherol degradation was reported as the major factor determining the frying stability of the oils. This was in agreement with the conclusion by Petukhov et al. [63] during frying of potato chips in regular, high oleic low linolenic, and hydrogenated canola oils.

The α- and γ-isomers of tocopherol are the most common tocochromanol components of frying oils, and whereas there is a general agreement on their relative stability under oxidative and storage conditions, the controversy on their relative stability during frying still remains. In Table 11.1, the major studies regarding the stability of tocopherol isomers under frying conditions are presented. For studies conducted using actual frying experiments, it is generally agreed that the loss of γ-tocopherol was more rapid than α-tocopherol [1, 3, 6, 9, 27, 48, 51]. Conversely, majority of the studies involving heating the oil in an oven or a Rancimat apparatus concluded that α-tocopherol was less stable than the γ- isomer [11, 37, 38]. However, the chemical reactions taking place during actual frying of food are different from those during continuous heating. Thus, the activity of tocochromanols can be better assessed during an actual frying or in a frying test that truly mimics actual deep fat frying conditions. Based on available data (Table 11.1), it can be deduced that the relative stability of α- and γ-tocopherol under frying conditions depends on: (1) The relative amounts of the isomers in the initial oil, with the isomer present in the higher amount degrading much faster than the one present in lesser amount Warner and Moser [83]; (2) frying temperature and period; and (3) experimental set up, that is, frying of food vs heating; heating with vs heating without air, water, or agitation, etc.

Tocochromanols are thermally unstable, and are known to be easily removed during frying by evaporation/distillation [44]. The loss of tocopherols has been attributed to both oxidative and thermal degradation, with the rate being significantly slower in unsaturated oils than in saturated oils [59, 87]. A number of tocopherol degradation products notably, α-tocopherolquinone, 4a,5-epoxy-α-tocopheroxyquinone, 5,6-epoxy-α-tocopheroxyquinone, and 7,8-epoxy-α-tocopheroxyquinone have been identified both in model systems and during actual deep fat frying[53, 68, 81]. However, despite the prevalence of oligomerization during frying, oligomers of tocopherols are rarely encountered, presumably because coupling of tocopheroxyl radicals with lipid peroxy and alkyl radicals predominates over the formation of tocopherol oligomers through self-coupling [31]. The pro-oxidant effect observed for tocopherols under thermos oxidative conditions has been attributed to their oxidation products. Rietjens et al. [70] suggested that increased levels of oxidized α-tocopherol could result in increased levels of intermediate radicals, which can initiate lipid oxidation.

11.2.2 PHYTOSTEROLS

Phytosterols are the major constituents of unsaponifiables present in edible oils [74]. They are triterpenic compounds, structurally different from cholesterols only in the side chain configuration. The most common phytosterols in edible oils are β-sitosterol, campesterol, stigmasterol, Δ5-avenasterol, and brassicasterol (Figure 11.4). In vegetable oils, phytosterols are the dominant class of minor components and occur primarily as free sterolsorsteryl fatty acid esters. There is a general agreement that phytosterols offer no protection to oils under storage conditions or low temperature applications (<120°C) [13]. However, under frying conditions, phytosterols have been reported to inhibit thermooxidative alterations of frying oils. Sims et al. [76] investigated the ability of some phytosterols to inhibit thermooxidation of safflower oil by heating it to 180°C, and the extent of thermo oxidation was assessed by iodine value. Fucosterol, Δ7-avenasterol, and vernosterol offered significant protection to the oil, while ergosterol, β-sitosterol, and sigmasterol were either ineffective or

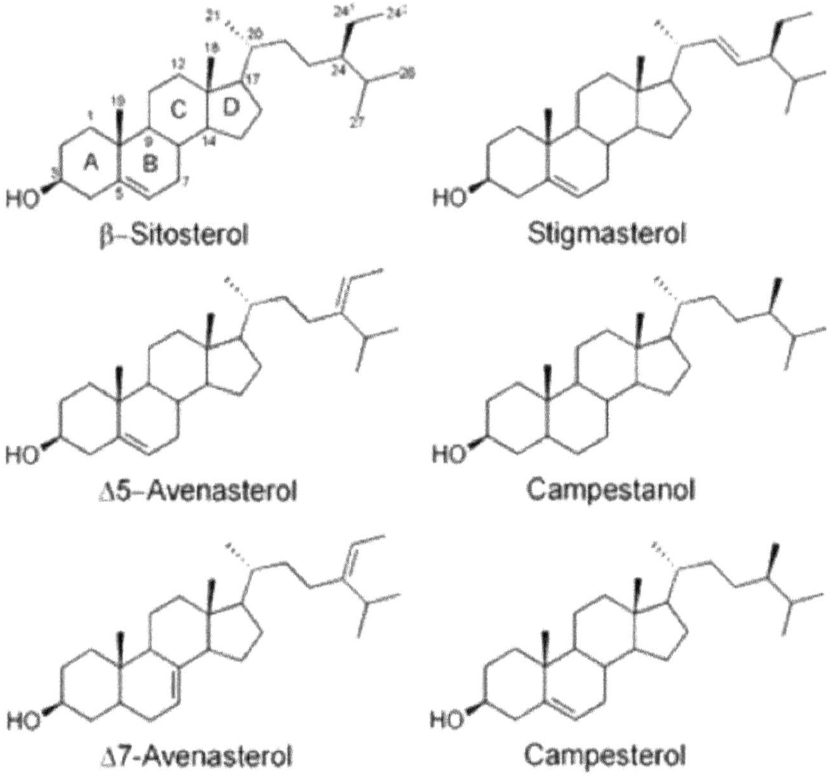

FIGURE 11.4 Structures of some common phytosterols and phytostanols.

slightly prooxidant. In a similar study, Gordon and Magos [25] reported that the addition of fucosterol and Δ5-avenasterol to technical triolein and heating it to 180°C inhibited thermo oxidation as measured by iodine value. According to Kochhar and Gertz [23], a mixture of phytosterols isolated from canola or sunflower oils significantly increased value of the Oxidative Stability at Elevated Temperature (OSET) index of canola oil heated at 170°C, indicating antioxidant activity. In a recent study, Winkler and Warner [82] observed an oil dependent activity of phytosterols. A mixture of phytosterols was added to purified soybean and high oleic sunflower oil triacylglycerols. The oils were heated at 180°C for up to 12 h, and formation of polymers was quantified by high performance size exclusion chromatography (HPSEC). The authors reported that the added phytosterols significantly decreased thermal polymerization of soybean oil triacylglycerols; however in high oleic sunflower oil triacylglycerols polymerization was significantly increased. Thus, the phytosterol mixture was effective in unsaturated oil but ineffective in the more saturated oil. A mixture of endogenous phytosterol isolated from canola (SCAN) and rice bran oil (SRBO) offered a concentration dependent protection for canola oil triacylglycerol during a model frying [4]. When 500 µg/g of SCAN was added, no protective activity was observed. On the contrary, at this low concentration the protection by SRBO was comparable to the observed for endogenous tocopherols isolated from canola or rice bran oils. This is presumably due to the presence of higher amount of sterols with known antioxidant activity such as avenasterol in RBO [5]. When 3,000 µg/g was added, however, the protection offered by SCAN was 20% better than tocopherols and comparable to SRBO as measured by the amount of TPC at the end of the frying period.

Like tocopherols, phytosterols can undergo thermos oxidative degradation under the conditions employed during deep frying, leading to a variety of polar and nonpolar compounds. The formation of phytosterol oxidation products have been studied both in model heating systems and under actual deep frying conditions [77, 79]. Dutta et al. [21] assessed the contents of phytosterol oxides in a hydrogenated canola/palm blend, sunflower oil, high oleic sunflower oils, and French fries fried in the various oils. 7α-, and 7β-hydroxysterols, 7-ketosterols, 5α,6α-epoxysterols, and dihydroxysterols were the major phytosterol oxides identified. Soupas et al. [77] evaluated the effects of the degree of unsaturation of both the phytosterols and the lipid medium on the formation of phytosteroloxides under different temperatures (60–180°C). Stigmasterol (unsaturated phytosterol) and sitostanol (saturated phytosterols) were added as model compounds to tripalmitin and canola triacylglycerols. The authors reported a significant influence of lipid matrices and temperature on the level and reaction pathway of phytosterol oxidation. For instance, after 3 h of heating at 180°C, the stigmasterol oxide contents were 24.2 and 7.4% in tripalmitin and purified canola oil, respectively. However, heating at 100°C yielded 0.3 and 26.5% stigmasterol oxide in tripalmitin and canola oil, respectively. It was also observed that the level of oxidation products from the unsaturated phytosterol was significantly higher than the level from the saturated phytosterol.

Whereas the biological activity of phytosterol oxidation products has been well studied and reviewed, their influence, antioxidative or prooxidative, on thermos oxidative stability of frying oil is yet to be studied.

11.2.3 GAMMA-ORYZANOL

Gamma-oryzanol, a mixture of ferulic acid steryl esters is a major antioxidant found in rice bran oil. At least 16 sterylferulates have been identified [55, 61]. Major components are presented in Figure 11.5. Studies on the activity of γ-oryzanol during frying are scarce, and are conducted by heating the oils rather than assessing during actual deep frying. It was reported that the addition of γ-oryzanol to refined canola and sunflower oils resulted in lower accumulation of dimers and polymers during an OSET test at 170°C [23]. Sitostanylferulate prevented polymerization in antioxidant stripped high oleic sunflower oil during heating at 180°C for 6 h [60]. The thiobarbituric acid-reactive substances (TBARs) in lard heated in the oven at 180°C for up to 10 days were significantly reduced in the presence of γ-oryzanol.

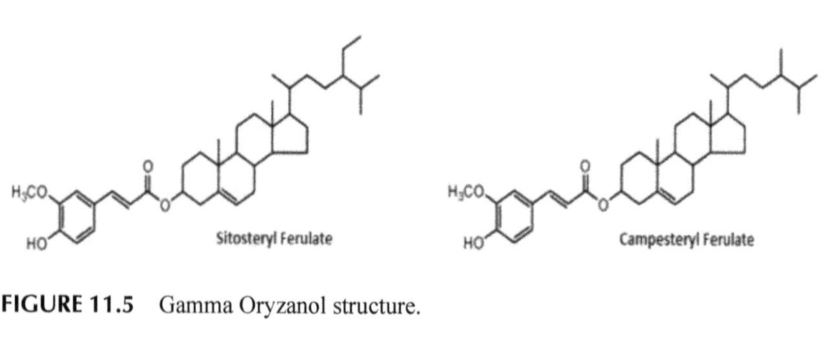

FIGURE 11.5   Gamma Oryzanol structure.

In a recent study, Winkler-Moser et al. [83] compared the ability of corn sterylferulate and γ-oryzanol on the thermos oxidative stability of refined, bleached, deodorized soybean oil in a 2-day frying experiment using a miniature frying protocol with potato cubes. The components were added at 0.5% concentration and the performance was assessed by the rate of polymerized triacylglycerol (PTAG) formation. Both corn sterylferulate and γ-oryzanol significantly delayed the formation of PTAG with corn sterylferulate offering a markedly superior performance lasting for the entire frying period. The protection by γ-oryzanol, on the other hand, was only significant on the first day of frying. The authors attributed the differences in the performance of corn sterylferulate and rice sterylferulate (γ-oryzanol) to structural differences and differential interactions with endogenous tocopherols; while corn sterylferulate had a protective effect on the tocopherols naturally present in soybean oil, γ-oryzanol increased their degradation.

11.2.4 LIGNANS

Lignans are compounds with a dibenzylbutane skeleton formed by coupling of two coniferyl alcohol residues that are present in the plant cell wall [80]. Sesamin, sesamol, sesamolin, sesaminol and sesamolinol are lignan compounds naturally present in sesame oil, and have been implicated in the oil's high stability [86]. The ability of sesamol, sesamin and sesamolin to inhibit lipid oxidation in model systems has been reported. Sunflower oil containing sesamol, sesamin, and sesamolin extracted from roasted sesame seed oil was heated at 180°C for 10 h, and thermos oxidative degradation assessed by conjugated diene contents, p-anisidine value, and fatty acid composition. Samples containing sesame lignans showed significantly higher stability compared to sunflower oil without them [39]. The effect of sesame lignans on thermo oxidation of methyl linoleate during heating at 180°C for 1 h in a Rancimat was assessed by Lee et al. [40, 41]. The contents of conjugated dienes and the p-anisidine value were significantly lower when lignans were added to oil than samples without lignans. The frying stability of soybean oil also increased after addition of sesamin and sesamolin [29].

11.2.5   CAROTENOIDS

Carotenoids are a group of naturally occurring tetraterpenoids, consisting of isoprenoid units [15]. They are lipid-soluble pigments that contribute to the yellow or deep orange color of oils. Depending on source and variety, crude palm oil may contain up to 0.5% carotenoids [72]. β-Carotene is the most widespread carotenoid present in vegetable oils. Although the antioxidant activity of carotenoids against photo-oxidation has been recognized, their antioxidant activity during storage without light exposure or at elevated temperature remains controversial. Yanishlieva et al. [84] observed a prooxidant effect when β-carotene was added to antioxidant free sunflower oil triacyglycerols during accelerated storage at 100°C. However, in the same study, an antioxidant activity was reported for regular sunflower oil. The observed activity was attributed to a synergistic action between β-carotene and the endogenous α-tocopherol in the sunflower oil [84]. According to Schroeder et al. (2006), the addition of 100–1000 μg/g β-carotene to antioxidant depleted palm olein did not extend the induction period in a Rancimat stability test at 120°C. In a recent study, Zeb and Murkovic [88] evaluated the effects of β-carotene on the oxidation of triacylglycerols. They observed that addition of β- carotene significantly increased the peroxide value of model triacylglycerols during oxidation at 110°C in a Rancimat apparatus. Procida et al. [65] reported that β-carotene inhibited the formation of some deleterious carbonyl compounds such as pentanal during frying in olive oil.

11.2.6   SQUALENE

Squalene is a triterpene hydrocarbon (Figure 11.6) widely distributed in vegetable oils, with olive (10–1200 mg/kg) and rice bran oils (100–330 mg/kg) containing the highest amounts [8]. The richest known source of squalene is shark liver oil. In vegetable oils, squalene is found over broad ranges. For example, in flaxseed, grape seed, and soybean oils it is not detected, but is quite prominent in peanut (1.28 g/kg), pumpkin (3.53 g/kg), and olive oils (5.99 g/kg). Squalene is the main component of skin surface polyunsaturated lipids and shows some advantages for the skin as an emollient and

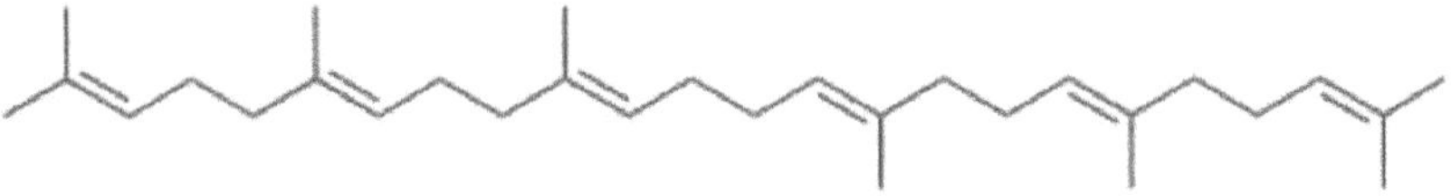

FIGURE 11.6   Chemical structure of Squalene.

antitumor compound. The triterpene has also been found to have protective activity against several carcinogens. The capacity of squalene to protect oils against oxidative degradation has been evaluated by Rao and Achaya [26]. Literature reports, however, on squalene antioxidant activity remain controversial, highly dependent on the model used [8, 36].

Conforti et al. [4] reported an antioxidant effect of squalene in a model of lipid peroxidation of liposomes; the IC50 value for Squalene was 0.023 mg/mL. An ethyl acetate extract of *Amaranthus caudatus* examined using the same method was 20-fold less active. The regeneration of a-tocopherol by Squalene in photo-oxidation studies was suggested in the past [67]. Squalene showed slight antioxidant activity when assayed by the crocin bleaching method [14]. In the same study, squalene demonstrated a synergistic effect with a-tocopherol and bsitosterol. The authors suggested that squalene could act as a competitive compound in the crocin bleaching reaction, thereby reducing the rate of oxidation.

Squalene, subjected to accelerated oxidation in a Rancimat apparatus at 100°C, showed a negligible antioxidant effect [47]. From experiments of Psomiadou and Tsimidou [66], a concentration-dependent moderate antioxidant activity of squalene – when stored at 40 and 62°C in the dark – was evident, which was stronger than the case of olive oil compared to that found for sunflower oil and lard. The authors concluded that the weak antioxidant efficacy of squalene in olive oil may be explained by competitive oxidation of the various lipids present, which leads to a reduction in the rate of oxidation.

11.2.7   PHOSPHOLIPIDS

Phospholipids, such as phosphatidylcholine (PC), phosphatidylethanolamine (PE), phosphatidylserine (PS), phosphatidylinositol (PI), and phosphatidic acid (PA) are endogenous minor components of oils. Unlike studies describing the antioxidant activity of phospholipids under accelerated storage conditions, rather fragmented information is available on their application under frying conditions, probably because of their adverse effects on color and foaming of oils [20]. The addition of 0.1% soy lecithin remarkably inhibited thermos oxidative alteration of oils during frying [16]. The thermos oxidative stability of salmon oil heated at 180°C was significantly improved in the presence of a phospholipid fraction isolated from bluefish [34]. The antioxidant effect of egg yolk during frying of flour dough containing different amounts of egg yolk powder was attributed to phospholipids present in this ingredient [33].

The observed antioxidant activity of phospholipids has been attributed to:

- their synergistic activity with phenolic antioxidants such as tocopherols;
- the ability of the phosphate group to chelate prooxidant metals;
- the formation of non-enzymatic browning reaction products between amino phospholipids and sugar or lipid oxidation products;
- the ability of phospholipids to form an oxygen barrier between the oil and air interface.

11.2.8 POLYPHENOLICS

A recent trend in the search for natural antioxidants is the application of extracts and isolates from different plants. The most prominent compounds present in those extracts are polyphenols [12, 41]. The isolation of Polyphenolics compounds, their antioxidant activities and applications in biological and food systems has been extensively studies by different authors in past [2, 28, 32, 69].

Spices and herbs such as rosemary, oregano, savory, marjoram, sage, thyme, basil, clove, cinnamon, nutmeg, turmeric, cumin, pepper, and garlic have long been recognized as important sources of potent antioxidants [14]. Active compounds present in most common spices and herbs include phenolic mono- and diterpenes (e.g., carnosic acid, carnosol, rosmanol, rosmadial, cavacrol, and thymol), phenolic acids and derivatives (e.g., rosemarinic, caffeic, gallic, ferulic, and protocatechuic), gingerol-related compounds (e.g., gingerol and shagoal) diarylheptanoids (e.g., curcumin, cassamunin A, B, C), phenolic amides (e.g., capsaicin and capsaicinol), and flavonoids (e.g., quercetin, luteolin, apigenin, kaempferol, and iso-harmnetin) (Figure 11.7). Whereas reports abound on the antioxidant activity of extracts and components from various spices and herbs in edible oils under ambient and storage conditions, investigations on their effectiveness during frying have received relatively less attentions.

Generally, extracts from different parts of spices, herbs and fruits studied so far indicated their efficacy in inhibiting thermos oxidative degradation and extend the fry life of vegetable oils [62]. In one of the recent work a comparative analysis was done using natural antioxidant derived from rosemary extract, mixed tocopherol, alpha tocopherol, Ascorbyl Palmitate along with synthetic antioxidant (TBHQ) during deep fat frying [19]. It was found in the study that the rosemary-based antioxidant (PRESOL^{TM}) was found out to be highly effective as compared to synthetic antioxidant with reference to the inhibition of palm oil degradation in deep-fried potato chips. Besides, the extracts and their isolated polyphenolic components effectively inhibited the formation of toxic thermos oxidative degradation products including heterocyclic amines during pan-frying of beef patties, and acrylamide during deep-frying of potato and bread sticks. Available data indicated that the addition of plants extracts to frying oil did not negatively affect the sensory attributes of the fried products, lending credence to their utilization as antioxidants for frying applications.

Although plant extracts and polyphenols show good potential as natural antioxidants for frying applications, nevertheless, a major set-back is their poor solubility in oils. However one of the recent patent suggest that herbal extract can be used as natural antioxidant without issues of solubility [52]. Additionally it has been suggested that for improved effectiveness, it is required of applied antioxidant to be soluble or optimally dispersed in the substrate. The observed improved efficiency of polyphenolic compounds with temperature may as well be related to increased solubility or dispersal at higher temperature. The solubility and dispersal of Polyphenolics compounds in oil is also a very important factor that can affect the level of performance observed in a model frying involving heating of oil to a frying temperature without food, compared to actual frying experiment. During heating of oil without food, water or aeration, 'caramelization' of

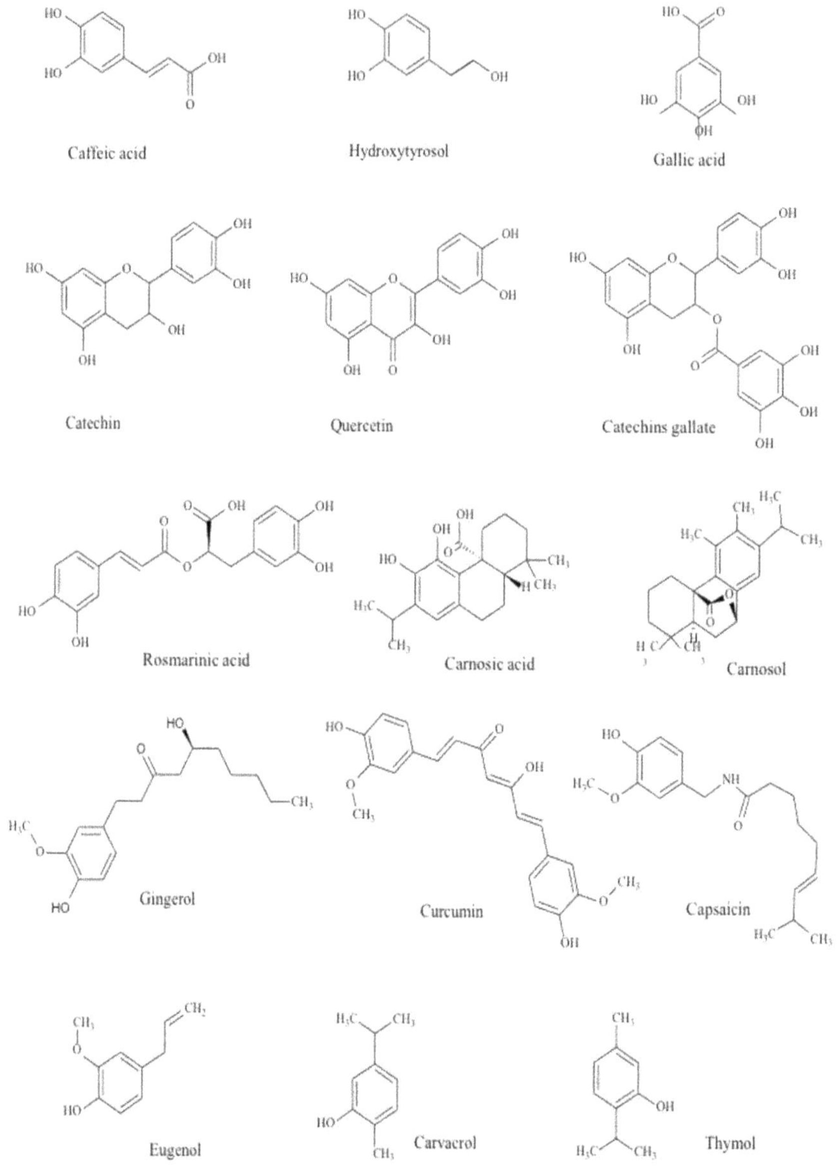

FIGURE 11.7 Representative antioxidant compounds in natural phenolic extracts.

extracts (forming black sediments), with consequent inactivation of poly-phenolic active ingredients is not unlikely. It is expected, however, that during frying of food, water from the food, agitation of the oil arising from escaping water vapor and aeration will improve polyphenolic solubility and dispersal. Investigations into methods to enhance the lipophilicity of polyphenolic compounds through appropriate modifications are therefore warranted.

11.3 CONCLUSIONS

Deep frying is a very complex phenomenon, and the harsh conditions employed places a huge demand on the applied antioxidants, resulting in common failure of traditional antioxidants. However, the formation of a host of potentially toxic degradation products during frying, coupled with an ever-increasing demand for fried food necessitates the development of safe, yet potent antioxidants for frying applications. Presently, only a very negligible fraction of the available (semi)synthetic antioxidants in the lit-erature has been evaluated under frying conditions presumably due to: (1) the presupposition that their radical scavenging activity in a chemical test or antioxidant activity under oxidative condition correlates with their per-formance during frying, which is not necessarily true because usually dif-ferent conditions are applied; (2) lack of an official fast and reliable frying test that truly mimics actual frying conditions to evaluate small amounts of developed antioxidants considering the large amounts required for a typical restaurant-type frying; (3) the time and resources required to assess toxicity of potential antioxidant for frying applications, considering the innumerable number of an antioxidant's decomposition products that are possible under frying conditions; and (4) the impact of antioxidants and their decomposition products on sensory attributes of fried food. Never-theless, the need to develop a safe, potent, and stable antioxidant for frying application still remains an appealing task. Based on current knowledge of the antioxidant mechanisms of activity, compounds with antioxidative properties generated from non-toxic natural precursors and possessing the following functional groups and properties could behave as effective anti-oxidants for frying:

302 Food Process Engineering

- at least two phenolic hydroxy functional groups, for enhanced radical scavenging activity;
- a catechol moiety, for possible metal chelating activity;
- an amino and/or amide functional group, for possible removal of carbonyl compounds through condensation reaction;
- relatively high molecular weight for reduced volatility under frying conditions;
- composed of independent small antioxidant units which may still exhibit some antioxidant activity consequent to thermos oxidative or hydrolytic degradation;
- made of components offering different mechanism of antioxidative activity; and
- at least moderate lipophilicity for better dispersal in oil.

Thus, the evaluation of performance during frying of lyophilized phenolic acid derivatives and a number of semi-synthetic antioxidant hybrids such as ascorbic acid-α-tocopherol, tocopherol-carotenoid, ascorbic-phenolic acid, amino phospholipid-chromanol, and tocopherol-sterol is warranted.

11.4 SUMMARY

Synthetic antioxidants like butylated hydroxytoluene (BHT), butylated hydroxyanisole (BHA), and tert–butylhydroquinone (TBHQ) are often added to processed oils to retard oxidative degradation during storage and frying; however, beside their poor performance under frying conditions, consumers' acceptance of synthetic antioxidants remains negative due to their perceived detrimental effect on human health. Consequently, there is a growing interest in the search for effective natural antioxidants for frying applications, notably, from phenolic components of common spices and herbs.

Natural antioxidants occurring in oils and those added to frying oils can play a prominent role in their protection against thermal and oxidative deterioration and hence in the manufacture of high quality fried products. Efficacy of antioxidants under frying conditions is far more difficult to be evaluated and defined because availability of air is lower and variable, and both oxidation and thermal reactions are simultaneously involved. This review is focused on the analysis and evaluation of efficacy of antioxidants in frying. Specific aspects of the action of natural and synthetic antioxidants at high temperature are discussed, and the most important methods used for the analysis of antioxidants and their efficacies are described.

This book chapter focused on the evaluation of efficacy of natural antioxidants during frying scenario. Specific aspects of the action of natural antioxidants at high temperature will be discussed, and the most important methods used for the analysis of antioxidants acting in frying and the evaluation of their efficacy will be described.

KEYWORDS

- Efficacy
- Extracts
- Frying
- Herbs
- Natural antioxidants
- Phenolic extracts
- Phytosterol
- Prooxidant
- Rancimat
- Spices
- Sunflower oil
- Synergistic
- Synthetic antioxidants
- Thermo oxidant
- Thermos oxidative stability

REFERENCES

1. Aggelousis, G., & Lalas, S. (1997). Quality changes of selected vegetable oils during frying of doughnuts. *Rivista Italiana delle Sostanze Grasse*. 74, 559–566.

2. Ahn, J., Grün, I., & Fernando, L. (2002). Antioxidant properties of natural plant extracts containing polyphenolic compounds in cooked ground beef. *Journal of Food Science*. 67(4), 1364–1369.

3. Al Khusaibi, M., et al. (2012). Frying of potato chips in a blend of canola oil and palm olein: changes in levels of individual fatty acids and tocols. *International Journal of Food Science & Technology*. 47(8), 1701–1709.

4. Aladedunye, F., Catel, Y., & Przybylski, R. (2012). Novel dihydrocaffeic acid amides: synthesis, radical scavenging activity, and evaluation as antioxidants under storage and frying conditions. *Food Chem*. 130, 945–952.

5. Aladedunye, F., & Przybylski, R. (2012). Frying Performance of Canola Oil Triacylglycerides as Affected by Vegetable Oils Minor Components. *Journal of the American Oil Chemists' Society*. 89(1), 41–53.

6. Aladedunye, F. A., & Przybylski, R. (2009). Degradation and nutritional quality changes of oil during frying. *Journal of the American Oil Chemists' Society*. 86(2), 149–156.

7. Aladedunye, F. A., & Przybylski, R. (2009). Protecting oil during frying: A comparative study. *European Journal of Lipid Science and Technology*. 111(9), 893–901.

8. Amarowicz, R. (2009). Squalene: A natural antioxidant. *European Journal of Lipid Science and Technology*. 111(5), 411–412.

9. Bansal, G., et al. (2010). Performance of palm olein in repeated deep frying and controlled heating processes. *Food Chemistry*. 121(2), 338–347.

10. Barrera–Arellano, D., et al. (1999). Loss of tocopherols and formation of degradation compounds in triacylglycerol model systems heated at high temperature. *Journal of the Science of Food and Agriculture*. 79(13), 1923–1928.

11. Barrera–Arellano, D., et al. (2002). Loss of tocopherols and formation of degradation compounds at frying temperatures in oils differing in degree of unsaturation and natural antioxidant content. *Journal of the Science of Food and Agriculture*. 82(14), 1696–1702.

12. Campos-Esparza, M. R., Sanchez-Gomez, M. V., & Matute, C. (2009). Molecular mechanisms of neuroprotection by two natural antioxidant polyphenols. *Cell Calcium*. 45(4), 358–368.

13. Cercaci, L., et al. (2007). Composition of total sterols (4-desmethyl-sterols) in extra-virgin olive oils obtained with different extraction technologies and their influence on the oil oxidative stability. *Food Chemistry*. 102(1), 66–76.

14. Charles, D. J. (2012); Antioxidant properties of spices, herbs and other sources. 2012, Springer Science & Business Media.

15. Choe, E., & Min, D. B. (2006). Mechanisms and factors for edible oil oxidation. *Comprehensive reviews in Food Science and Food Safety*. 5(4), 169–186.

16. Chu, Y.-H. (1991). A comparative study of analytical methods for evaluation of soybean oil quality. *Journal of the American Oil Chemists' Society*. 68(6), 379–384.

17. Conforti, F., et al. (2005). In Vitro Antioxidant Effect and Inhibition of. ALPHA.-Amylase of Two Varieties of *Amaranthus caudatus* Seeds. *Biological and Pharmaceutical Bulletin*. 28(6), 1098–1102.

18. Corsini, M. S., Silva, M. G., & Jorge, N. (2009). Loss in tocopherols and oxidative stability during the frying of frozen cassava chips. Grasas y aceites. 60(1), 77–81.

19. Deora, N. S., Aastha Deswal, & S. Madhavan, (2015). Comparative Analysis of Natural vs. Synthetic Antioxidant during Deep-Fat Frying of Potato Chips. Research & Reviews. *Journal of Food Science and Technology*. 4(2), 1–12.

20. Dobarganes, C., Márquez–Ruiz, G., & Velasco, J. (2000). Interactions between fat and food during deep frying. *European Journal of Lipid Science and Technology.* 102(8–9), 521–528.
21. Dutta, P. (1997). Studies on phytosterol oxides. II: Content in some vegetable oils and in French fries prepared in these oils. *Journal of the American Oil Chemists' Society.* 74(6), 659–666.
22. Fisnar, J., Dolezal, M., & Reblova, Z. (2014). Tocopherol losses during pan frying. *European Journal of Lipid Science and Technology.* 116(12), 1694–1700.
23. Gertz, C., Klostermann, S., & Kochhar, S. P. (2000). Testing and comparing oxidative stability of vegetable oils and fats at frying temperature. *European Journal of Lipid Science and Technology.* 102(8–9), 543–551.
24. Gordon, M. H., & Kourimská, L. (1995). Effect of antioxidants on losses of tocopherols during deep-fat frying. *Food Chemistry.* 52(2), 175–177.
25. Gordon, M. H., & Magos, P. (1983). The effect of sterols on the oxidation of edible oils. *Food chemistry.* 10(2), 141–147.
26. Govind Rao, M., & Achaya, K. (1968). Antioxidant activity of squalene. *Journal of the American Oil Chemists' Society.* 45(4), 296–296.
27. Gruczynska, E., Przybylski, R., & Aladedunye, F. (2015). Performance of structured lipids incorporating selected phenolic and ascorbic acids. *Food Chemistry.* 173(0), 778–783.
28. Haslam, E. (1996). Natural polyphenols (vegetable tannins) as drugs: possible modes of action. *Journal of Natural Products.* 59(2), 205–215.
29. Hemalatha, S. (2007). Sesame lignans enhance the thermal stability of edible vegetable oils. *Food Chemistry.* 105(3), 1076–1085.
30. Juárez, M. D., et al. (2011). Degradation in soybean oil, sunflower oil and partially hydrogenated fats after food frying, monitored by conventional and unconventional methods. *Food Control.* 22(12), 1920–1927.
31. Kamal-Eldin, A., & Appelqvist, L.-Å. (1996). The chemistry and antioxidant properties of tocopherols and tocotrienols. *Lipids.* 31(7), 671–701.
32. Kamal-Eldin, A., & Budilarto, E. (2015). Tocopherols and tocotrienols as antioxidants for food preservation. *Handbook of Antioxidants for Food Preservation,* 141.
33. Kim, H., & Choe, E. (2008). Effects of egg yolk powder addition to the flour dough on the lipid oxidation development during frying. *LWT-Food Science and Technology.* 41(5), 845–853.
34. King, M., Boyd, L., & Sheldon, B. (1992). Effects of phospholipids on lipid oxidation of a salmon oil model system. *Journal of the American Oil Chemists Society.* 69(3), 237–242.
35. Kmiecik, D., et al. (2015). Stabilization of phytosterols by natural and synthetic antioxidants in high temperature conditions. *Food Chemistry.* 173, 966–971.
36. Ko, T. F., Weng, Y. M., & Chiou, R. Y. Y. (2002). Squalene content and antioxidant activity of Terminalia catappa leaves and seeds. *Journal of Agricultural and Food Chemistry.* 50(19), 5343–5348.
37. Lampi, A.-M., & Kamal-Eldin, A. (1998). Effect of α-and γ-tocopherols on thermal polymerization of purified high-oleic sunflower triacylglycerols. *Journal of the American Oil Chemists' Society.* 75(12), 1699–1703.
38. Lampi, A.-M., et al. (2009). Distribution of monomeric, dimeric and polymeric products of stigmasterol during thermo-oxidation. *European Journal of Lipid Science and Technology.* 111(10), 1027–1034.

39. Lee, J., Kim, M., & Choe, E. (2007). Antioxidant activity of lignan compounds extracted from roasted sesame oil on the oxidation of sunflower oil. *Food Science and Biotechnology.* 16(6), 981–987.

40. Lee, J., Lee, Y., & Choe, E. (2008). Effects of sesamol, sesamin, and sesamolin extracted from roasted sesame oil on the thermal oxidation of methyl linoleate. *LWT-Food Science and Technology.* 41(10), 1871–1875.

41. Leopoldini, M., Russo, N., & Toscano, M. (2011). The molecular basis of working mechanism of natural polyphenolic antioxidants. *Food Chemistry.* 125(2), 288–306.

42. Li, W. (1997). Phytosterol and tocopherol changes in modified canola oils during frying and storage of fried products.

43. Madhavi, D., Deshpande, S., & Salunkhe, D. K. (1995); Food antioxidants: Technological. *Toxicological and Health Perspectives.* 1995, CRC Press.

44. Marmesat, S., et al. (2010). Action and fate of natural and synthetic antioxidants during frying. *Grasas Y Aceites.* 61(4), 333–340.

45. Marmesat, S., et al. (2008). Thermostability of genetically modified sunflower oils differing in fatty acid and tocopherol compositions. *European Journal of Lipid Science and Technology.* 110(8), 776–782.

46. Márquez–Ruiz, G., Ruiz–Méndez, M., & Velasco, J. (2014). Antioxidants in frying: Analysis and evaluation of efficacy. *European Journal of Lipid Science and Technology.* 116(11), 1441–1450.

47. Mateos, R., et al. (2003). Antioxidant effect of phenolic compounds, α-tocopherol, and other minor components in virgin olive oil. *Journal of Agricultural and Food Chemistry.* 51(24), 7170–7175.

48. Matthäus, B. (2006). Utilization of high-oleic rapeseed oil for deep-fat frying of French fries compared to other commonly used edible oils. *European Journal of Lipid Science and Technology.* 108(3), 200–211.

49. Matthäus, B. (2006). Utilization of high–oleic rapeseed oil for deep–fat frying of French fries compared to other commonly used edible oils. *European Journal of Lipid Science and Technology.* 108(3), 200–211.

50. Messina, V., et al. (2009). Effect of pan–frying in extra–virgin olive oil on odour profile, volatile compounds and vitamins. *International Journal of Food Science & Technology.* 44(3), 552–559.

51. Miyagawa, K., Hirai, K., & Takezoe, R. (1991). Tocopherol and fluorescence levels in deep-frying oil and their measurement for oil assessment. *Journal of the American Oil Chemists' Society.* 68(3), 163–166.

52. Mooppil, S., & Francis, P. (2013). Herbal extract composition and a process thereof. Google Patents.

53. Murkovic, M., Wiltschko, D., & Pfannhauser, W. (1997). Formation of α-Tocopherolquinone and α-Tocopherolquinone Epoxides in Plant Oil. *Lipid/Fett.* 99(5), 165–169.

54. Nadeem, M., Abdullah, M., & Ellahi, M. (2010). Effect of incorporating rape seed oil on quality of ice cream. *Mediterranean Journal of Nutrition and Metabolism.* 3(2), 121–126.

55. Nakayama, S., et al. (1987). Comparative Effects of Two Forms of γ-Oryzanol in Different Sterol Compositions on Hyperlipidemia Induced by Cholesterol Diet in Rats. *The Japanese Journal of Pharmacology.* 44(2), 135–143.

56. Neff, W. E., Warner, K., & Eller, F. (2003). Effect of γ-tocopherol on formation of nonvolatile lipid degradation products during frying of potato chips in triolein. *Journal of the American Oil Chemists' Society*. 80(8), 801–806.
57. Niness, K. R. (1999). Inulin and oligofructose: what are they? *Journal of Nutrition*. 129(7), 1402S–1406s.
58. Nogala-Kalucka, M., et al. (2005). Effect of α- and δ-tocopherol on the oxidative stability of a mixed hydrogenated fat under frying conditions. *European Food Research and Technology*. 221(3–4), 291–297.
59. Normand, L., Eskin, N., & Przybylski, R. (2001). Effect of tocopherols on the frying stability of regular and modified canola oils. *Journal of the American Oil Chemists' Society*. 78(4), 369–373.
60. Nyström, L., et al. (2007). A comparison of the antioxidant properties of steryl ferulates with tocopherol at high temperatures. *Food Chemistry*. 101(3), 947–954.
61. Parrado, J., et al. (2003). Prevention of brain protein and lipid oxidation elicited by a water-soluble oryzanol enzymatic extract derived from rice bran. *European Journal of Nutrition*. 42(6), 307–314.
62. Perumalla, A. V. S., & Hettiarachchy, N. S. (2011). Green tea and grape seed extracts—Potential applications in food safety and quality. *Food Research International*. 44(4), 827–839.
63. Petukhov, I., et al. (1999). Frying performance of genetically modified canola oils. *Journal of the American Oil Chemists' Society*. 76(5), 627–632.
64. Pokorný, J. (2007). Are natural antioxidants better–and safer–than synthetic antioxidants? *European Journal of Lipid Science and Technology*. 109(6), 629–642.
65. Procida, G., et al. (2009). Influence of chemical composition of olive oil on the development of volatile compounds during frying. *European Food Research and Technology*. 230(2), 217–229.
66. Psomiadou, E., & Tsimidou, M. (1999). On the role of squalene in olive oil stability. *Journal of agricultural and food chemistry*. 47(10), 4025–4032.
67. Psomiadou, E., & Tsimidou, M. (2002). Stability of virgin olive oil. 2. Photo-oxidation studies. *Journal of Agricultural and Food Chemistry*. 50(4), 722–727.
68. Rennick, K. A., & Warner, K. (2006). Effect of Elevated Temperature on Development of Tocopherolquinones in Oils. *Journal of Agricultural and Food Chemistry*. 54(6), 2188–2192.
69. Rice-Evans, C. A., Miller, N. J., & Paganga, G. (1996). Structure-antioxidant activity relationships of flavonoids and phenolic acids. *Free Radical Biology and Medicine*. 20(7), 933–956.
70. Rietjens, I. M. C. M., et al. (2002). The pro-oxidant chemistry of the natural antioxidants vitamin C, vitamin E, carotenoids and flavonoids. *Environmental Toxicology and Pharmacology*. 11(3–4), 321–333.
71. Romero, N., et al. (2007). Effect of α-tocopherol, α-tocotrienol and Rosa mosqueta shell extract on the performance of antioxidant-stripped canola oil (*Brassica* sp.) at high temperature. *Food Chemistry*. 104(1), 383–389.
72. Rossell, J. (2001). Frying: Improving Quality. Vol. 56. 2001, Woodhead Publishing.
73. Rossi, M., Alamprese, C., Ratti, S. (2007). Tocopherols and tocotrienols as free radical-scavengers in refined vegetable oils and their stability during deep-fat frying. *Food Chemistry*. 102(3), 812–817.

74. Rudzińska, M., Przybylski, R., & Wąsowicz, E. (2009). Products Formed During Thermo-oxidative Degradation of Phytosterols. *Journal of the American Oil Chemists' Society.* 86(7), 651–662.

75. Simonne, A., & Eitenmiller, R. (1998). Retention of vitamin E and added retinyl palmitate in selected vegetable oils during deep-fat frying and in fried breaded products. *Journal of Agricultural and Food Chemistry.* 46(12), 5273–5277.

76. Sims, R. J., Fioriti, J. A. & Kanuk, M. J. (1972). Sterol additives as polymerization inhibitors for frying oils. *Journal of the American Oil Chemists Society.* 49(5), 298–301.

77. Soupas, L., et al. (2004). Effects of Sterol Structure, Temperature, and Lipid Medium on Phytosterol Oxidation. *Journal of Agricultural and Food Chemistry.* 52(21), 6485–6491.

78. Steel, C. J., Dobarganes, M. C., & Barrera-Arellano, D. (2005). The influence of natural tocopherols during thermal oxidation of refined and partially hydrogenated soybean oils. Grasas y aceites. 56(1), 46–52.

79. Tabee, E., et al. (2008). Effects of α-Tocopherol on Oxidative Stability and Phytosterol Oxidation During Heating in Some Regular and High-Oleic Vegetable Oils. *Journal of the American Oil Chemists' Society.* 85(9), 857–867.

80. Torres, P., et al. (2008). Enzymatic modification for ascorbic acid and alpha-tocopherol to enhance their stability in food and nutritional applications.

81. Verleyen, T., et al. (2001). Identification of α-Tocopherol Oxidation Products in Triolein at Elevated Temperatures. *Journal of Agricultural and Food Chemistry.* 49(3), 1508–1511.

82. Warner, K., & Moser, J. (2009). Frying stability of purified mid-oleic sunflower oil triacylglycerols with added pure tocopherols and tocopherol mixtures. *Journal of the American Oil Chemists' Society.* 86(12), 1199–1207.

83. Winkler-Moser, J., et al. (2012). Comparison of the Impact of γ-Oryzanol and Corn Steryl Ferulates on the Polymerization of Soybean Oil During Frying. *Journal of the American Oil Chemists' Society.* 89(2), 243–252.

84. Yanishlieva, N., Raneva, V., & Marinova, E. (2001). β-carotene in sunflower oil oxidation. Grasas y aceites. 52(1), 10–16.

85. Yanishlieva, N. V., & Marinova, E. M. (2001). Stabilization of edible oils with natural antioxidants. *European Journal of Lipid Science and Technology.* 103(11), 752–767.

86. Yoshida, H. (1994). Composition and quality characteristics of sesame seed (*Sesamum indicum*) oil roasted at different temperatures in an electric oven. *Journal of the Science of Food and Agriculture.* 65(3), 331–336.

87. Yuki, E., & Ishikawa, Y. (1976). Tocopherol contents of nine vegetable frying oils, and their changes under simulated deep-fat frying conditions. *Journal of the American Oil Chemists' Society.* 53(11), 673–676.

88. Zeb, A., & Murkovic, M. (2010). Characterization of the effects of β-carotene on the thermal oxidation of triacylglycerols using HPLCESIMS. *European Journal of Lipid Science and Technology.* 112(11), 1218–1228.

PART V

FOOD HAZARDS AND THEIR CONTROLS

CHAPTER 12

HAZARD ANALYSIS CRITICAL CONTROL POINT PROGRAM

RUPESH CHAVAN,[1] ANIT KUMAR,[2] SHRADDHA BHATT,[3] VAIBHAV VYAS,[4] SOMYA TEWARI,[5] and TANMAY NALAWADE[6]

[1]Quality Assessment Analyst, Mother Dairy, Gandhinagar, Gujarat. Mailing Address:B 501 Nobel Platinum, Rayjibaug Near Motibaug Circle, Junagadh 362001, Gujarat, India. Mobile: +91-9992525438, E-mail: rschavanb_tech@rediffmail.com

[2]Department of Food Science and Technology, National Institute of Food Technology Entrepreneurship and Management (NIFTEM), Kundli, Sonepat, Haryana, India

[3]Assistant Professor, Department of Biotechnology, Junagadh Agricultural University, Junagadh, Gujarat, India

[4]M.Tech. Scholar, Department of Food Science and Technology, National Institute of Food Technology Entrepreneurship and Management (NIFTEM), Kundli, Sonepat, Haryana, India

[5]Senior Research Fellow, Department of Food Process Engineering, National Institute of Food Technology Entrepreneurship and Management (NIFTEM), Kundli, Sonepat, Haryana, India

[6]Assistant Professor, Department of Bachelor of Food Technology and Management, CNCVCW, CSIBER, Kolhapur, India

CONTENTS

12.1   INTRODUCTION

Hazard Analysis Critical Control Point (HACCP) is considered as a systematic and preventive approach to address biological, chemical and physical hazards in food products by anticipating and implementing preventive actions rather than having postproduction analysis. To ensure the food safety, food operators and processors worldwide implement HACCP which safe guards the product by identifying, monitoring, verifying, and controlling critical-processing steps for hazards. HACCP is implemented at each and every step during the food production i.e., raw materials procurement, handling production, distribution, and consumption of finished food. Briefly HACCP follows the following steps [2]:

   a.   Conduct the hazard analysis.
   b.   Determine the critical control points (CCPs).
   c.   Establish the critical limit for each step.
   d.   Establish a system to monitor control of each CCP.
   e.   Establish the corrective action to be taken when monitoring indicates that a particular CCP is not under control.
   f.   Establish the procedures for verification to confirm that the HACCP system is working effectively.
   g.   Establish the record keeping and documentation procedures.

12.2   HISTORY AND BACKGROUND

In 1960s, HACCP concept was pioneered by the Pillsbury Company, the United States Army and the United States National Aeronautics and Space

Administration (NASA) as a collaborative development for the production of safe foods for the United States space program. In the early 1980s, the HACCP approach was adopted by other major food companies in the world. Other than NASA, International Commission on Microbiological Specifications for Foods (ICMSF) and the International Association of Milk, Food and Environmental Sanitarians (IAMFES), have recommended the broad application of HACCP for ensuring food safety.

12.3 HAZARD ANALYSIS

Hazard analysis is one of the most important tasks in the whole process of HACCP. According to Codex 2003 [2], hazard is defined as "*a biological, chemical or physical agent in, or condition of, food with the potential to cause an adverse health effect.*" Hazard analysis is a critical step because it makes the food processor to understand the hazard in the food product and the steps which will eliminate or reduce the hazards to acceptable levels to make the food safe. The hazards' in the foods are classified into three categories, i.e., biological, chemical and physical.

12.3.1 BIOLOGICAL HAZARDS

Biological hazards including bacteria, viruses, fungi and parasites can enter the food cycle through raw materials and also during the processing of food. Food products contaminated by pathogens or by toxic microbial by-products may not look, smell or taste bad, but can make a person sick (Table 12.1). Food spoilage or decomposition, which can create food safety problems and these problems are usually prevented or controlled by a HACCP program. During the processing of foods, most of the microorganisms are killed or inactivated and numbers of them can be minimized by adequate control of handling and storage practices (hygiene, temperature and time).

Bacterial hazards are defined as those bacteria that, if they occur in food, may cause illness in humans, either by infection or intoxication. Bacterial hazards can also be grouped into spore formers and non-spore formers. Certain types of bacteria (e.g., *Clostridium* and *Bacillus* spp.)

may pass through adormant stage in their life cycle called a spore. When the bacterium exists as aspore, it is very resistant to chemicals, heat and other treatments that would normally be lethal to non-spore forming bacteria. Although they are alive, viruses differ from other microorganisms in what they need to live and how they multiply. Viruses exist in foods without growing, so they need no food, water or air to survive and can cause illness by infection. Parasites are organisms that need a host to survive, living on or within it. There are two types of parasites that can infect people through food or water: protozoa and parasitic worms.

TABLE 12.1   Bacterial Hazards Associated with Food Products

Bacterial hazards	Reasons to be a hazard
Clostridium botulinum (sporeformer)	Causes intoxication which affects the central nervous system and can cause shortness of breath, blurred vision, loss of motor capabilities and death.
Cryptosporidium parvum	Causes watery diarrhea, coughing, low grade fever, and severe intestinal distress.
Cyclospora cayetanensis	Causes disease symptoms that mimic those of cryptosporidiosis. Other symptoms may include loss of appetite, fatigue, and weight loss.
Entamoeba histolytica	Causes dysentery (severe, bloody diarrhea).
Giardia lamblia	Causes diarrhea, abdominal cramps, fatigue, nausea, flatulence (intestinal gas) and weight loss. Illness may last for one to two weeks, but chronic infections can last months to years.
Hepatitis A virus	Causes fever and abdominal discomfort, followed by jaundice.
Listeria monocytogenes (nonsporeformer)	Causes an infection with mild flu-like symptoms. Severe forms of listeriosis are possible in people with weakened immune systems, causing septicemia, meningitis, encephalitis and still births
Norwalk virus	Headache and low-grade fever may also occur.
Rotavirus	Causes nausea, vomiting, diarrhea and abdominal pain (gastroenteritis).
Salmonella spp. (nonsporeformer)	Causes an infection with symptoms like nausea, vomiting, abdominal cramps, diarrhea, fever and headache.

12.3.2 CHEMICAL HAZARDS

Chemical hazards include chemical contaminants and in food they may be naturally occurring or may be added during the processing of food. Harmful chemicals at high levels can cause food borne illnesses and can be responsible for chronic illness at lower levels (Table 12.2). Chemical hazards can be separated into three categories:
   a.   Naturally-occurring chemicals.
   b.   Intentionally-added chemicals.
   c.   Unintentional or incidental chemical additives.

TABLE 12.2   Chemical Hazards Associated With Food Products

Chemical hazards	Reasons to be a hazard
A. Naturally Occurring chemical	
Apple, nuts, cereal grains	Certain molds can form mycotoxins, e.g., Aflatoxin, Alternaria toxins, Fumonisin, Ochratoxin, Patulin and Vomitoxin
Egg, milk, nuts, seafood, soy	Certain varieties or species produce an allergic reaction in sensitive people
B. Intentionally-added chemicals	
FD&C Yellow No. 5 (food coloring)	Can produce an adverse reaction in sensitive people.
Sodium nitrite (preservative)	Toxic in high concentrations.
Sulphur dioxide (preservative)	Can cause an intolerance reaction in sensitive individuals.
Vitamin A (nutritional supplement)	Toxic in high concentrations.
C. Incidental Contaminants	
Agricultural chemicals (e.g., pesticides, herbicides)	Many are approved for use on food. However, if improperly used or applied, some can be acutely toxic or may cause health risks with long-term exposure.
Cleaning chemicals (e.g., acids, caustics)	Can cause chemical burns if present in the food at high levels.
Equipment components (e.g., copper pipe fittings)	Acidic foods can cause leaching of heavy metals from pipes and joints (e.g., copper and lead).
Maintenance chemicals (e.g., lubricants)	Some chemicals that are not approved for food use may be toxic.
Packaging materials (e.g., tin)	High nitrite levels in food can cause excessive detinning of uncoated cans resulting in excessive levels of tin in food.

12.3.3 PHYSICAL HAZARDS

Physical hazards are due to presence of foreign objects in the food and can result from contamination and/or poor practices in the food chain. When a consumer mistakenly eats the foreign material or object, it is likely to cause choking, injury or other adverse health effects. Physical hazards are the most commonly reported consumer complaints because the injury occurs immediately or soon after consumption, and the source of the hazard are often easy to identify. Examples of physical hazards include: glass, wood, metals, etc.

12.4 PRE-REQUISITE PROGRAMS (PRPS)

Hazards of a low probability of occurrence and a low severity should not be addressed under the HACCP system but can be addressed through the good manufacturing practices (GMPs) contained in the Codex General Principles of Food Hygiene. HACCP systems must be built upon a firm foundation of compliance with (GMPs) [1] acceptable sanitation standard pertaining procedures (SSOPs) and appropriate industry practices. GMPs and sanitation procedures affect the processing environment and should be considered pre-requisite programs to HACCP.

The GMPs define measures of general hygiene as well as measures that prevent food from becoming adulterated due to unsanitary conditions. SSOPs are procedures used by food processing firms to help accomplish the overall goal of maintaining GMPs in the production of food. SSOPs can help reduce the likelihood of occurrence of a hazard by: preventing product cross-contamination, providing hand-washing and sanitizing stations near the processing area to facilitate proper employee hygiene, and ensuring appropriate equipment maintenance and cleaning and sanitizing procedures.

12.5 PRINCIPLES OF THE HACCP SYSTEM

Principle 1: Conduct a hazard analysis.
Principle 2: Determine the Critical Control Points (CCPs).

Principle 3: Establish critical limit.

Principle 4: Establish a system to monitor control of the CCP.

Principle 5: Establish the corrective action to be taken when monitoring indicates that a particular CCP is not under control.

Principle 6: Establish procedures for verification to confirm that the HACCP system is working effectively.

Principle 7: Establish documentation concerning procedures appropriate to these principles and their application.

12.6   GUIDELINES FOR THE APPLICATION OF THE HACCP SYSTEM

Prior to application of HACCP, PRPs must be in place including training, should be well established, fully operational and verified in order to facilitate the successful application and implementation. Apart from system the food business management must be well aware and committed towards implementation of HACCP system and they must also deploy employees with a proficiency and skill in the systems. HACCP should be applied to each specific operation separately and should be reviewed and necessary changes made when any modification is made in the product, process, or any step. The application of the HACCP principles should be the responsibility of each individual in business. The application of HACCP principles consists of the following tasks as identified in the Logic Sequence (Figure 12.1).

12.6.1   ASSEMBLE HACCP TEAM

The HACCP team must preferably comprising of people having knowledge and expertise in the food product and operations and is usually advised that the team is multidisciplinary i.e., comprising personnel of quality, production, management, maintenance and other departments if required. Before implementation of HACCP, the scope of the HACCP plan must be identified and it must be able to describe, which segment of the food chain is involved and the general classes of hazards to be addressed.

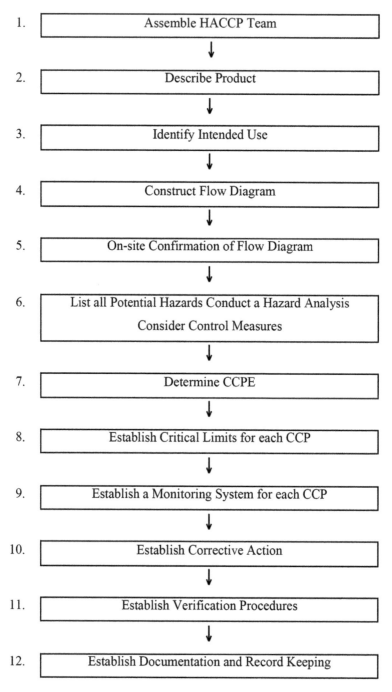

FIGURE 12.1   Logic sequence for application of HACCP.

12.6.2   PRODUCT DESCRIPTION

The product to be covered under HACCP plan must be described in detail. The details of a product must cover the information pertaining to composition, physical/chemical structure (including a_w, pH, etc.), matricidal/static treatments (heat-treatment, freezing, brining, smoking, etc.), packaging, durability and storage conditions and method of distribution. The food processing sector in which multiple products are manufactured it is advised to the group the products with similar characteristics or processing steps.

12.6.3   INTENDED USE OF THE PRODUCT

The intended use of the food product must be based on the expected uses by the end user or consumer. In specific cases, vulnerable groups of the population, e.g., institutional feeding, may have to be considered.

12.6.4   CONSTRUCTION OF FLOW DIAGRAM

The flow diagram should be constructed by the HACCP team and must cover all the steps involved during manufacturing of a food product. Products using similar processing steps, same flow diagram can be used. While applying HACCP to a given operation, consideration should be given to steps preceding and following the specified operation.

12.6.5   ON-SITE CONFIRMATION OF FLOW DIAGRAM

Once the flow diagram for a product is constructed, it must be confirmed on-site y the HACCP team and amended if required. For on-site confirmation of flow diagram or persons with sufficient knowledge of the processing operation must be deployed.

12.6.6   CONTROL IDENTIFIED HAZARDS

List all potential hazards associated with each step, conduct a hazard analysis, and consider any measures to control identified hazards (See Principle 1)

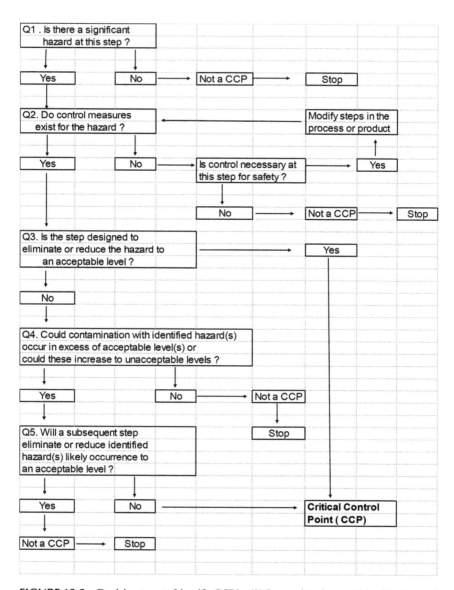

FIGURE 12.2   Decision tree to Identify CCP's. (*) Proceed to the next identified hazard in the described process. (**) Acceptable and unacceptable levels need to be defined within the overall objectives in identifying the identifying the CCP of HACCP plan.

The HACCP team should list all of the hazards that may be reasonably expected to occur at each step according to the scope from primary production, processing, manufacture, and distribution until the point of consumption. The HACCP team should next conduct a hazard analysis to identify the hazards for which elimination or reduction to acceptable levels is essential to produce safe food.

In conducting the hazard analysis, wherever possible, the following should be included:

- The likely occurrence of hazards and severity of their adverse health effects;
- The qualitative and/or quantitative evaluation of the presence of hazards;
- Survival or multiplication of micro-organisms of concern;
- Production or persistence in toxins, chemicals or physical agents; and
- Conditions leading to the above.

Consideration should be given to what control measures, if any exist, can be applied to each hazard. More than one control measure may be required to control a specific hazard(s) and more than one hazard may be controlled by a specified control measure.

12.6.7   DETERMINE CRITICAL CONTROL POINTS (SEE PRINCIPLE 2)

Critical Control Points (CCP) can be determined by the application of a decision tree (e.g., Figure 12.2), which indicates a logic reasoning approach. Application of a decision tree should be flexible, given whether the operation is for production, slaughter, processing, storage, distribution or other. This example of a decision tree may not be applicable to all situations and other approaches may be used and for effective usage training must be provided for the HACCP team. If a hazard has been identified at step where control is necessary for safety, and no control measure exists at that step, or any other, then the product or process should be modified at that step, or at any earlier or later stage, to include a control measure.

12.6.8 ESTABLISH CRITICAL LIMITS FOR EACH CCP (SEE PRINCIPLE 3)

Critical limits must be specified and validated for each CCP and must be measurable. In some cases of food manufacturing more than one critical limit will be elaborated at a particular step. Criteria often used include measurements of temperature, time, moisture level, pH, a_w, available chlorine, and sensory parameters such as visual appearance and texture.

12.6.9 ESTABLISH A MONITORING SYSTEM FOR EACH CCP (SEE PRINCIPLE 4)

Monitoring is the scheduled measurement or observation of a CCP relative to its critical limits and must be able to report any deviation. Where possible, process adjustments should be made when monitoring results which indicate a trend towards loss of control at a CCP and adjustments must be made in the process to tackle the deviation. Data derived from monitoring must be evaluated by a designated person with knowledge and authority to carry out corrective actions when indicated. If monitoring is not continuous, then the amount or frequency of monitoring must be sufficient to guarantee the CCP is in control. All the records and documents associated with monitoring of CCPs must be signed by the person(s) doing the monitoring and by a responsible reviewing official(s) of the company.

12.6.10 ESTABLISH CORRECTIVE ACTIONS (SEE PRINCIPLE 5)

To deal with the deviations, corrective actions must be developed for each CCP which must ensure that the CCP has been brought under control. Actions taken must also include proper disposition of the affected product. Deviation and product disposition procedures must be documented in the HACCP record keeping.

12.6.11 VERIFICATION (SEE PRINCIPLE 6)

Verification and auditing methods, procedures and tests, including random sampling and analysis, can be used to determine if the HACCP system is working correctly. The frequency of verification should be sufficient to confirm that the HACCP system is working effectively.

It is always advisable that verification should be carried out by those who are not responsible for performing the monitoring and corrective actions. Examples of verification activities include: review of the HACCP system and plan and its records; review of deviations and product dispositions; confirmation that CCPs are kept under control.

12.6.12 DOCUMENTATION AND RECORD KEEPING (SEE PRINCIPLE 7)

Documentation and record keeping should be appropriate to the nature and size of the operation and sufficient to assist the business to verify that the HACCP controls are in place and being maintained. Documentation examples are: hazard analysis; CCP determination; critical limit determination. Examples of records in HACCP system are CCP monitoring activities; deviations and associated corrective actions; verification procedures performed; and modifications to the HACCP plan. An example of a HACCP worksheet for the development of a HACCP plan is attached as Figure 12.3.

1. Describe Product							
2. Flow Process Diagram							
3. List							
Steps	Hazard (s)	Control measure(s)	CCPs	Critical limit(s)	Monitoring procedure(s)	Corrective action(s)	Record(s)
4. Verification							

FIGURE 12.3 Example of a HACCP worksheet.

12.7   COMMERCIAL PROCESSING EXAMPLE

To facilitate our discussion of HACCP, the ABC Dairy is introduced. With this fictitious company as a base, evolution of a HACCP plan for pouch milk will be discussed and illustrated. Since HACCP plans are very product, process and plant specific, ABC Dairy's plan may not be suitable for other firms. Processing narratives can help explain the current processing steps needed to produce a product covered by a particular HACCP plan. They offer a historical, working reference for the processor and facilitate communication with the staff and inspectors. For these reasons, a written narrative should accompany a HACCP plan.

1. **Company: ABC Dairy.**
2. **Product name: Pouch Packed Milk**
3. **Raw Materials, Ingredients and Product contact materials characteristics:**
    3.1. **Raw material and ingredients:** Raw milk, skim milk powder, cream, water.
        3.1.1. **Milk**
            a. **Biological, chemical and physical characteristic:** Milk is in continuous phase in which lactose, apportion of minerals and salts are in phase of true solution and protein and fat are in suspended form. Good quality of raw milk less acidity milk is yellowish to creamy white (cow milk) and creamy white (Buffalo milk). The flavor of milk is blend of a sweet taste of lactose and salty taste of minerals. The phospholipids, fatty acid and fat are also contributed to the flavor. Milk serves as a nutritional rich complex material for the optimum growth of micro-organisms.
            b. **Composition:** % Fat: NLT 3.0, SNF: NLT 8.5% for cow milk and % Fat: NLT 5.0, SNF: NLT 9.0% for buffalo milk.
            c. **Origin:** Obtain from healthy milch animal.
            d. **Method of production:** By milking healthy milch animal.

e. **Packaging and delivery method:** Raw milk is received from District cooperative Society through SS cans.

f. **Storage condition and shelf life:** After chilling raw milk is stored in silo at less than 10°C.

g. **Preparation and/or handling before use or processing:** Raw milk is stored at less than 10°C before use for not more than 24 hrs.

h. **Food safety–related acceptance criteria or specifications of materials and ingredients appropriate to the intended use:** Raw milk when received should be free from visible dirt and dust. It should have clean, clean flavor and free from any objectionable flavor. Clot on boiling (COB) must be negative. Raw milk should have 3% fat and 8.5% SNF in case of cow milk and 5% fat and 9% SNF for buffalo milk.

i. **Statutory and Regulatory food safety requirements:** Milk is the normal mammary secretion derived from complete milking of healthy animal without either addition thereto or extraction there from. It shall be free from colostrum.

3.1.2. SMP

a. **Biological chemical and physical characteristics:** The milk powder has characteristic to absorb moisture when proper packaging technique not applied and same stored in humid atmosphere. On prolong storage at room temperature leads to rancid taste in powder. Due to less than 4% of moisture, its microbiological activity is very less and product remains safe for about 18 months when stored in cool and dry place. To ensure maximum keeping quality, the dried product should be stored in a vapor proof, moisture proof, and sealed package in a dark, cool, dry place. Color: normally yellowish-white for cow and chalky white for buffalo milk. The ideal dry milk is one which rapidly recombines with water without agitation to give the characteristics of regular milk. One of the factors affecting the reconstitution is the temperature of water and dried milk, the nature and extent of hardness in water, and the time and nature of agitation. The flavor of dried milk should be clean, rich, sweet and very pleasant. A more common defect in appearance is the presence of burnt particles. Generally the deterioration occurring during storage involves flavors, color and solubility index.

b. **Composition:** Total milk solids (%wt) 95.0% (min.) and Moisture (%wt) 5.0% (max.)

c. **Origin:** Skimmed milk

d. **Method of Production:** It is prepared by condensing and spray drying of skim milk.

e. **Packaging and delivery method:** Packing material: primary packing material is Polyethylene pouch; secondary packing material is plastic bag. Pack size: 25 kg. Deliver through local transport in covered truck.

f. **Storage condition and shelf life:** Stored at room temperature in cool and dry place and product remains safe for 18 months.

g. **Food safety-related acceptance criteria:** Skimmed milk powder Should not have more than 4% of Moisture, Not more than 1.5% of Fat, Not more than 1.5% lactic acid, brown and scorched particles should be absent and Coli form should be absent in 0.1 g.

h. **Statutory and Regulatory food safety requirements:**
Moisture: NMT 5.0% m/m; Milk Fat: NMT 1.5% m/m; Milk protein in MSNF: NLT 34.0% m/m; Titratable acidity: NMT 0.18 lactic acid; Insolubility index: NMT 2 mL; Total ash on dry weight basis: NMT 8.2%.

3.1.3. Cream

a. **Biological chemical and physical characteristics:** Available in liquid form with high viscosity. Cream used contain generally % fat > 40% and should be properly pasteurized and chilled <10°C. And should be stored at <10°C to avoid microbial deterioration. Cream stored at more than 12°C may develop flavor defects like Cheese, acidic, sour, fruity, and yeasty and leads to COB positive test. Generally the acidity of fresh cream remains in the range of (0.06 to 0.09% lactic acid).

b. **Composition:** Fat: min. 40% and SNF: max. 6%

c. **Origin and method of production:** Cream is separated from milk by centrifugal separation.

d. **Storage condition and shelf life:** Should be stored below 10°C.

e. **Preparation and/or handling before use or processing:** Separated cream must be pasteurized at 80 °C with no hold and transferred into cream storage tank at less than 12 °C. The processed cream should not be stored more than 24 hrs. Fresh cream (raw or pasteurized) can be used for the preparation of pouch milk.

f. **Food safety-related acceptance criteria appropriate to the intended:** COB should be negative for cream.

3.1.4. Water

The quality of water directly affects the safety of product and care should be taken during production and processing that potable water is only used. To minimize the risk of contamination of product through water, it is important to minimize the chances of water contamination during handling steps.

3.2. Product contact material

Film, hand, SS Equipment and pipes, plastic crate, SS can.

3.2.1. Film

a. Biological chemical and physical characteristics: Milk pouch film is manufactured from virgin polymer of Low-density polyethylene(LDPE) and linear low-density polyethylene (LLDPE). It must have a good heat sealing property. Milk pouch is added with Titanium dioxide (TiO_2) which makes the film opaque and prevents the milk from getting oxidized flavor when exposed to sunlight.

b. Composition: LDPE and LLDPE (Food grade)

c. Method of production: Co-extrusion and blowing

d. Packaging and delivery method: Approximately 18–25 kg roll pack in corrugated box/plastic covered and delivered in covered vehicle at ABC Dairy from supplier.

e. Storage condition and shelf life: Stored in clean and dry place at room temperature.

f. Food safety-related acceptance criteria or specifications of purchased materials: Film should be clean and should be free from off smell. Film width: 350±2 mm, Thickness: 40 micron±7%.

g. Statutory and Regulatory food safety requirements: Should be food grade.

3.2.2. Hand Wash

Germs are transmitted from unclean hands into food; hand washing can prevent the transfer of germs. It is recommended to vigorously scrub the hands with soapy water for at least 15 s. Sanitizer helps to prevent the spread of germs and infection, which is extremely important in food contact and health care industries. To impede the spread of bacteria, AHD is a powerful germ-fighting agent.

3.2.3. Stainless Steel

In metallurgy, stainless steel is defined as a steel alloy with a minimum of 10% chromium content by mass.

Stainless steel does not stain, corrode, or rust as easily as ordinary steel (it stains less), but it is not stain-proof. It is also called corrosion-resistant. Acidic food products like milk can be stored, processed, transported and handled through stainless steel pipes, tanks, cans or any kind of storage or transporting equipments due to its corrosion resistance and antibacterial properties.

3.2.4. Plastic Crates

Plastic crates are used in dairy industry for handling of milk pouches as they are light weight and economical. Almost all plastic form of petroleum are melted and formed into polyethylene. Polyethylene is easily molded to any shape and holds together when stretched thin, it is water-proof, nontoxic, break-resistant and is resistant to bacteria a and other microbial growth. Plastic crates are manufactured by injection molding.

4. Labeling Instructions

Each and every milk pouch must carry the following information:
1. Name of milk
2. Manufacturer name: ABC Dairy, XYZ
3. FSSAI License Number: XXXXXXXXXX
4. Customer care No: 0000000000
5. Market name: XYZ.
6. MRP:
7. Batch no:
8. Use by date:
9. Volume of milk:
5. Characteristics, Intended Use and Consumer of End Products

Product: Pasteurized standardized milk

Biological, chemical and physical characteristics: Water is in continuous phase, in which lactose, minerals and salts are in phase of true solution and protein and fat are in suspended form. Good quality of milk has acidity less than 0.135% lactic acid and pH is 6.6 to 6.8. Color of milk is yellowish to creamy white (cow milk) and creamy white (buffalo milk).The flavor of milk is a blend of a sweet taste of lactose and salty taste of minerals. The phospholipids, fatty acid and fat are also contributes to the flavor. Milk serves as a nutritional rich complex material for the optimum growth of micro-organisms. Milk is considered as good source of fat, protein, carbohydrate, vitamins and minerals. Further processing of milk leads to increase in shelf-life and makes the milk safe for human consumption. Processed and packed milk is required to store below 8°C to prevent growth of microorganisms and storage above 8°C will lead to increased acidity due to microbial activity causing souring of milk resulting into spoilage of milk.

6. **Intended Shelf life and Storage Conditions**
   Use-by-date will be printed on every pouch indicated the shelf-life of the product when stored under refrigeration below 8°C.

7. **Intended use**
   To be consumed as milk, to be made into products by the consumers and then consumed.

8. **Packaging**
   Item: Polythene film

9. **Material**
   LDPE & LLDPE

10. **Pack size**
    200 mL and 500 mL.

11. **Method of Distribution**
    Pasteurized standardized milk will be dispatched through insulated vehicles to the retailers and distributors. During the transportation of milk the temperature of the milk must be maintained well below 8°C.

12. **Flow Diagram of Pouch Milk**

13. **Risk Rating Criteria**
    P – Probability of contamination
    Rating = N (1) = Negligible: The P of contamination is extremely low or negligible given the combination of factors previously described like once in a year
    Rating = L (2) = Low: The P of contamination is low but clearly possible given the combination of factors previously described like once in quarter

Rating = M (3) = Medium: The P of contamination is likely given the combination of factors previously described like every month

Rating = H (4) = High: The P of contamination is very likely or certain given the combination of factors previously described like every day.

S – Severity of contamination

Rating = N (1) = Negligible: Disease occurs in a limited group and/or has negligible impact on health and quality of life like feeling uneasiness.

Rating = L (2) = Low: There is a limited range of people who suffer disease, with minor impact on health like headache.

Rating = M (3) = Medium: There is a moderately broad range of people who suffer disease and/or moderate impact on health like vomiting.

Rating = H (4) = High: Disease occurs in a broad range of people and/or with severe impact on health and quality of life like death or poisoning.

RV = Risk Value: P-Physical contamination, C-Chemical contamination, M-Microbiological contamination.

Method of arriving at risk value: Probability x Severity = Risk value

Hazards having Risk value "3 or above" is considered "Significant Hazards to be controlled." After that the classification of all the "Significant Hazards to be controlled" is done in the form that whether they are going to be controlled by OPRP or going to be considered as CCP. Classification of CCPs and OPRP is done by categorization of control measures. The reason behind considering the *Risk Value* greater that equal to 3:

If RV = 16 ------> 100% chances of any risk involved.

If RV = 12 ------> 75% chances of any risk involved.

If RV = 08 ------> 50% chances of any risk involved.

If RV = 04 ------> 25% chances of any risk involved.

Hence for any identified Hazard we have taken RV near about 4, i.e., considering RV ≥3 for any "Significant Hazards to be controlled," which shows we have taken only 25% chances and considering each and every identified hazard whose RV ≥3.

14. List of Food Safety Hazards–Hazard Analysis

Process steps		Source of hazards	Risk Rating			Control Measures Implemented
			P	S	RV	
Receiving of Raw Milk in cans from society at Room Temp and Milk Received in Tanker	P	Contamination of dust, dirt and foreign materials may be there from tanker	2	1	2	Provision of filter during receiving
	C	Chances of neutralizer, preservative and adulterant availability in raw milk. More temperature & acidity will lead to COB positive.	1	3	3	As per schedule checking of neutralizer, preservative and any adulterant is done and also regular testing of all tanker in Q.A. for temperature of milk, Acidity, OT and COB is done.
	M	Presence of harmful toxins like Aflatoxins, Ocratoxin, Patulintoxin, Botulinum toxin, Stapylococcal toxin, endotoxins, exotoxins etc. Presence of Coliform, Salmonella spp., Shigella spp., Listeria spp., Escherichia coli, Clostridia spp., Vibrio spp., Alkaligens spp, Pseudomonas spp., Staphylococcus spp., Streptococcus spp., Bacillus spp, etc. Corynebacterium spp, Clostridia spp. Mycobacterium spp., Coxiella spp., Brucella spp., and certain Yeast and Moulds certain viruses like Adenovirus, Enterovirus etc. and certain protozoa like Toxoplasma, Amoeba, etc. Which can lead to acid production and hence COB can be positive.	4	3	12	Proper checking of milk temperature during receiving for preventing the growth and checking the quality of milk for any objectionable flavor, acidity and COB in Q.A.

Plungering ~20 times for proper mixing

Type	Hazard	P	S		Corrective measure
P	Contamination of dust and dirt may be possible of improper cleaning of plunger	1	1	1	Proper cleaning of plunger by following Code of Practice to clean plunger.
C	Contamination of teepol and iodophor residue left after improper cleaning.	1	1	1	Proper cleaning of plunger by following Code of Practice to clean plunger.
M	Improper sanitization leads to contamination of harmful micro organism like	1	2	2	Proper cleaning of plunger by following Code of Practice to clean plunger.

Sampling by Q.A. Personnel

Type	Hazard	P	S		Corrective measure
P	Not Any	NA	NA	NA	
C	Not Any	NA	NA	NA	
M	Improper sanitization of sampling vessels leads to contamination of harmful micro-organism like Coliforms and Pathogens like *Salmonella* spp., *Listeria* spp., *Escherichia coli, Shigella* spp., *Staphylococcus* spp., *Yersinia* spp., *Vibrio* spp. various milk spoilage bacteria, and certain yeast and molds.	1	2	2	Proper cleaning of plunger and sampling vessels by following Code of Practice.

Process steps	Source of hazards	Risk Rating			Control Measures Implemented
		P	S	RV	
Q.A. Clearance (Basis of fat, SNF, Acidity, COB&OT)and Grading		P	S		
	P	Not Any	NA	NA	NA	
	C	Not Any	NA	NA	NA	
	M	Not Any	NA	NA	NA	
Hose Pipe Connection & Dump Tank		P	S		
	P	Contamination of dust, dirt or foreign material if hose pipe is not cleaned and stored properly	1	1	1	Proper monitoring and cleaning and arrangement of hose pipe at Receiving dock
	C	Contamination of lye residue if CIP is not done properly	1	1	1	Proper Monitoring and cleaning
	M	Improper sanitization leads to contamination of harmful micro-organism like Coliforms and Pathogens like *Salmonella* spp., *Listeria* spp. *Escherichia coli*, *Shigella* spp., *Staphylococcus* spp., *Yersinia* spp., *Vibrio* spp. various milk spoilage bacteria, and certain yeast and molds.	1	2	2	Proper monitoring and cleaning and arrangement of hose pipe at receiving dock

Addition of sweet cream & skim milk Powder.			P	S		
	P	Improper cleaning of trolley can or dumping tank may lead to dust, dirt or other foreign material contamination.	1	1	1	Proper monitoring, recording of cleaning of can, trolley, dumping tank daily as per schedule and code of practice
	C	Contamination of lye residue if CIP of dumping line is not done properly. Also contamination of iodophor and teepol may be possible in case manual cleaning of can and trolley is not done properly.	1	1	1	Proper CIP of dumping line with monitoring and recording. Proper monitoring, recording of cleaning of can and trolley by following code of practices.
	M	Improper CIP or storage at unhygienic condition may lead to harmful microbial contamination Coliforms and Pathogens like *Salmonella* spp., *Listeria* spp. *Escherichia coli, Shigella* spp., *Staphylococcus* spp., *Yersinia* spp., *Vibrio* spp. various milk spoilage bacteria, and certain yeast and molds. Also increase in temperature during storage may lead to growth of Psychrophilic or cryophilic bacteria like *Pseudomonas* spp., *Acromobacter* spp., *Flavobacterium* spp., *Alcaligens* spp., *Aerobacter* spp., *Serretia* spp., etc. also yeast and molds, etc. leading to acid production and hence COB positive results can be possible.	NA	NA	NA	CIP, cleaning, recording and monitoring of Can, trolley and dumping line. Also doing COB test, organoleptic test, acidity test in Q.A. before transferring the milk to process section, which is recorded, monitored and analyzed. Until Q.A. clearance addition of sweet cream butter milk or any dumping milk is not done in process section.

Process steps	Source of hazards	Risk Rating			Control Measures Implemented
		P	S	RV	
Chilling to <=10°C through plate chiller	P Not Any	P NA	S NA	NA	
	C Contamination of lye residue if CIP is not done properly	1	1	1	PT should be negative at the end of CIP.
	M Increase in chilling temperature of milk, i.e., >6°C in chiller may lead to harmful microbial proliferation of Psychrophilic or cryophilic bacteria like *Pseudomonas spp., Acromobacter spp., Flavobacterium spp., Alcaligens spp., Aerobacter spp., Serretia spp.,* etc. *also yeast and molds.*	1	3	3	Keeping chilling temp. of milk always <=10°C and is monitored and recorded.
	Generation of metabolites like acids, esters, alcohols, carbon dioxide, NH_3 and certain toxins is possible.				

Process Step		Hazard	P	S	NA	Control Measures
Transferred and stored in silo at <= °10 C	P	Not Any	NA	NA	NA	
	C	Contamination of lye residue if CIP is not done properly	1	1	1	Ensure proper cleaning and PT should be negative at the end of CIP.
	M	Improper CIP of silo may lead to further addition of microbial contamination of *Pseudomonas* spp, *Acromobacter* spp., *Flavobacterium* spp., *Alcaligens* spp., *Aerobacter* spp., *Serretia* spp., etc. *also yeast and molds.* and increase in milk temperature to above 6 C can cause microbial multiplication of Psychrophilic or cryophilic bacteria *Pseudomonas* spp., *Acromobacter* spp., *Flavobacterium* spp., *Alcaligens* spp., *Aerobacter* spp., *Serretia* spp., etc. *also yeast and molds.*, etc. Generation of metabolites like acids, esters, alcohols, carbon dioxide, NH3 & certain toxins is possible.	1	2	2	Proper implementation of Code of Practice for CIP of silo as per that the time, concentration, and temperature which is properly maintained, monitored and recorded. Insulated silo is used so that increase in temperature of milk inside is silo is prevented.
Processing of Milk from Silo for Pasteurization	P	Not Any	NA	NA	NA	
	C	Not Any	NA	NA	NA	
	M	Not Any	NA	NA	NA	

Process steps		Source of hazards	Risk Rating			Control Measures Implemented
			P	S	RV	
Milk in balance tank	P	Not Any	NA	NA	NA	
	C	Contamination of lye residue if CIP is not done properly	1	1	1	Ensure proper cleaning and PT should be negative at the end of CIP.
	M	Not Any	NA	NA	NA	
Regeneration -1 Temp≈45°C	P	Not Any	NA	NA	NA	
	C	Contamination of lye residue if CIP is not done properly	1	1	1	Ensure proper cleaning and PT should be negative at the end of CIP.
	M	Not Any	NA	NA	NA	
Filtration	P	If filter is not clean and done properly than there is chances of contamination of dust and dirt or any other foreign material.	1	1	1	Proper cleaning before starting and end of operation.
	C	Not Any	NA	NA	NA	
	M	Not Any	NA	NA	NA	

			P	S	
Cream and skim milk separation	P	If separator is not carried out properly than chances of contamination of dust or dirt or sludge is possible.	1	1	Separator is cleaned every day after completion of the processing and it is recorded.
	C	Contamination of teepol residue if manual cleaning is not done properly.	1	1	Separator is cleaned every day after completion of the processing and it is recorded.
	M	Improper cleaning may lead to microbial contamination like Coliforms and Pathogens like *Salmonella* spp., *Listeria* spp., *Escherichia coli*, *Shigella* spp., *Staphylococcus* spp., *Yersinia* spp., *Vibrio* spp. various milk spoilage bacteria, and certain yeast and molds.	1	1	Separator is cleaned every day after completion of the processing and it is recorded.
			P	S	
Regeneration −2 temp ≈ 67°C	P	Not Any	NA	NA	NA
	C	Contamination of lye residue if CIP is not done properly.	1	1	Ensure proper cleaning and PT should be negative at the end of CIP.
	M	Not Any	NA	NA	NA

Process steps	Source of hazards	Risk Rating			Control Measures Implemented
		P	S	RV	
Heating section temp. at 76 to 82°C for holding min. 16 s.					
	P Not Any	NA	NA	NA	
	C Not Any	NA	NA	NA	
	M Improper temperature and time maintenance will lead to microbial survival of pathogenic micro organisms like *Bacillus cerus, Clostridium perfringenes, Coxiellaburnetti, Mycobacterium tuberculosis, Salmonella* spp. etc. and multiplication of Psychrophilic bacteria like *Pseudomonas* spp., *Acromobacter* spp., *Flavobacterium* spp., *Alcaligens* spp., *Aerobacter* spp., *Serretia* spp., etc. *also yeast and molds.*	2	3	6	Proper pasteurization time and temperature combination is maintained, monitored and recorded and proper working of temperature controller and FDV is ensured, i.e., if temperature of milk reaches to 71.9°C, FDV opens and milk is again transferred to balance tank. Holding time is ensured by the length of our holding tube and flow rate of milk passing through it.
Regeneration –2 temp≈54°C					
	P Not Any	NA	NA	NA	
	C Contamination of lye residue if CIP is not done properly.	1	1	1	Ensure proper cleaning and PT should be negative at the end of CIP.
	M Not Any	NA	NA	NA	

			P	S	NA	
Regeneration -1 temp≈15°C	P	Not Any	NA	NA	NA	
	C	Contamination of lye residue if CIP is not done properly.	1	1	1	Ensure proper cleaning and PT should be negative at the end of CIP.
	M	Not Any	NA	NA	NA	
Chilling section temp.<=8°C	P	Not Any	P	S	NA	
	C	Contamination of lye residue if CIP is not done properly	1	1	1	PT should be negative at the end of CIP.
	M	Not Any	NA	NA	NA	

Process steps	Source of hazards	Risk Rating			Control Measures Implemented
		P	S	RV	
Pasteurized Milk is Taken in Pre Selected Sterilized Silo		P	S		
	P	NA	NA	NA	
	C Contamination of lye residue if CIP is not done properly	1	1	1	PT should be negative at the end of CIP.
	M Improper CIP of silo will lead to microbial contamination in pasteurized milk Coliforms and Pathogens like *Salmonella* spp., *Listeria* spp. *Escherichia coli*, *Shigella* spp., *Staphylococcus* spp., *Yersinia* spp., *Vibrio* spp. various milk spoilage bacteria, and certain yeast and molds., and various milk spoilage bacteria, yeast and moulds. And if temperature inside silo is increase than growth of Psychrophilic or cryophilic bacteria like *Pseudomonas* spp., *Acromobacter* spp., *Flavobacterium* spp., *Alcaligens* spp., *Aerobacter* spp., *Serretia* spp., etc. also yeast and molds, etc. is possible.	1	2	2	Proper time, temp. and conc. Of the CIP solution to clean silo is monitored and recorded and PT should be negative. Also insulated silo is there which will not allow increase in temperature of pasteurized milk inside silo.
Q.C. Clarence		P	S		
	P Not Any	NA	NA	NA	
	C Not Any	NA	NA	NA	
	M Not Any	NA	NA	NA	

		P	S	NA		
Transfer to HMST through Chiller <=8°C	P	NA	NA	NA	Not Any	
	C	1	1	1	Contamination of lye residue if CIP is not done properly.	PT should be negative at the end of CIP.
	M	1	2	2	Improper CIP of silo will lead to microbial contamination in pasteurized milk Coliforms and Pathogens like *Salmonella* spp., *Listeria* spp. *Escherichia coli*, *Shigella* spp., *Staphylococcus* spp., *Yersinia* spp., *Vibrio* spp. various milk spoilage bacteria, and certain yeast and molds. and various milk spoilage bacteria, yeast and moulds. And if temperature inside silo is increase than growth of Psychrophilic or cryophilic bacteria like *Pseudomonas* spp., *Acromobacter* spp., *Flavobacterium* spp., *Alcaligens* spp., *Aerobacter* spp., *Serretia* spp., etc. also yeast and molds. etc. is possible.	Since silos are insulated temp. of stored milk does not rises and then transferred to storage tanks and packed immediately. Proper time, temp., and conc. of the CIP solution is monitored and recorded.
		P	S	NA		
S.S. Line Filter	P	NA	NA	NA	Not Any	
	C	1	1	1	Contamination of lye residue if CIP is not done properly.	PT should be negative at the end of CIP.
	M	NA	NA	NA	Not Any	

Process steps	Source of hazards	Risk Rating			Control Measures Implemented
		P	S	RV	
Pouch Packing		P	S		
	P	Not Any	NA	NA	NA	
	C	Contamination of lye residue if packing machine CIP is not done properly	1	1	1	PT should be negative
	M	Improper cleaning of heads can lead to microbial contamination like Coliforms and Pathogens like *Salmonella* spp., *Listeria* spp., *Escherichia coli*, *Shigella* spp., *Staphylococcus* spp., *Yersinia* spp., *Vibrio* spp. various milk spoilage bacteria, and certain yeast and molds.	1	2	2	Proper time, temp. and conc. of the CIP solution is maintained, monitored and recorded and.
Pouch Film		P	S		
	P	Poor quality of packing material and improper testing of pouch film can lead to contamination of dust, dirt or any foreign material.	1	1	1	Ensure quality of Packing film as per testing according to specifications by QA.
	C	Not Any	NA	NA	NA	
	M	Air flora contamination may be possible by *Bacillus* species, *Clostridium* species, *Micrococcus* species, *Corynebacterium* etc., and certain yeast and moulds.	1	2	2	UV light in milk packaging machine ensures sterilization of milk film.

			P	S		
Pouch Filling in Crates	P	Improper cleaning of crates might spoil the outer surface of milk pouches.	1	1	1	Ensure proper cleaning of milk crates and segregation of unclean crates by visual observation at the time of packing.
	C	Not Any	NA	NA	NA	
	M	Not Any	NA	NA	NA	
Separation of leaky pouch	P	Leaky pouch can contaminate the outer surface of other Milk pouch and hence spoil the good pouches.	1	1	1	Segregate the leaky pouches
	C	Not Any	NA	NA	NA	
	M	Not Any	NA	NA	NA	

Process steps	Source of hazards	Risk Rating			Control Measures Implemented
		P	S	RV	
Cold Storage <=8°C	P	P	S		
	C	NA	NA	NA	
		NA	NA	NA	
	M	1	3	3	Proper temp. of milk cold storage must be maintained, monitored and recorded.

Not Any

Not Any

Increase in storage temperature more than 6 °C will lead to microbial multiplication of Psychrophilic or cryophilic bacteria like *Pseudomonas* spp., *Acromobacter* spp., *Flavobacterium* spp., *Alcaligens* spp., *Aerobacter* spp., *Serretia* spp., etc. also yeast and molds, etc. and if temperature increases to room temperature may lead to multiplication of post pasteurization contaminants coliforms and pathogens like *Salmonella* spp., *Listeria* spp. *Escherichia coli*, *Shigella* spp., *Staphylococcus* spp., *Yersinia* spp., *Vibrio* spp. various milk spoilage bacteria, and certain yeast and molds.

Dispatch			P	S		
P	Leaky pouches. Improper transportation and handling may lead to puncture in pouches of milk		1	1	1	Ensure that leaky pouch should not be dispatched and segregate the leaky pouches for further processing. Also ensure proper stacking of crates and handling as per code of practice of dispatch of milk.
C	Not Any		NA	NA	NA	
M	If milk is kept for longer period of time outside cold store for dispatch than increase in temperature may lead to multiplication of Psychrophilic or cryophilic bacteria like *Pseudomonas* spp, *Acromobacter* spp., *Flavobacterium* spp., *Alcaligens* spp., *Aerobacter* spp., *Serretia* spp., etc. also yeast and molds.		1	3	3	Proper temp. of milk cold storage room, maintained, monitored and recorded. Also implementing code of practices for dispatch of milk effectively by monitoring and maintaining temperature of milk pouches.

15. List of Significant Hazards to be Controlled

Process steps		Source of hazards	Risk value	Control measures implemented	CCP/OPRP
Receiving of raw milk in S.S. cans from village cooperative society	C	Chances of neutralizer, preservative and adulterant availability in raw milk. More temperature and acidity will lead to COB positive.	6	As per schedule checking of neutralizer, preservative and any adulterant is done and also regular testing of all tanker in Q.A. for milk Grading	OPRP-01
	M	Presence of harmful toxins like Aflatoxins, Ocratoxin, Patulintoxin, Botulinum toxin, Stapylococcal toxin, endotoxins, exotoxins etc. Presence of Coliform, Salmonella spp., Shigella spp., Listeria spp., Escherichia coli, Clostridia spp., Vibrio spp., Alkaligens spp, Pseudomonas spp., Staphylococcus spp., Streptococcus spp., Bacillus spp, etc. Corynebacterium spp, Clostridia spp. Mycobacterium spp., Coxiella spp., Brucella spp., and certain Yeast and Moulds certain viruses like Adenovirus, Enterovirus etc. and certain protozoa like Toxoplasma, Amoeba, etc., which can lead to acid production and hence COB positive.	8	Proper checking of milk grade during receiving for preventing the growth and checking the quality of milk for any objectionable flavor, acidity and COB in Q.A.	OPRP-01
Chilling to <10 °C through plate chiller	M	Increase in chilling temperature, i.e., >10°C in chiller may lead to microbial proliferation of Psychrophilic or cryophilic bacteria like Pseudomonas spp, Acromobacter spp., Flavobacterium spp., Alcaligens spp., Aerobacter spp., Serretia spp., etc. i. Generation of metabolites like acids, esters, alcohols, carbon dioxide, NH_3 and certain toxins is possible.	4	Keeping chilling temp. of milk always <=10°C and is monitored and recorded.	OPRP-02

Addition of sweet cream/ SMP	M	4	Improper CIP or storage at unhygienic condition may lead to harmful microbial contamination Coliforms and Pathogens like *Salmonella* spp., *Listeria* spp. *Escherichia coli, Shigella* spp., *Staphylococcus* spp., *Yersinia* spp., *Vibrio* spp. various milk spoilage bacteria, and certain yeast and molds. Also increase in temperature during storage may lead to growth of Psychrophilic or cryophilic bacteria like *Pseudomonas* spp, *Acromobacter* spp., *Flavobacterium* spp., *Alcaligens* spp., *Aerobacter* spp., *Serretia* spp., etc. also yeast and molds, etc. leading to acid production and hence COB positive results can be possible.	CIP, cleaning, recording and monitoring of Can, trolley and dumping line. Also doing COB test, organoleptic test, acidity test in Q.A. before transferring the milk to process section, which is recorded, monitored and analyzed. Until Q. A. clearance addition of sweet cream butter milk or any dumping milk is not done in process section.	OPRP-03
Pasteurization temperature is 76 to 82°C for 16 s. and chilling below 10°C.	M	4	Improper heating chilling will lead to higher temperature of milk as a result chances of multiplication of micro organisms is possible of psychrophilic or cryophilic bacteria like *Pseudomonas* spp, *Acromobacter* spp., *Flavobacterium* spp., *Alcaligens* spp., *Aerobacter* spp., *Serretia* spp. also yeast and molds, etc.	Proper heating chilling of milk to <= 10°C is maintained. This temperature is monitored and recorded.	CCP-01 CCP-02
S.S. line filter	P	2	Dust particle filtered and also foreign particles are filtered.	Proper filtration due to on line filter.	OPRP-04

Process steps	Source of hazards		Risk value	Control measures implemented	CCP/OPRP
Cold storage at temp. <= 8°C	M	Increase in storage temperature more than 8°C will lead to microbial multiplication of Psychrophilic or cryophilic bacteria like *Pseudomonas* spp., *Acromobacter* spp., *Flavobacterium* spp., *Alcaligens* spp., *Aerobacter* spp., *Serretia* spp. also yeast and molds, etc. And if temperature increases to room temperature may lead to multiplication of post pasteurization contaminants Coliforms and Pathogens like *Salmonella* spp., *Listeria* spp., *Escherichia coli, Shigella* spp., *Staphylococcus* spp., *Yersinia* spp., *Vibrio* spp. various milk spoilage bacteria, and certain yeast and molds.	2	Proper temperature of milk cold storage room, maintained, monitored and recorded.	CCP-03

16. Operational Pre-requisite Programs

Process Steps/OPRP	Source of Hazards	Control Measures	Monitoring Procedures	Correction/Corrective Action	Monitoring Record	Responsibility/ Authority
Receiving of raw milk and grading by Q.A Personnel	Refer to hazard analysis of milk.	Checking of neutralizer, preservative and any adulterant is done and also regular testing of all tanker and Cans of some selected society in Q.A. for temperature of milk, Acidity, OT and COB is also done.	**What:** Milk flavor (O.T.), Acidity, Temperature of milk in tanker, COB test, and As per schedule checking of neutralizer, preservative and any adulterant./milk flavor, grading for good, sour and unacceptable quality **How:** O.T., COB Test, Acidity Check, Temperature Check, Adulteration or Preservative Check. **When:** On each tanker arrival/each society cans **Where:** Q.A. Lab./RMRD **Who:** Chemist/Lab Attendant	**Correction Action:** Rejection of tanker and cans if result is not found as per standards and specification. **Corrective Action:** 1. Receiving of tanker/Cans after Q.A. clearance. 2. Ensuring testing of each and every tanker/cans and recording of results tanker wise society wise. 3. Regularly ensuring the specification our test results. **Who:** Chemist/Shift In charge	**Record No:** xx **Record name:** yy **Who:** Chemist	**Responsibility:** Chemist. **Authority:** Q.A. In-charge.

Process Steps/OPRP	Source of Hazards	Control Measures	Monitoring Procedures	Correction/Corrective Action	Monitoring Record	Responsibility/ Authority
Addition of sweet cream/ Skim Milk Powder	Refer to Hazard Analysis of Milk.	CIP, cleaning, recording and monitoring of Can, trolley and dumping line. Also doing COB test, organoleptic test, acidity test in Q.A. before transferring the milk to process section, which is recorded, monitored and analyzed. Until Q. A. clearance addition of sweet cream and Skim Milk Powder or any dumping milk is not done in process section.	**What:** Phenolphthalein test negative and COB is negative. **How:** By doing PT Test and COB test. **When:** PT negative after CIP and COB negative for every lot of dumping milk. **Where:** In Production PT Test and In Q.A. Lab COB test. **Who:** Section In-charge Packing/Operator Packing for PT test. Chemist for COB test.	**Correction Action:** 1. If COB is positive than ensure that the dumping milk should be drain. 2. Flushing of water up till PT is negative. **Corrective Action:** Always proper ensuring of COB results of each and every dumping milk and not clear by Q.A. it should not be allotted to Processing section. **Who:** In-charge Packing/ In charge Q.A. and FSFL.	**Record No:** xxxxxxxx **Record name:** yyyyyyyyyy **Who:** In-Charge/ Chemist./ Operator.	**Responsibility:** In-Charge Q.A/ Packing. **Authority:** Assistant Manager (Production)/ Q.A. In-charge.

			Monitoring	Corrective Action	Record / Responsibility
Chilling to <10 °C through plate chiller	Refer to hazard analysis of milk.	Keeping chilling temp. of milk always <=10 °C and is monitored and recorded.	**What**: Temperature of Chilled milk. **How**: Manually from digital temperature indicator. **When**: Continuously and recording after every half an hour. **Where**: Production **Who**: In-charge/Operator.	**Correction Action:** Increase Chilled water flow if temp. is not achieved. **Corrective Action:** 1. Always proper ensuring of IBT temperature and proper working of Chilled water pump. 2. Calibration of Temp. Indicator. **Who**: In-charge.	**Record No:** xxxxxxxxxxx **Record name:** yyyyyyyyy **Who:** In-Charge/Operator. **Responsibility:** In-Charge **Authority:** Assistant Manager (Production).
S.S. Line Filter	Refer to hazard analysis of milk.	Proper cleaning. of the S.S. Line Filter is to be maintained and recorded each shift.	**What**: Proper cleaning of the S.S. Line Filter **How:** Manually cleaning of the S.S. Line Filter **When**: Continuously and recording after every shift **Where**: Production **Who**: In-charge/Operator.	**Correction Action:** Each shift the S.S. line filter dismantling and cleaning **Corrective Action:** 1. Always proper ensuring of proper working. 2. Every shift dismantling and checking **Who**: In-charge/Operator.	**Record No:** xxxxxxxxxxx **Record name:** yyyyyyyyyy **Who:** In-Charge/Operator. **Response.** In-Charge **Authority:** Assistant Manager (Production).

17. Critical Control Points

Process steps/ CCP	Critical limit of CCP	Monitoring procedure	Acceptable Level of Possible Hazards	Correction/ Corrective action on CCP violation	Monitoring Record	Verification Schedule	Responsibility/ authority
Pasteurization of milk (Heating at 76 to 82° C for Min. 16 Sec.)	Temp. >= 76°C. (Heating at 76 to 82 °C for min. 16 s.)	**What:** Pasteurization temperature of milk **How:** Visual observation of temp. indication on pasteurizer & through Data Logger system **When:** During continuously processing of milk at 30 min for manually entry and by Data logger at every min **Where:** At pasteurizer & Process Data logger **Who:** S.I./Operator	AP test −ve Coliform: Nil/0.1 mL & SPC: <30,000/ mL.	**Correction:** Reprocessing of milk. **Corrective Action:** Ensure proper operation of FDV and Calibration of temp. Controller.	**Record No:** xxxxxxxxx **Record name:** yyyyyyyyy **Who:** Sift In-Charge/ Operator	**By:** In-charge **When:** On daily basis	**Responsibility:** In-charge **Authority:** Asst. Manager (Production)

CCP	Critical Limit	Monitoring	Critical Limits	Correction / Corrective Action	Records	By / When	Responsibility / Authority
Chilling section (Temp. 8°C).	<= 10°C	**What:** Chilling temperature of milk. **How:** Maintaining proper flow rate by chilled water pump. **When:** During continuously processing of milk at 30 min for manually entry and by Data logger **Where:** At pasteurizer & Process Data logger. **Who:** S. I./Operator.	AP test −ve Coliform: Nil/0.1 mL & SPC: <30,000 mL.	**Correction:** Increasing the flow rate of chilled water by running additional pump. **Corrective Action:** Proper maintaining the temperature of Ice Bank Tank and working of Chilled water pump.	**Record No:** xxxxxxxx **Record name:** yyyyyyyy **Who:** Sift In-Charge/Operator/In-Charge engineering	**By:** In-charge **When:** On daily basis	**Responsibility:** In-charge **Authority:** Asst. Manager (Production) Asst. Manager (Engineering)
Temperature of Cold Storage Min <=8°C	Cold storage temp. below 8°C	**What:** Cold storage temp. **How:** Visual observation of temp. indication on Cold Storage after every hour **When:** During transfer of milk from Silo to HMST and HMST to Packing. **Where:** At Silo or Milk Storage Tank **Who:** S.I./Operator	AP test −ve Coliform: Nil/0.1 mL & SPC: <30,000/ mL.	**Correction:** To control the liquid refrigerant supply **Corrective Action:** Proper maintaining the temperature of IBT and working of liquid refrigerant supply to expansion valve.	**Record No:** xxxxxxxx **Record name:** yyyyyyyy **Who:** Sift In-Charge/Operator/In-Charge engineering.	**By:** In-charge **When:** On daily basis	**Responsibility:** In-charge **Authority:** Asst. Manager (Production) Asst. Manager (Engineering)

18. Food Safety Management Verification Schedule

Implementation of PRP(S).	To ensure in controlling food safety related hazards.	By following statutory and regulatory requirements, any recognized guidelines, CODEX guidelines, etc.	Once in Six Months	Food Safety Team/In charge/Food Safety Team leader (FSTL)/M.D.
Review of all the product Flow Chart	To check any change in process flow chart resulting in addition or reduction of processing steps.	By following the ISO 22000 standards and Food Safety Team will verify the accuracy of the flow diagrams by on site checking.	Once in Six Months	Food Safety Team/In charge/FSTL
Review of Hazard Analysis	To check any new source of hazards within and outside the process.	Conducting once again the study of Hazard Analysis for the entire product and listing any new significant source of hazards.	Once in a year.	Food Safety Team/In-charge/FSTL/M.D.
Effectiveness of OPRP Placed.	To check the food safety management system and effectiveness of system implemented.	Monitoring and reviewing of any violation of OPRP. Any Correction/ Corrective action and control over the possible hazards.	Once in six months.	Food Safety Team/ Section In-charge/FSTL.
Presence and Correctness of CCP monitoring	To check whether all the critical control points are under control and under specified limit.	Number of product conformity and non conformity observed. Records verification and any correction/corrective action undertaken.	Once in six months	Food Safety Team/FSTL/ Internal Auditor.

Item	Objective	Details	Frequency	Responsibility
Validation of CCP	To remain update with the latest research and literatures and searching for any possibility of hazard. Any new CCP development.	To collect the literatures and any information or recent research data relating to our requirements whether it is legal requirement or any other. Concerning with the concern bodies and new amendments implemented with respect to food safety.	Once in year	Food Safety Team/FSTL.
Monitoring Equipment/ Calibration/ Operating.	To ensure whatever results we get with the help of any equipment or measuring devices is accurate, resulting into total food safety.	Authorized calibration of each and every equipment and measuring devices with the help of authorized body and devices.	Once in six months	Food Safety Team/ FSTL./Calibrator.
Review of identified acceptable levels of Hazard.	To get assured by considering and analyzing each and every aspect of acceptable levels of hazards.	To analysis results regularly and constantly study the process related to product chemically as well as microbiologically.	Once in six months	Food Safety Team/FSTL.
Review of correction/ corrective action (If any).	To ensure whatever action taken is valid and verification of action leading to bringing the control back to get acceptable level of hazards.	To analyze the results after correction and planning for implementing corrective action and providing whatever resource required.	Once in six months	Food Safety Team/FSTL/ Internal Auditor.
Review of Customer related process	To get familiar about customer requirement, opinion, complaints and comments. Market requirements and related competition.	Field review, any withdrawal, root cause study of any complaints etc.	Once in six months	Food Safety Team/FSTL/ Marketing Head

12.8   SUMMARY

To ensure that milk and milk products are processed, stored and distributed are safe and wholesome for human consumption implementation of HACCP turns to be vital step. The food industries including dairy must be committed for application of HACCP for food safety prior to application of HACCP, it has been ensured that the general principles of food hygienic cleaning and sanitation of equipment, personnel hygiene, housekeeping and good manufacturing practices are in place and food safety legislations are implemented. During hazard identification, evaluation and subsequent operations in designing and application of food safety management system, considerations must be given to the impact of raw materials ingredients, contact surfaces, food manufacturing practices and role of manufacturing process to control hazard. The implementation and execution of HACCP must be done by a multi-disciplinary and experienced team.

KEYWORDS

- Codex Aliment Arius Commission
- Critical control point
- Critical limits
- Food and Drug Administration
- Food safety team
- HACCP
- Hazard analysis
- Hazard identification
- Hazards
- Management systems
- Milk products
- Risk and safety
- Validation
- Verification

REFERENCES

1. 21CFR110 (2014). Current good manufacturing practice in manufacturing, packing, or holding human food. CFR – Code of Federal Regulations, Title 21 – Food and Drugs, Chapter 1. Subchapter B: Food for Human Consumption. *Food and Drug Administration, US Department of Health and Human Services.*
2. Codex Aliment Arius Commission (2003). *Hazard Analysis and Critical Control Point (HACCP) System and Guidelines for its Application.* Annex to CAC/RCP 1, 1969, Rev. 4.

CHAPTER 13

ANTIBIOTICS IN FOOD PRODUCING ANIMALS: PRESENT STATUS OF RESISTANCE HAZARDS

SHAHID PRAWEZ,[1] AZAD AHMAD AHANGER,[2]
MANISH KUMAR,[3] PRIYA RANJAN KUMAR,[4] and
RAMADEVI NIMANNAPALLI[5]

[1]Associate Professor, Veterinary and Animal Sciences, Institute of Agricultural Sciences, Banaras Hindu University, Varanasi – 221005 (UP), India. E-mail: Shahidprawez@gmail.com

[2]Associate Professor, Division of Pharmacology and Toxicology, Sher-e-Kashmir University of Agricultural Sciences and Technology, Kashmir, Shuhama, Alustang, Srinagar–190006, India. E-mail: azadpharm@rediffmail.com

[3]Assistant Professor, Veterinary and Animal Sciences, Institute of Agricultural Sciences, Banaras Hindu University, Varanasi – 221005 (UP), India. E-mail: manish.vet82@gmail.com

[4]Assistant Professor, Veterinary and Animal Sciences, Institute of Agricultural Sciences, Banaras Hindu University, Varanasi – 221005 (UP), India. E-mail: dr.pranjan007@gmail.com

[5]Professor, Veterinary and Animal Sciences, Institute of Agricultural Sciences, Banaras Hindu University, Varanasi – 221005 (UP), India. E-mail: ramadevi.nimannapalli@gmail.com

CONTENTS

13.1   INTRODUCTION

Since the existence of earth, microorganisms also became an indispensable part of biosphere. Till today, microorganisms are still surviving by making some changes on their own so as adapt to the changing surroundings. After the discovery of antibiotics in 1940, a golden era started that influenced the world by reducing illness as well as mortality. Antibiotics are used to prevent and treat infectious diseases caused by bacteria. Profitable enterprise in rearing food-producing animals depends upon high productivity. Use of antibiotics as growth promoters in food producing animals is a trend to achieve maximum growth within short period of time. The rampant use of antibiotics, either for therapeutics or for enhancing production in food producing animals has led to the development of antibiotic resistance among various species of bacteria [7].

Term Antibiotic resistance connotes to the capacity of bacteria to survive the onslaught of antibiotics. Bacteria develop this resistance against the antibiotic by adopting small changes to survive the deadly effect of antibiotic [38]. More an antibiotic is used; more are the chances of developing resistant population amongst pathogenic as well as commensal organisms. Day after day development of bacterial resistance against antibiotics is propagating at a rapid pace, which is of grave concern for developing country like India where enforcement agencies are inadequate to check the indiscriminate use of antibiotics [8]. Further, in the face chronic diseases like tuberculosis, the antibiotic resistance among bacteria poses tremendous challenges as the diseases become practically incurable. Infectious organisms in which resistance is encountered most commonly include

Escherichia coli, Salmonella, Mycobacterium tuberculosis, Staphylococcus aureus [18, 30]. The pace of development of antibiotic resistance in bacteria far exceeds the pace at which new antibiotics are discovered [11]. A healthy person can get infected with resistant bacteria through carriers: food, water, soil, environment, and contact with diseased animals or persons [11]. Besides, indiscriminate use of antibiotics in animals, antibiotics resistance also arises from hospital and community settings, e.g., Methicillin resistant *Staphylococcus aureus* (MRSA) [16]. In present scenario, antibiotics resistance presents a critical situation, as diseases from multiple drugs resistant bacteria are difficult to treat [13].

13.2 ANTIBIOTICS AS GROWTH PROMOTERS

The name "antibiotics as growth promoters" means antibiotics administered at low or sub-therapeutic dose which helps to destroys or check bacterial growth. There is no doubt about efficacy of growth promoters as it improves daily growth of animals (1 to 10 percentage), producing meat of better quality, however the effect is more appreciable for sick and animals reared under poor management condition [29]. Worldwide use of antibiotics varies from country to country as Sweden say no to antibiotics however in USA a wide range of antibiotics are being used as growth promoter [20]. Antibiotics increases feed efficiency and keep the growing animal healthy resulting increase in productivity [23]. Mostly growth promoter antimicrobials are being used in species like broiler chicken, pig, turkeys and beef producing cattle in which pig stand on the top. Its period of administration vary from two weeks to entire period of production cycle. Interesting fact the amount of antibiotics is being in use for food producing animals is comparable to human medicines [40]. Presently global curiosity arises because of overuse of growth promoters constituting emergence of global problem known as bacterial resistant.

13.3 ANTIBIOTICS RESISTANCE HAZARDS

Antibiotics resistance came first in news through Swann report [34] and spread worldwide like a wild fire. Report presumes that antibiotics as animal growth promoter are prescribed at sub-therapeutic dose to which the bacteria get accustomed and finally negate its effects. The committee proposed that the antibiotics used as growth promoters should never be used for therapeutics in animals or human. Aarestrup et al. [1] demonstrated that use of savoparcin in pigs and chicken as growth promoter produced resistant enterococci. Further, scientific community correlated the finding of Aarestrup et al. [1] and concluded that antibiotic resistance developed in human basically came from animals source. Injudicious and rampant use of antibiotics in food producing animals produces colonization of resistant bacteria which is a good recipe for selection of resistant bacteria. Resistant bacteria of animal origin act as reservoir of resistance gene, which is hazardous in nature and propagates to human population via water, food etc. [44].

13.4 OVERUSE OF ANTIBIOTICS IN FOOD PRODUCING ANIMALS

Golden era of antibiotics arrived with the discovery of first antibiotics named Penicillin-G. After Penicillin-G, many new antibiotics were discovered each one with a unique property. Antibiotics at therapeutic dosage kill or halt the growth of infectious microorganisms, whereas at sub-therapeutic dose, the growth and productivity of food producing animals is enhanced.

To prevent spread of zoonotic diseases and to improve the quality of life in humans, antibiotics are used quite often. Antibiotics are even sprayed over crops to reduce bacterial load. In food producing animals, antibiotics are not only used for prophylaxis and therapeutics of infectious diseases but for growth promotion, as well [23]. Uncontrolled proliferation of antibiotics resistant bacteria can increase the chances of their transmission to susceptible patient, which can prove detrimental to human health [14]. Antimicrobials widely used in poultry, swine, goats and sheep include beta-lactams, tetracyclines, lincosamides, sulfonamides and fluroquinolones, therefore, resistances against these classes are widely encountered in the field [17, 21].

The menace of antibiotic resistance is further compounded when same antibiotic classes are prescribed for both animal as well as human clinical cases [31]. World Health Organization [41] published a report which attributes the rise in antimicrobials resistance in bacteria afflicting human population, to misuse or overuse of antibiotics. It has been noted that even a susceptible human commensals such as enterococcus became resistant as a consequence of antimicrobials misuse. These resistant bacteria may be transmitted to humans by various means in which food of animal origin is one of most important medium.

13.5 MECHANISM OF BACTERIAL RESISTANCE AGAINST ANTIBIOTICS

Mechanism of antibiotics resistance development is invariably determined by the mechanism of action with which an antibiotic acts against microorganism. Antibiotic when present in adequate quantity can kill or inhibit a bacterium by acting on a specific target on the susceptible bacterium. This implies that for effective action:

1. Antibiotic should be present in adequate quantity at the site of action.
2. Bacterium should be susceptible, i.e., the target for the antibiotic action is present in the bacterium.
3. Interaction between antibiotic and susceptible bacteria should take place.

A bacterium can develop resistance either by modifying the structure of target (cell membrane, ribosomal subunit, nucleic acid, metabolic enzyme) or by inducing some changes in the pharmacophore of antibiotic. Some bacteria can also reduce the effective concentration of antibiotic at the site of action by pumping the drug molecules out of the cell [3, 6, 33, 35]. Some bacteria are naturally resistant to antibiotics and some may acquire resistant through changes in their genome [38]. Gram-negative bacteria are naturally resistant to some antibiotics meant for gram positive due the presence of outer membrane lipo-polysaccharides. Some gram-negative bacteria acquire resistance either by modifying structural proteins or extrude the drug molecules by inducing extrusion pumps in the cell membrane [26, 32]. Some bacteria acquire resistance by developing enzymes, which inactivate the antibiotics [15].

Bacteria can acquire resistance either with fast pace or slow pace. Fast pace of development of resistance can occur by modification in its structure, e.g., lipopolysaccharides and porin protein or by changing the orientation of extrusion pumps so that drug is always extruded out of the cell [12, 25, 26]. However, slow pace of resistance acquisition is through modification in genes [10]. Numerous genetic loci are present in infectious microorganisms (bacteria) that participate in gene associated antibiotics resistance [4]. The acquisition of slow paced resistance depends on type of bacterial species [43].

Changes in antibiotic permeability across the membrane of bacterial cell have been attributed mostly to variation in lipopolysaccharides, porin proteins. Porins are basically channels constituted of protein through which different hydrophilic antibiotics cross the bacterial membrane and enter into the bacterial cell, e.g., penicillin, cephalosporin, carbopenems [9, 26]. Efflux pumps also manage to pass the drug across bacterial membrane but in reverse direction, i.e., movement is outward in direction, which helps to remove the antibiotic and endogenous substance out of cell [22, 28]. Resistant bacteria posses activated efflux pumps that pump out antibiotics, therefore, insufficient concentration of antibiotic remains at target to produce the action [37].

13.6 ENZYMATIC INACTIVATION OF ANTIBIOTICS

Antibiotic either kills or inhibits the growth of sensitive bacteria. However, chronic administration of antibiotics induces bacteria to develop protective measure to save own life. Antibiotics susceptible bacteria get converted into resistant one because of long-term use of antibiotics in food producing animals. Resistant bacteria keep their life intact by inactivating active moiety of antibiotics by various means including release of enzymes. The β-lactam antibiotics such as penicillins, and cephalosporins possessing β-lactam ring, which require for its antibacterial activity. Enzyme β-lactamase hydrolyzes the β-lactam ring of penicillins and cephalosporins leads to loss of antibacterial activity resulting into treatment failure [27]. Similarly, bacterial enzymes from resistant bacteria inactivate aminoglycosides (streptomycin, gentamicin, neomycin, amikacin) by modifying their structure [42]. Enzyme chloramphenicol-acetyltransferase inactivates the antibiotic chloramphenicol, whereas streptogramin-B inactivation occurs through enzymes released by resistant bacteria [19, 36].

13.7 HUMAN AND ENVIRONMENTAL CONCERNS OF ANTIBIOTICS

Antibiotic resistance is an emerging global threat and if it continues to progresses at such a rate, then in near future, a number of antibiotics will lose therapeutic efficacy. In present scenario best examples of food borne antibiotic resistant microbes that are transmitted from animals into the human food chain are Campylobacter and Salmonella [16]. Antibiotics like penicillins, cephlosporins, sulfonamides, dihydropyrimidines, fluroquinolones, aminoglycosides, and tetracyclines are prescribed for both humans and animals [2]; therefore, bacterial populations colonizing them are exposed to similar drugs, which are a trigger point for the development of antimicrobial resistance. This trigger is when unlocked among zoonotic organisms, the antibiotic resistance raises its head far and wide when these microbes get transmitted back and forth [5, 24, 39].

Thus indiscriminate use of antibiotic is a major threat to human population and environment, which commences with treatment failure and ends with the recalcitrant diseases. Consumers continuously are exposed to sub-therapeutic concentration of antibiotics in the form of residues present in the food products from food producing animals for longer duration of time result in toxicities like reproductive toxicity, carcinogenicity and teratogenicity micro-flora imbalance and secondary infections. In addition to these, antibiotic reduces the beneficial gut micro-flora of consumers, which favors selection of dormant resistant bacteria.

13.8 SUMMARY

Indiscriminate use of antibiotics creates resistance to the extent that very few antibiotics are clinically effective and to control infections due to resistant bacteria, even fewer options are left with the clinician. The best example of antibiotic resistant microbes is Mycobacterium. To prevent further the situation from worsening, various measures are to be undertaken to like:

a. Antibiotics should be used only against susceptible infectious organism, for which selection of an antibiotic should be made on the basis of culture and sensitivity test.
b. Dose of the antibiotic should be meticulously worked out so as to avoid under or over dosage of the antibiotic.
c. Antibiotics being used against a susceptible infection should not be withdrawn before the completion of treatment regimen.
d. Antibiotics should be strictly sold on prescription.
e. Educational Institute should regularly conduct programs.
f. Use of antibiotics as growth promoter should be avoided.

KEYWORDS

- Antibiotics
- Antibiotics inactivation
- Bacterial resistance
- Food producing animals
- Growth promoters
- Human and environment concern
- Mechanism of resistance
- Overuse

REFERENCES

1. Aarestrup, F. M., Ahrens, P., Madsen, M., Pallesen, L. V., Poulsen, R. L., & Westh, H. (1995). Glycopeptide susceptibility among Danish *Enterococcus faecium* and *Enterococcus faecalis* isolates of animal and human origin and PCR identification of genes within the VanA cluster. *Antimicrob. Agents Chemother*, 40, 1938–1940.
2. Aarestrup, F. M., Wegener, H. C., & Collignon, P. (2008). Resistance in bacteria of the food chain: Epidemiology and control strategies. *Expert Review of Anti-Infective Therapy*, 6, 733–750.

3. Aminov, R. I. (2009). The role of antibiotics and antibiotic resistance in nature. *Environ. Microbiol,* 11, 2970–2988.
4. Aminov, R. I., & Mackie, R. I. (2007). Evolution and ecology of antibiotic resistance genes, FEMS Microbiol. *Lett,* 271, 147–161.
5. Anderson, S. A., Yeaton, Woo, R. W., & Crawford, L. M. (2001). Risk assessment of the impact on human health of resistant *Campylobacter jejuni* from fluoroquinolone use in beef cattle. *Food Control,* 12, 13–25.
6. Benton, B., Breukink, E., Visscher, I., Debabov, D., Lunde, C., Janc, J., Mammen, M., & Humphrey, P. (2007). Telavancin inhibits peptidoglycan biosynthesis through preferential targeting of transglycosylation: Evidence for a multivalent interaction between telavancin and lipid II. *Int. J. Antimicrob. Agents* (Suppl.), 29, 51–52.
7. Blackman, B. T. (2002). Resistant bacteria in retail meats and antimicrobial use in animals. *New Engl. J. Med,* 346, 777–779.
8. Byarugaba, D. K. (2005). Antimicrobial resistance and its containment in developing countries. In: *Antibiotic Policies: Theory and Practice,* I. Gould & V. Meer, (Ed.) New York: Springer, pp. 617–646.
9. Ceccarelli, M., & Ruggerone, P. (2008). Physical insights into permeation of and resistance to antibiotics in bacteria. *Curr Drug Targets,* 9(9), 779–88.
10. Chopra, I., O'Neill, A. J., & Miller, K. (2003). The role of mutators in the emergence of antibiotic-resistant bacteria. *Drug Resist Updat,* 6(3), 137–45.
11. Coates, A., & Halls, G. (2012). Antibiotics in Phase II and II Clinical Trials. *Handbook of Experimental Pharmacology,* 211, 167–83.
12. Delcour, A. H. (2009). Outer membrane permeability and antibiotic resistance. *Biochim. Biophys. Acta,* 1794(5), 808–16.
13. Dessen, A., Di-Guilmi, A. M., Vernet, T., & Dideberg, O. (2001). Molecular mechanisms of antibiotic resistance in gram-positive pathogens. *Curr. Drug Targets Infect. Dis,* 1, 63–77.
14. Duval-Iflah, Y., Gainche, I., Ouriet, M. F., Lett, M. C., & Hubert, J. C. (1994). Recombinant DNA transfer to *Escherichia coli* of human faecal origin in vitro and in digestive tract of gnotobiotic mice. *FEMS Microbiol Ecol,* 15, 79–88.
15. Dzidic, S., Suskovic, J., & Kos, J. (2008). Antibiotic Resistance Mechanisms in Bacteria: Biochemical and Genetic Aspects. *Antibiotic Resistance in Bacteria, Food Technol. Biotechnol,* 46(1), 11–21.
16. Gold, H. S., & Moellering, R. C. (1996). Antimicrobial-drug resistance. *New Engl. J. Med,* 335, 1445–1453.
17. Gorbach, S. L. (2001). Antimicrobial use in animal feed-time to stop. *N. Engl. J. Med,* 345, 1202–1203.
18. Hidron, A. I., Edward, J. R., Patel, J., Horan, T. C., Sievert, D. M., Pollock, D. A., & Fridkin, S. K. (2008). NHSN Annual Update: Antimicrobial-resistant pathogens associated with healthcare-associated infections: Annual summary of data reported to the national healthcare safety network at the centers for disease control and prevention, 2006–2007. *Infect Control Hosp Epidemiol,* 29(11), 996–1011.
19. Johnston, N. J., de-Azavedo, J. C., Kellner, J. D., & Low, D. E. (1998). Prevalence and characterization of the mechanisms of macrolide, lincosamide and streptogramin resistance in isolates of Streptococcus pneumoniae. *Antimicrob. Agents Chemother,* 42, 2425–2426.

20. Joint Expert Technical Advisory Committee on Antibiotic Resistance (JETACAR). The use of antibiotics in food producing animals: antibiotic resistant bacteria in animals and humans. Commonwealth Department of Health and Aged Care, Commonwealth Department of Agriculture, Fisheries and Forestry-Australia. Prepared for JETACAR by Biotex Canberra, 1999, Commonwealth of Australia. Available from: http://www.health.gov.au/pubs/ jetacar.htm. Accessed: July 2002.

21. Jones, F. T., & Ricke. S. C. (2003). Observations on the history of the development of antimicrobials and their use in poultry feeds. *Poult. Sci,* 82, 613–617.

22. Li, X. Z., & Nikaido, H. (2009). Efflux-mediated drug resistance in bacteria: An update. *Drugs,* 69(12), 1555–623.

23. McEwen S. A., & Fedorka-Cray, P. J. (2002). Antimicrobial use and resistance in animals. *Clin Infect Dis,* 34(Suppl 3), S93–106.

24. Morrell, V. (1997). Antibiotic resistance: the road of no return. *Science,* 278, 575–6.

25. Nikaido, H. (2003). Molecular basis of bacterial outer membrane permeability revisited. *Microbiol Mol Biol Rev,* 67(4), 593–656.

26. Pagés, J. M., James, C. E., & Winterhalter, M. (2008). The porin and the permeating antibiotic: A selective diffusion barrier in Gram-negative bacteria. *Nat Rev Microbiol,* 6(12), 893–903.

27. Poole, K. (2004). Resistance to β-lactam antibiotics. *Cell. Mol. Life Sci,* 61, 2200–2223.

28. Poole, K. (2008). Bacterial multidrug efflux pumps serve other functions. *Microbe,* 3(4), 179–85.

29. Prescott J. F., & Baggot J. D. 1993. *Antimicrobial Therapy in Veterinary Medicine,* 2nd edition: Iowa State University Press. pp. 564–565.

30. Rice, L. B. (2012). Mechanisms of resistance and clinical relevance of resistance to beta-lactams, glycopeptides and Fluoroquinolones. *Mayo Clinic Proc,* 87(2), 198–208.

31. Sarmah, A. K., Meyer, M. T., & Boxall, A. B. (2006). A global perspective on the use, sales, exposure pathways, occurrence, fate and effects of veterinary antibiotics (VAs) in the environment. *Chemosphere,* 65, 725–59.

32. Stephan, J., Mailaender, C., Etienne, G., Daffé, M., & Niederweis, M. (2004). Multidrug resistance of a porin deletion mutant of *Mycobacterium smegmatis. Antimicrob Agents Chemother,* 48(11), 4163–70.

33. Sutcliffe, J. A., Mueller, J. P., & Utt, E. A. (1999). Antibiotic Resistance Mechanisms of Bacterial Pathogens. In: *Manual of Industrial Microbiology and Biotechnology, A. L. Demain, J. E. Davies (Eds.).* ASM Press, Washington, USA. pp. 759–775.

34. Swann, M. M. (1969). *Use of Antibiotics in Animal Husbandry and Veterinary Medicines.* UK Joint Committee Report.

35. Tenover, F. C. (2006). Mechanisms of antimicrobial resistance in bacteria. *Am. J. Med.* (Suppl.), 119, 3–10.

36. Traced, P., de-Cespedes, G., Bentorcha, F., Delbos, F., Gaspar, E., & Horaud, T. (1993). Study of heterogeneity of chloramphenicol acetyltransferase (CAT) genes in streptococci and enterococci by polymerase chain reaction: characterization of a new CAT determinant. *Antimicrob. Agents Chemother,* 37, 2593–2598.

37. Vila, J., Fabrega, A., Roca, I., Hernandez, A., & Martinez, J. L. (2011). Efflux pumps as an important mechanism for quinolone resistance. *Adv Enzymol Related Areas Mol Biol,* 77, 167–235.

38. Walsh, C. (2000). Molecular mechanisms that confer antibacterial drug resistance. *Nature*, 406, 775–781.
39. Witte, W. (1998). Medical consequences of antibiotic use in agriculture. *Science*, 279, 996–997.
40. World Health Organization. (1997). Division of Emerging and other Communicable Diseases Surveillance and Control, The Medical impact of the use of antimicrobials in food animals: report of a WHO meeting, Berlin, Germany, 13–17 October, Geneva: World Health Organization. p. 24.
41. World Health Organization Report. (2002). Use of antimicrobials outside human medicine and resultant antimicrobial resistance in humans.
42. Wright, G. D. (1999). Aminoglycoside-modifying enzymes. *Curr Opin Microbiol*, 2, 499–503.
43. Wright, G. D. (2005). Bacterial resistance to antibiotics: Enzymatic degradation and modification. *Adv. Drug Deliv. Rev*, 57, 1451–1470.
44. Yan, J. J., Hong, C. Y., Ko, W. C., Chen, Y. J., Tsai, S. H. Chuang, C. L., & Wu, J. J. (2004). Dissemination of Bla$_{CMY-2}$ among *Escherichia coli* isolates from food animals, retail ground meats, and humans in Southern Taiwan. *Antimicrob Agents Chemother*, 48, 1353–1356.

FOOD ADDITIVES

INTERNATIONAL NUMBERING SYSTEM (INS) AND FOOD ADDITIVES

This appendix is only for identifying the INS Numbers of respective food additives. This food additive list is based on Codex; and the food additives allowed under the Food Safety and Standards Regulations 2010 are listed in these regulations. The list given below as published by Codex updated on November 23, 2015 by Indian Food Safety and Standards Regulations 2010.

List of food additives arranged as per INS number

INS Number	Food Additive Name	Technical functions
100	Curcumins	Color
100(i)	Curcumin	Color
100(ii)	Turmeric	Color
101	Riboflavins	Color
102	Tartrazine	Color
103	Alkanet	Color
104	Quinoline yellow	Color
107	Yellow 2G	Color
110	Sunset yellow FCF	Color
120	Carmines	Color
121	Citrus red 2	Color
122	Azorubine/Carmoisine	Color
123	Amaranth	Color
124	Ponceau 4R	Color
125	Ponceau SX	Color
127	Erythrosine	Color

INS Number	Food Additive Name	Technical functions
128	Red 2G	Color
129	Allurared AC/Fast Red E	Color
130	Manascorubin	Color
131	Patent blue V	Color
132	Indigotine	Color
133	Brilliant blue FCF	Color
140	Chlorophyll	Color
141	Copper chlorophylls	Color
142	Green S	Color
143	Fast green FCF	Color
150a	Caramel I-plain	Color
151	Brilliant black PN	Color
152	Carbon black (hydrocarbon)	Color
153	Vegetable carbon	Color
154	Brown FK	Color
155	Brown HT	Color
160a	Carotenes	Color
161a	Flavoxanthin	Color
162	Beet red	Color
163	Anthocyanins	Color
164	Gardenia yellow	Color
166	Sandalwood	Color
170	Calcium carbonates	Surface Colorant, Anticaking agent, Stabilizer
171	Titanium dioxide	Color
172	Iron oxides	Color
173	Aluminum	Color
174	Silver	Color
175	Gold	Color
180	Lithol rubine BK	Color
181	Tannins, food grade	Color, Emulsifier, Stabilizer, Thickener
182	Orchil	Color
200	Sorbic acid	Preservative

INS Number	Food Additive Name	Technical functions
201	Sodium sorbate	Preservative
202	Potassium sorbate	Preservative
203	Calcium sorbate	Preservative
209	Heptyl p-hydroxybenzoate	Preservative
210	Benzoic acid	Preservative
211	Sodium benzoate	Preservative
212	Potassium benzoate	Preservative
213	Calcium benzoate	Preservative
214	Ethyl p-hydroxybenzoate	Preservative
215	Sodium ethyl p-hydroxybenzoate	Preservative
216	Propyl p-hydroxybenzoate	Preservative
217	Sodium propyl p-hydroxybenzoate	Preservative
218	Methyl p-hydroxybenzoate	Preservative
219	Sodium methyl p-hydroxybenzoate	Preservative
220	Sulphur dioxide	Preservative, Antioxidant
221	Sodium sulphite	Preservative, Antioxidant
222	Sodium hydrogen sulphite	Preservative, Antioxidant
223	Sodium metabisulphite	Preservative, Bleaching agent, Antioxidant
224	Potassium metabisulphite	Preservative, Antioxidant
225	Potassium sulphite	Preservative, Antioxidant
226	Calcium sulphite	Preservative, Antioxidant
227	Calcium hydrogen sulphite	Preservative, Antioxidant
228	Potassium bisulphate	Preservative, Antioxidant
230	Diphenyl	Preservative
231	Ortho-phenylphenol	Preservative
232	Sodium o-phenylphenol	Preservative
233	Thiabendazole	Preservative
234	Nisin	Preservative
235	Pimaricin (natamycin)	Preservative
236	Formic acid	Preservative
237	Sodium formate	Preservative

INS Number	Food Additive Name	Technical functions
238	Calcium formate	Preservative
239	Hexamethylene tetramine	Preservative
240	Formaldehyde	Preservative
241	Gum guaicum	Preservative
242	Dimethyl dicarbonate	Preservative
249	Potassium nitrite	Preservative, Color fixative
250	Sodium nitrite	Preservative, Color fixative
251	Sodium nitrate	Preservative, Color fixative
252	Potassium nitrate	Preservative, Color fixative
260	Acetic acid, glacial	Preservative, Acidity regulator
261	Potassium acetates	Preservative, Acidity regulator
262	Sodium acetates	Preservative, Acidity regulator, Sequestrant
263	Calcium acetate	Preservative, Stabilizer, Acidity Regulator
264	Ammonium acetate	Acidity regulator
265	Dehydroacetic acid	Preservative
266	Sodium dehydroacetate	Preservative
270	Lactic acid (L-, D- and Dl-)	Acidity regulator
280	Propionic acid	Preservative
281	Sodium propionate	Preservative
282	Calcium propionate	Preservative
283	Potassium propionate	Preservative
290	Carbon dioxide	Carbonating agent, Packing agent
296	Malic acid (DL-L-)	Acidity regulator, Flavoring agent.
297	Fumaric acid	Acidity regulator
300	Ascorbic acid (L)	Antioxidant
301	Sodium ascorbate	Antioxidant
302	Calcium ascorbate	Antioxidant
303	Potassium ascorbate	Antioxidant
304	Ascorbyl palmitate	Antioxidant
305	Ascorbyl stearate	Antioxidant
306	Mixed tocopherols	Antioxidant
307	Alpha-tocopherol	Antioxidant

INS Number	Food Additive Name	Technical functions
308	Synthetic gamma-tocopherol	Antioxidant
309	Synthetic delta-tocopherol	Antioxidant
310	Propyl gallate	Antioxidant
311	Octyl gallate	Antioxidant
312	Dodecyl gallate	Antioxidant
313	Ethyl gallate	Antioxidant
314	Guaiac resin	Antioxidant
315	Isoascorbic acid	Antioxidant
316	Sodium isoascorbate	Antioxidant
317	Potassium isoascorbate	Antioxidant
318	Calcium isoascorbate	Antioxidant
319	Tertiary butylhydroquinone	Antioxidant
320	Butylated hydroxyanisole	Antioxidant
321	Butylated hydroxytoluene	Antioxidant
322	Lecithins	Antioxidant, Emulsifier
323	Anoxomer	Antioxidant
324	Ethoxyquin	Antioxidant
325	Sodium lactate	Antioxidant, Synergist, Humectant, Bulking agent
326	Potassium lactate	Antioxidant, Synergist, Acidity Regulator
327	Calcium lactate	Acidity regulator, Flour treatment agent
328	Ammonium lactate	Acidity regulator, Flour treatment agent
329	Magnesium lactate (D-,L-)	Acidity regulator, Flour treatment agent
330	Citric acid	Acidity regulator, Synergist for Sequestrant
331	Sodium citrates	Acidity regulator, Emulsifier stabilizer
332	Potassium citrates	Acidity regulator, Sequestrant, Stabilizer
333	Calcium citrates	Acidity regulator, Firming agent, Sequestrant

INS Number	Food Additive Name	Technical functions
334	Tartaric acid [L(+)-]	Acidity regulator, Antioxidant synergist
335	Sodium tartrates	Stabilizer, Sequestrant,
336	Potassium tartrate	Stabilizer, Sequestrant
337	Potassium sodium tartrate	Stabilizer, Sequestrant
338	Orthophosphoric acid	Acidity regulator, Antioxidant Synergist
342	Ammonium phosphates	Acidity regulator, Flour treatment agent
343	Magnesium phosphates	Acidity regulator, Anticaking Agent
344	Lecithin citrate	Preservative
345	Magnesium citrate	Acidity regulator
349	Ammonium malate	Acidity regulator
350	Sodium malates	Acidity regulator, Humectant
351	Potassium malates.	Acidity regulator
352	Calcium malates	Acidity regulator
353	Metatartaric acid	Acidity regulator
354	Calcium tartrate	Acidity regulator
355	Adipic acid	Acidity regulator
356	Sodium adipates	Acidity regulator
357	Potassium adipates	Acidity regulator
359	Ammonium adipates	Acidity regulator
363	Succinic acid	Acidity regulator
364	(i) Monosodium succinate	acidity regulator, Flavor Enhancer
365	Sodium fumarates	Acidity regulator
366	Potassium fumarates	Acidity regulator
367	Calcium fumarates	Acidity regulator
368	Ammonium fumarates	Acidity regulator
370	1,4-Heptonolactone	Acidity regulator, Sequestrant
375	Nicotinic acid	Color retention agent
380	Ammonium citrates	Acidity regulator
381	Ferric ammonium citrate	Anticaking agent
383	Calcium glycerophosphate	Thickener, Gelling agent, Stabilizer

INS Number	Food Additive Name	Technical functions
384	Isopropyl citrates	Antioxidant, Preservative, Sequestrant
388	Thiodipropionic acid	Antioxidant
389	Dilauryl thiodipropionate	Antioxidant
390	Distearyl thiodipropionate	Antioxidant
391	Phytic acid	Antioxidant
399	Calcium lactobionate	Stabilizer
400	Alginic acid	Thickener, Stabilizer
401	Sodium alginate	Thickener, Stabilizer, Gelling Agent
402	Potassium alginate	Thickener, Stabilizer
403	Ammonium alginate	Thickener, Stabilizer
404	Calcium alginate	Thickener, Stabilizer, Gelling Agent, Antifoaming
405	Propylene glycol alginate	Thickener, Emulsifier
406	Agar	Thickener, Gelling agent, Stabilizer
407	Carrageenan and its Na, K, NH_4 salts	Thickener, Gelling agent, Stabilizer
407a	Processed Euchema Seaweed (PES)	Thickener, Stabilizer
408	Bakers yeast glycan	Thickener, Gelling agent, Stabilizer
409	Arabinogalactan	Thickener, Gelling agent, Stabilizer
410	Carob bean gum	Thickener, Stabilizer
411	Oat gum	Thickener, Stabilizer
412	Guar gum	Thickener, Stabilizer, Emulsifier
413	Tragacanth gum	Thickener, Stabilizer, Emulsifier
414	Gum arabic (acacia gum)	Thickener, Stabilizer
415	Xanthan gum	Thickener, Stabilizer, Emulsifier, Foaming agent
416	Karaya gum	Thickener, Stabilizer
417	Tara gum	Thickener, Stabilizer
418	Gellan gum	Thickener, Stabilizer, Gelling Agent

INS Number	Food Additive Name	Technical functions
419	Gum ghatti	Thickener, Stabilizer, Emulsifier
421	Mannitol	Sweetener, Anticaking agent
422	Glycerol	Humectant, Bodying agent
424	Curd lan	Thickener, Stabilizer
425	Konjac flour	Thickener
429	Peptones	Emulsifier
430	Polyoxyethylene (8) stearate	Emulsifier
431	Polyoxyethylene (40) stearate	Emulsifier
432	Polyoxyethylene (20) sorbitan	Monolaurate Emulsifier, Dispersing agent
433	Polyoxyethylene (20) sorbitan	Monoleate Emulsifier, Dispersing agent
434	Polyoxyethylene (20) sorbitan	Monopalmitate Emulsifier, Dispersing agent
435	Polyoxyethylene (20) sorbitan	Monostearate Emulsifier, Dispersing agent
436	Polyoxyethylene (20) sorbitan	Tristearate Emulsifier, Dispersing agent
440	Pectins Thickener, Emulsifier	Stabilizer, Gelling agent
441	Superglycerinated hydrogenated rapeseed oil	Emulsifier
442	Ammonium salts of phosphatidic	Acid Emulsifier
443	Brominated vegetable oil	Emulsifier, Stabilizer
444	Sucrose acetate isobutyrate	Emulsifier, Stabilizer
445	Glycerol esters of wood resin	Emulsifier, Stabilizer
446	Succistearin	Emulsifier
451	Triphosphates	Sequestrant, Acidity regulator Texturizer
458	Gamma Cyclodextrin	Stabilizer, Binder
459	Beta-cyclodextrin	Stabilizer, Binder
460	Cellulose	Emulsifier, Dispersing, Anticaking, Texturizer
461	Methyl cellulose	Thickener, Emulsifier, Stabilizer
462	Ethyl cellulose	Binder, Filler
463	Hydroxypropyl cellulose	Thickener, Emulsifier, Stabilizer

INS Number	Food Additive Name	Technical functions
464	Hydroxypropyl methyl cellulose	Thickener, Emulsifier, Stabilizer
465	Methyl ethyl cellulose	Thickener antifoaming, Emulsifier, Stabilizer
466	Sodium carboxymethyl cellulose	Thickener, Emulsifier, Stabilizer
467	Ethyl hydroxyethyl cellulose	Thickener, Emulsifier, Stabilizer
468	Croscaramellose	Stabilizer, Binder
470	Salts of fatty acids	Emulsifier, Stabilizer, Anticaking agent
471	Mono-and di-glycerides of fatty acids	Emulsifier, Stabilizer
472a	Acetic and fatty acid esters of glycerol	Emulsifier, Stabilizer Sequestrant
473	Sucrose esters of fatty acids	Emulsifier, Stabilizer, Sequestrant
474	Sucroglycerides	Emulsifier, Stabilizer, Sequestrant
475	Polyglycerol esters of fatty acid	Emulsifier, Stabilizer, Sequestrant
477	Propylene glycol esters of fatty	Acids Emulsifier, Stabilizer, Sequestrant
480	Dioctyl sodium sulphosuccinate	Emulsifier, Wetting agent
481	Sodium lactylate	Emulsifier, Stabilizer
482	Calcium lactylates	Emulsifier, Stabilizer
483	Stearyl tartrate	Flour treatment agent
484	Stearyl citrate	Emulsifier, Sequestrant
485	Sodium stearoyl fumarate	Emulsifier
486	Calcium stearoyl fumarate	Emulsifier
487	Sodium laurylsulphate	Emulsifier
488	Ethoxylated mono-and di-glycerides	Emulsifier
491	Sorbitan monostearate	Emulsifier
492	Sorbitan tristearate	Emulsifier
493	Sorbitan monolaurate	Emulsifier
494	Sorbitan monooleate	Emulsifier
495	Sorbitan monopalmitate	Emulsifier
496	Sorbitan trioleate	Stabilizer, Emulsifier

INS Number	Food Additive Name	Technical functions
500	Sodium carbonates	Acidity regulator, Raising agent, Anticaking agent
501	Potassium carbonates	Acidity regulator, Stabilizer
503	Ammonium carbonates	Acidity regulator, Raising agent
504	Magnesium carbonates	Acidity regulator, Anticaking, Color retention
505	Ferrous carbonate	Acidity regulator
507	Hydrochloric acid	Acidity regulator acid
508	Potassium chloride	Gelling agent
509	Calcium chloride	Firming agent
510	Ammonium chloride	Flour treatment agent
511	Magnesium chloride	Firming agent
512	Stannous chloride	Antioxidant, Color retention Agent
513	Sulphuric acid	Acidity regulator
514	Sodium sulphates	Acidity regulator
515	Potassium sulphates	Acidity regulator
516	Calcium Sulphate	Dough conditioner, Sequestrant, Firming agent
517	Ammonium sulphate	Flour treatment agent, Stabilizer
518	Magnesium sulphate	firming agent
519	Cupric sulphate	Color fixative, Preservative
520	Aluminum sulphate	firming agent
521	Aluminum sodium	Sulphate firming agent
522	Aluminum potassium	Sulphate Acidity regulator, Stabilizer
523	Aluminum ammonium	Sulphate Stabilizer, Firming agent
524	Sodium hydroxide	Acidity regulator
525	Potassium hydroxide	Acidity regulator
526	Calcium hydroxide	Acidity regulator, Firming agent
527	Ammonium hydroxide	Acidity regulator
528	Magnesium hydroxide	Acidity regulator, Color retention agent
529	Calcium oxide	Acidity regulator, Color retention agent
530	Magnesium oxide	Anticaking agent

INS Number	Food Additive Name	Technical functions
535	Sodium ferrocyanide	Anticaking agent
536	Potassium ferrocyanide	Anticaking agent
537	Ferrous hexacyanomanganate	Anticaking agent
538	Calcium ferrocyanide	Anticaking agent
539	Sodium thiosulphate	Antioxidant, Sequestrant
541	Sodium aluminum phosphate	Acidity regulator, Emulsifier
541(i)	Sodium aluminum phosphate-acidic	Acidity regulator, Emulsifier
542	Bone phosphate	Emulsifier, Anticaking agent, Water retention
550	Sodium silicates	Anticaking agent
551	Silicon dioxide, amorphous	Anticaking agent
552	Calcium silicate	Anticaking agent
553	Magnesium silicates	Anticaking agent, Dusting Powder
554	Sodium aluminosilicate	Anticaking agent
555	Potassium aluminum silicate	Anticaking agent
556	Calcium aluminum silicate	Anticaking agent
557	Zinc silicate	Anticaking agent
558	Bentonite	Anticaking agent
559	Aluminum silicate	Anticaking agent
560	Potassium silicate	Anticaking agent
570	Fatty acids	Foam stabilizer, Glazing agent, Antifoaming agent
574	Gluconic acid (D-)	Acidity regulator, Raising agent
575	Glucono delta-lactone	Acidity regulator, Raising agent
576	Sodium gluconate	Sequestrant
577	Potassium gluconate	Sequestrant
578	Calcium gluconate	Acidity regulator, Firming agent
579	Ferrous gluconate	Color retention agent
580	Magnesium gluconate	Acidity regulator, Firming agent
585	Ferrous lactate	Color retention agent
586	4-Hexylresorcinol	Color retention agent, Antioxidant
620	Glutamic acid (L (+)-)	Flavor enhancer
621	Monosodium glutamate	Flavor enhancer

INS Number	Food Additive Name	Technical functions
622	Monopotassium glutamate	Flavor enhancer
623	Calcium glutamate	Flavor enhancer
624	Monoammonium glutamate	Flavor enhancer
625	Magnesium glutamate	Flavor enhancer
626	Guanylic acid	Flavor enhancer
627	Disodium 5'-guanylate	Flavor enhancer
628	Dipotassium 5'-guanylate	Flavor enhancer
629	Calcium 5'-guanylate	Flavor enhancer
630	Inosinic acid	Flavor enhancer
631	Disodium 5'-inosinate	Flavor enhancer
632	Potassium inosinate	Flavor enhancer
633	Calcium 5'-inosinate	Flavor enhancer
634	Calcium 5'-ribonucleotides	Flavor enhancer
635	Disodium 5'-ribonucleotides	Flavor enhancer
636	Maltol	Flavor enhancer
637	Ethyl maltol	Flavor enhancer
638	Sodium L-Aspartate	Flavor enhancer
639	DL-Alanine	Flavor enhancer
640	Glycine	Flavor enhancer
641	L-Leucine	Flavor enhancer
642	Lysin hydrochloride	Flavor enhancer
900a	Polydimethylsiloxane	Antifoaming agent, Anticaking agent, Emulsifier
900b	Methylphenylpolysiloxane	Antifoaming agent
901	Beeswax	White and yellow glazing agent, Release
903	Carnaubawax	Glazing agent
904	Shellac	Glazing agent
905a	Mineral oil	Food grade glazing agent, Release agent sealing
905b	Petroleum jelly	Glazing, Release, Sealing agent
905c	Petroleum wax	Glazing agent, Release agent, Sealing agent
906	Benzoin gum	Glazing agent
907	Hydrogenated poly-1-decene	Glazing agent

INS Number	Food Additive Name	Technical functions
908	Rice bran wax	Glazing agent
909	Spermaceti wax	Glazing agent
910	Wax esters	Glazing agent
911	Methyl esters of fatty acids	Glazing agent
913	Lanolin	Glazing agent
916	Calcium iodate	Flour treatment agent
917	Potassium iodate	Flour treatment agent
918	Nitrogen oxide	Flour treatment agent
919	Nitrosyl chloride	Flour treatment agent
922	Potassium persulphate	Flour treatment agent
923	Ammonium persulphate	Flour treatment agent
924a	Potassium bromate	Flour treatment agent
924b	Calcium bromate	Flour treatment agent
925	Chlorine	Flour treatment agent
926	Chlorine dioxide	Flour treatment agent
927a	Azodicarbonamide	Flour treatment agent
927b	Carbamide (urea)	Flour treatment agent
929	Acetone peroxide	Flour treatment agent
930	Calcium peroxide	Flour treatment agent
938	Argon	Packing gas
939	Helium	Packing gas
940	Dichlorodifluoromethane	Propellant, Liquid freezing
941	Nitrogen	Packing gas, Freezing
942	Nitrous oxide	Propellant
943a	Butane	Propellant
943b	Isobutane	Propellant
944	Propane	Propellant
945	Chloropentafluoroethane	Propellant
946	Octafluorocyclobutane	Propellant
948	Oxygen	Packing gas
950	Acesulfame potassium	Sweetener, Flavor enhancer
951	Aspartame	Sweetener, Flavor enhancer
952	Cyclamic acid (and Na, K, Ca Salts)	Sweetener

INS Number	Food Additive Name	Technical functions
954	Saccharin (and Na, K, Ca salts)	Sweetener
955	Sucralose (trichlorogalactosucrose)	Sweetener
956	Alitame	Sweetener
957	Thaumatin	Sweetener, Flavor enhancer
958	Glycyrrhizin	Sweetener, Flavor enhancer
960	Stevioside	Sweetener
964	Polyglycitol syrup	Sweetener
965	Maltitol and maltitol Syrup	Sweetener, Stabilizer, Emulsifier
966	Lactitol	Sweetener, Texturizer
967	Xylitol Sweetener	Humectant, Stabilizer, Emulsifier, Thickener
968	Erythritol Sweetener	Flavor enhancer, Humectant
999	Qulillaia extracts	Foaming agent
1000	Cholic acid	Emulsifier
1001	Choline salts and esters	Emulsifier
1100	Amylases	Flour treatment agent
1101	Proteases flour treatment agent	Stabilizer, Tenderizer, Flavor enhancer
1102	Glucose oxidase	Antioxidant
1103	Invertases	Stabilizer
1104	Lipases flavor	Enhancer
1105	Lysozyme	Preservative
1503	Castor oil	Release agent
1505	Triethyl citrate	Foam stabilizer
1518	Triacetin	Humectant
1521	Polyethylene glycol	Antifoaming agent

REFERENCES

1. Codex Aliment Arius International Food Standards (2015). http://www.codexalimentarius.net
2. Food Safety and Standards Regulations (2010). http://www.fssai.gov.in/
3. http://www.codexalimentarius.net/web/jecfa.jsp

INDEX